VOLUME SIXTY EIGHT

EMBEDDED DISCRETE FRACTURE MODELING AND APPLICATION IN RESERVOIR SIMULATION

DEVELOPMENTS IN PETROLEUM SCIENCE

VOLUME SIXTY EIGHT

EMBEDDED DISCRETE FRACTURE MODELING AND APPLICATION IN RESERVOIR SIMULATION

KAMY SEPEHRNOORI
Hildebrand Department of Petroleum and Geosystems Engineering, The University of Texas at Austin

YIFEI XU
Hildebrand Department of Petroleum and Geosystems Engineering, The University of Texas at Austin, now with ExxonMobil Upstream Research Company

WEI YU
Sim Tech LLC

Series Editors

BAOJUN BAI AND ZHANGXIN CHEN

ELSEVIER

Elsevier
Radarweg 29, PO Box 211, 1000 AE Amsterdam, Netherlands
The Boulevard, Langford Lane, Kidlington, Oxford OX5 1GB, United Kingdom
50 Hampshire Street, 5th Floor, Cambridge, MA 02139, United States

Notices
Knowledge and best practice in this field are constantly changing. As new research and experience broaden our understanding, changes in research methods, professional practices, or medical treatment may become necessary.

Practitioners and researchers must always rely on their own experience and knowledge in evaluating and using any information, methods, compounds, or experiments described herein. In using such information or methods they should be mindful of their own safety and the safety of others, including parties for whom they have a professional responsibility.

To the fullest extent of the law, neither the Publisher nor the authors, contributors, or editors, assume any liability for any injury and/or damage to persons or property as a matter of products liability, negligence or otherwise, or from any use or operation of any methods, products, instructions, or ideas contained in the material herein.

Library of Congress Cataloging-in-Publication Data
A catalog record for this book is available from the Library of Congress

British Library Cataloguing-in-Publication Data
A catalogue record for this book is available from the British Library

ISBN: 978-0-12-821872-3
ISSN: 0376-7361

Publisher: Candice Janco
Acquisitions Editor: Amy Shapiro
Editorial Project Manager: Andrea Dulberger
Production Project Manager: Sruthi Satheesh
Designer: Greg Harris

Typeset by Thomson Digital

To my sisters Mahin Dokht and Parvin Dokht Sepehrnoori

Kamy Sepehrnoori

To my parents Zhongzheng Xu and Lihong Song and my wife Yue Zhang

Yifei Xu

To my parents Zhongzhao Yu and Ruie Sun and my in-laws Guilao Wu and Chunyun Tang and my wife Kan Wu

Wei Yu

Contents

Authors biography

Dr. Kamy Sepehrnoori is a professor in the Hildebrand Department of Petroleum and Geosystems Engineering at The University of Texas at Austin, where he holds the Texaco Centennial Chair in Petroleum Engineering. His research interest and teaching include computational methods, reservoir simulation, simulation of unconventional reservoirs, enhanced oil recovery modeling, flow assurance modeling, naturally fractured reservoirs, high-performance computing, and CO_2 sequestration. He has been teaching at The University of Texas for over 30 years and has graduated more than 200 MS and PhD students under his supervision working mainly in the areas of reservoir simulation and enhanced oil recovery modeling. For the last several years, he has been supervising a research group in simulation of unconventional reservoirs (shale gas and tight oil reservoirs). Sepehrnoori's research team along with his colleagues have been in charge of development of several compositional reservoir simulators (i.e., UTCOMPRS, UTCHEMRS, and UTGEL). Also, more recently, he supervised the development of a software package for embedded discrete fracture modeling for application in naturally and hydraulically fractured reservoirs. He has published over 600 articles in journals and conference proceedings in his research areas. He has also coauthored two books, which have been published by Elsevier. Sepehrnoori is the director of the Reservoir Simulation Joint Industry Project in the Center for Subsurface Energy and the Environment. He holds a PhD degree in petroleum engineering from The University of Texas at Austin.

Dr. Yifei Xu is a senior research engineer at ExxonMobil Upstream Research Company. His research interests include reservoir simulation, reservoir modeling, gridding, naturally and hydraulically fractured reservoirs, and unconventional resources. He has published more than 15 technical papers in his research area. His contributions to this book come from his research as a PhD candidate at The University of Texas at Austin. He worked on the embedded discrete fracture model (EDFM) for more than 5 years, and he was one of the developers of an EDFM preprocessing code. He holds a PhD degree in petroleum engineering from The University of Texas at Austin.

Dr. Wei Yu is the chief technology officer for Sim Tech LLC. His research interests include EDFM (Embedded Discrete Fracture Model) technology for modeling any complex fractures, shale gas and tight oil reservoir simulation, EDFM-AI for automatic history matching and well spacing optimization. He is an Associate Editor for SPE Journal, Journal of Petroleum Science and Engineering, and International Journal of Oil, Gas and Coal Technology. Yu authored or coauthored more than 150 technical papers and holds five patents and one book ("Shale Gas and Tight Oil Reservoir Simulation"). He holds a PhD degree in petroleum engineering from The University of Texas at Austin. Yu is an active member of SPE.

Preface

The simulation of fractured reservoirs is a challenging topic in reservoir simulation owing to the complexity of fracture geometry and recovery processes related to fractured reservoirs. Reliable and efficient numerical models are required for the representation of hydraulic and natural fractures in conventional and unconventional reservoirs.

This book is intended for engineers as well as graduate students studying fractured reservoir simulation. The objective of this book is to introduce a numerical approach for simulating complex fractures and complex recovery processes in fractured reservoirs using various types of computational grids. The method is called the embedded discrete fracture model (EDFM). This approach was recently developed to honor the accuracy of discrete fracture models while maintaining the ease of use and computational efficiency offered by non-conforming meshes. In recent years, the EDFM has gained much attention from both academia and industry. This book summarizes much of the research in this area and introduces the latest progress in the field.

In writing this book, we have sought to include both the fundamentals and advanced topics of the EDFM. We cover the application of the EDFM in various types of computational grids and discuss the challenges with respect to each type of grid. The geometrical algorithms and their implementation are also discussed to provide guidance for the implementation of this model in different types of reservoir simulators. We present many verification and application case studies using different reservoirs with various recovery mechanisms to compare the EDFM with other numerical methods for fractured reservoir simulation. Case studies in real fields are also presented to give examples of using the EDFM to solve practical engineering problems.

For the audience new to the area of fractured reservoir simulation, this book serves as an introductory book to the EDFM; for experienced researchers and engineers who have worked on this topic, this book can serve as a useful reference.

We would also like to thank previous graduate students at The University of Texas at Austin who worked on this topic: Ali Moinfar, Mahmood Shakiba, and Jose Sergio de Araujo Cavalcante Filho. This book would not be possible without their solid work. We would like to extend our thanks to many other researchers who have helped us in various ways, direct and indirect, during the writing of this book. We also would like to thank Ms. Jessica Yeager for her careful editing of the manuscript.

CHAPTER ONE

Introduction

1.1 Conventional reservoirs

Most hydrocarbon reservoirs are, to some extent, fractured (Bratton et al., 2006; Fernø, 2012). A naturally fractured reservoir is defined as a reservoir where natural fractures have a significant influence on the fluid flow (Nelson, 2001). A large portion of the world's oil and gas reserves are trapped in naturally fractured reservoirs (Bourbiaux, 2010). For example, carbonate reservoirs hold more than 60% of the proven oil reserves and 40% of the proven gas reserves (Total, 2017). Most carbonate reservoirs (around 80%) are naturally fractured (Total, 2017), including the Ghawar Field of Saudi Arabia, the largest conventional oil field in the world.

Natural fractures are typically formed along with certain geological processes, such as faulting, folding, and weathering (Fernø, 2012). An example of natural fractures is shown in Fig. 1.1. The patterns of natural fractures depict the local stress state at the time when the fractures were created. The length of natural fractures ranges from micrometers to several miles. Fractures can have permeability much higher than the surrounding matrix, or they can act as flow barriers. In different types of naturally fractured reservoirs, fractures might provide essential reservoir porosity, greatly enhance permeability, or create significant permeability anisotropy (Nelson, 2001).

1.2 Unconventional reservoirs

Besides natural fractures, induced fractures are also created using techniques such as hydraulic fracturing to improve the well performance. In the past, vertical wells were hydraulically fractured to enhance the fluid flow from the reservoir to the wellbore. In recent years, the combination of horizontal drilling and multi-stage hydraulic fracturing has made it possible to economically develop shale gas and tight oil resources with extremely low permeability and low porosity, as shown in Fig. 1.2. In 2016, hydraulically fractured horizontal wells contributed to 69% of all oil and natural gas wells drilled in the United States and account for about 83% of the total linear footage drilled, as shown in Fig. 1.3.

Embedded Discrete Fracture Modeling and Application in Reservoir Simulation.
http://dx.doi.org/10.1016/B978-0-12-821872-3.00001-0

Figure 1.1 ***Photograph of fractured limestone (Geiger et al., 2009).***

On the basis of US Energy Information Administration's (EIA's) Annual Energy Outlook (AEO) 2019, tight oil production in the United States was 6.5 million barrels per day in 2018, which accounted for 61% of total US oil production, as shown in Fig. 1.4. In addition, the tight oil production will continue to increase through 2030 and will reach more than 10 million barrels per day in the early 2030s. Furthermore, the improvement of drilling efficiency and reduction of cost will make tight oil resource development less sensitive to oil prices than in the past. As shown in Fig. 1.5, three major tight oil plays, Permian Basin, Bakken, and Eagle Ford, are expected to contribute half of the cumulative tight oil production through 2050. They account for approximately 41%, 19%, and 17% of US tight oil production in 2018, respectively.

In addition, in 2018, US shale gas production was about 65 billion cubic feet per day, which accounted for 70% of total US dry gas production (EIA, 2019b), as shown in Fig. 1.6. However, in 2008, shale gas production only accounted for 16% of total US gas production. The most productive shale gas play is Marcellus. On the basis of projections in the EIA's AEO 2016, shale gas production is expected to reach about 80 billion cubic feet per day by 2040, as shown in Fig. 1.7. Shale oil and gas resources are distributed all over the world (as shown in Fig. 1.8), and they are expected to be increasingly important in the next several decades, as shown in Fig. 1.9.

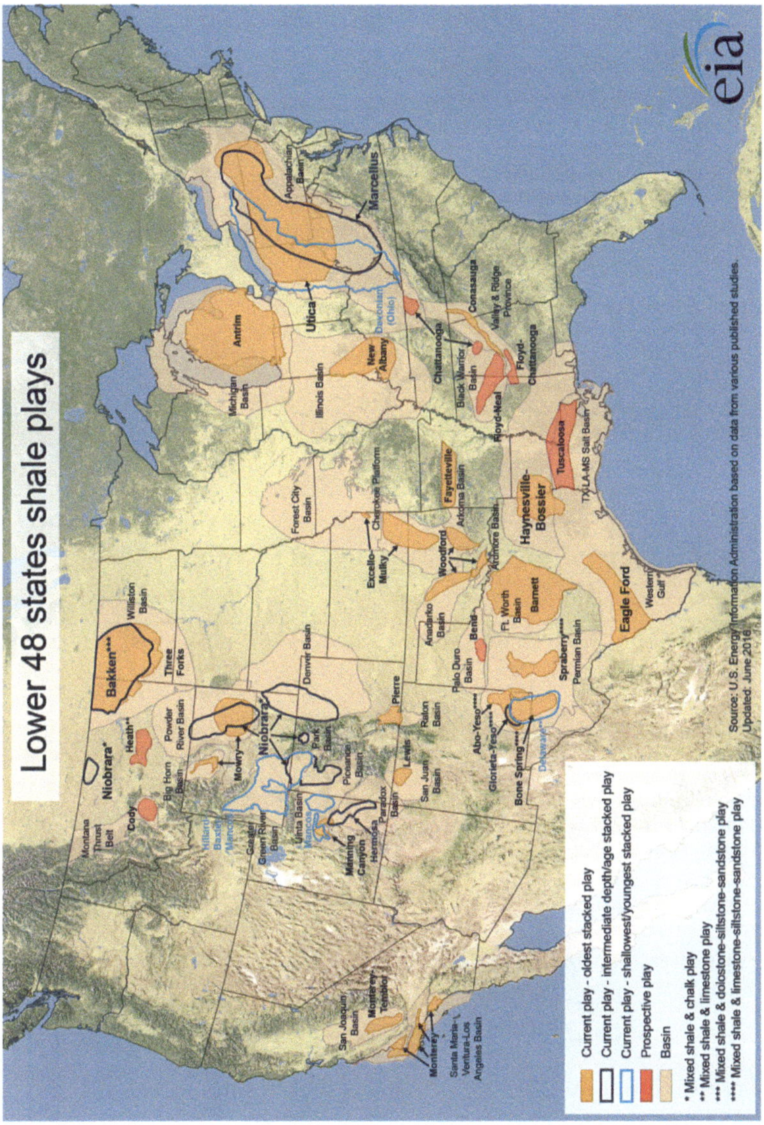

Figure 1.2 *Shale gas and oil plays, lower 48 states (EIA, 2016a).*

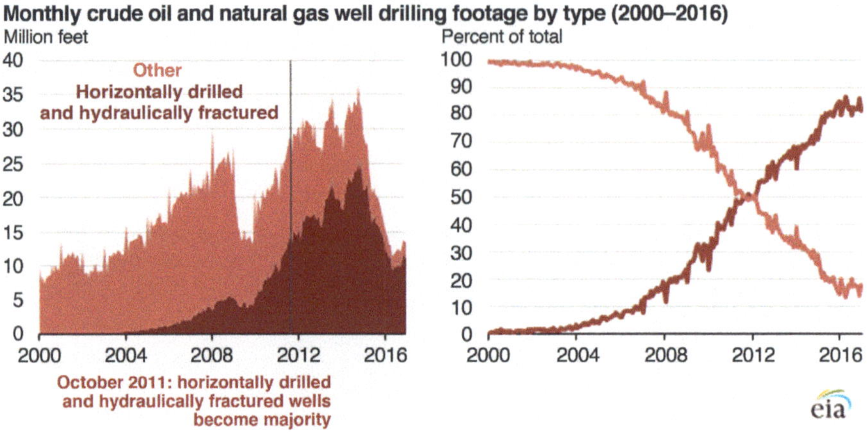

Figure 1.3 *Monthly crude oil and natural gas well drilling footage by type (EIA, 2018).*

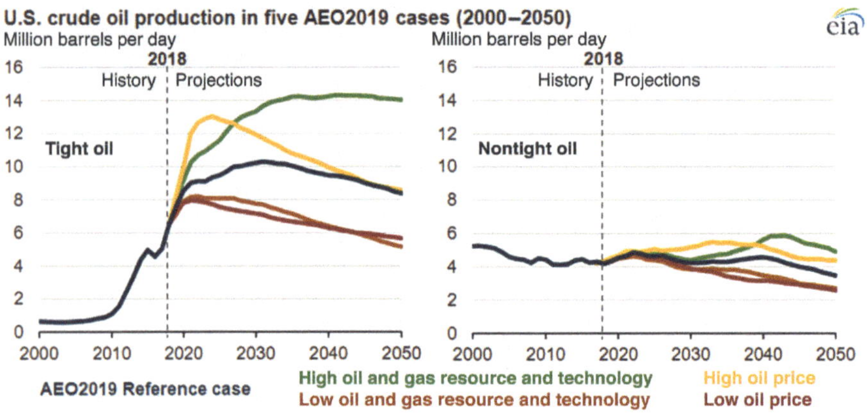

Figure 1.4 *US crude oil production in five AEO 2019 cases (EIA, 2019a).*

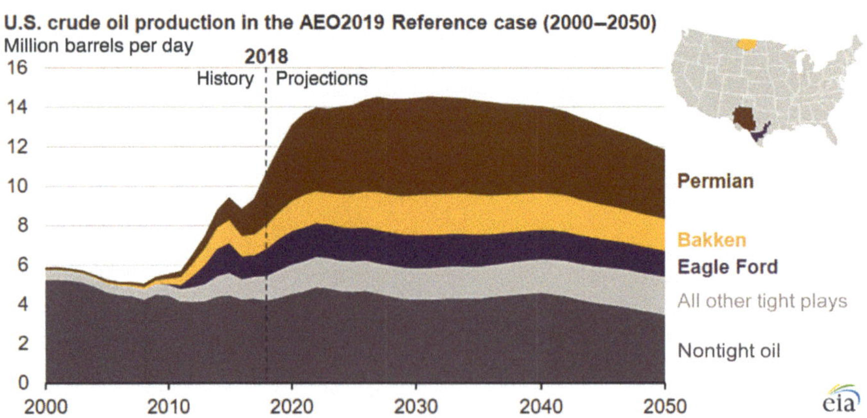

Figure 1.5 *Prediction of tight oil production of three major US tight oil plays in the AEO2019 Reference case (EIA, 2019a).*

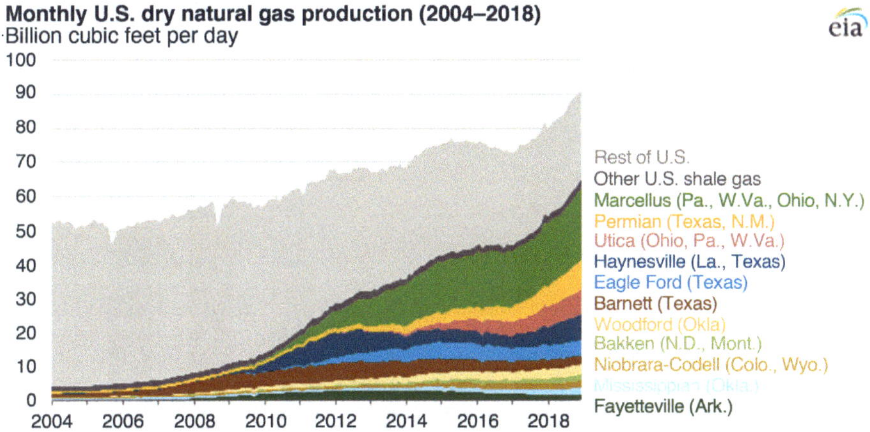

Figure 1.6 *Monthly US dry natural gas production from 2004 to 2018 (EIA, 2019b).*

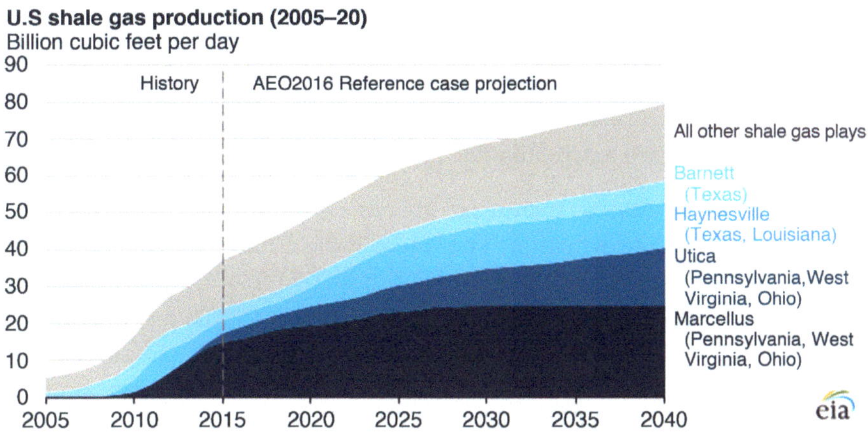

Figure 1.7 *US shale gas production from 2005 to 2040 (EIA, 2016b).*

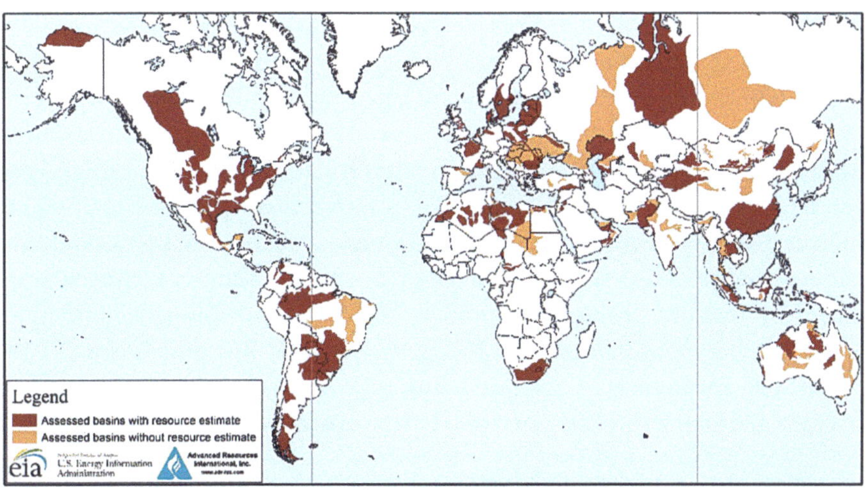

Figure 1.8 *Basins with assessed shale oil and shale gas formations in the world as of May 2013 (EIA, 2015).*

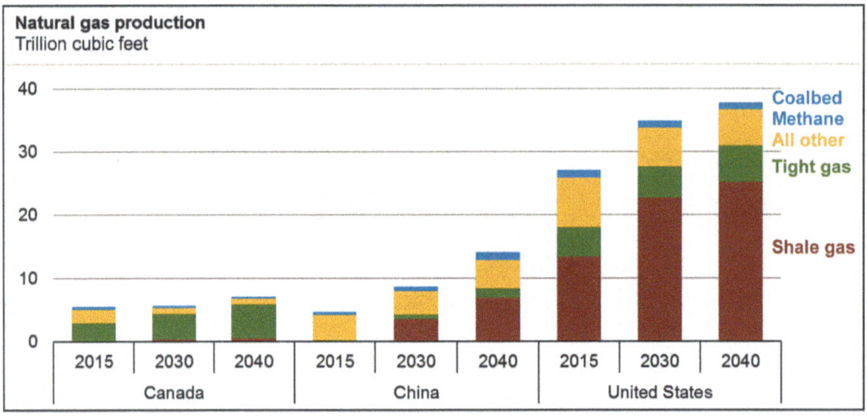

Figure 1.9 ***Outlook of natural gas supplies in Canada, China, and the United States (EIA, 2017).***

1.3 Fracture complexity

During a hydraulic fracturing operation, water, sand, and chemicals are injected into the formation. The injected fluid increases the pore pressure to break apart the rock and create conductive fractures in the formation. The created fractures vastly increase the contact surface area between the reservoir and the wellbore. The pre-existing natural fractures in the formation may also be connected to hydraulic fractures, forming complex fracture networks and further improving the reservoir contact (Maxwell et al., 2002; Fisher et al., 2004; Gale et al., 2007; Warpinski et al., 2009), as shown in Fig. 1.10.

Generally, there is high complexity in fracture geometry. In naturally fractured reservoirs, the orientation of natural fractures is quite complex, and complicated fracture networks are often created. In hydraulically fractured reservoirs, as fractures play a key role in reservoir fluid flow, a detailed description of fracture geometry becomes vital. The fracture dimensions and fracture geometry can be measured using fracture diagnostic techniques such as tiltmeter fracture mapping, microseismic fracture mapping, and distributed temperature sensing (Barree et al., 2002). They can also be predicted by fracture propagation models (Sneddon and Elliot, 1946; Settari and Cleary, 1986; Olson, 2008; Weng et al., 2011; Wu and Olson, 2014, 2015). The measurement and modeling results indicate the complexity of fracture geometry in both horizontal and vertical directions (Barree et al., 2002; Wu and Olson, 2014).

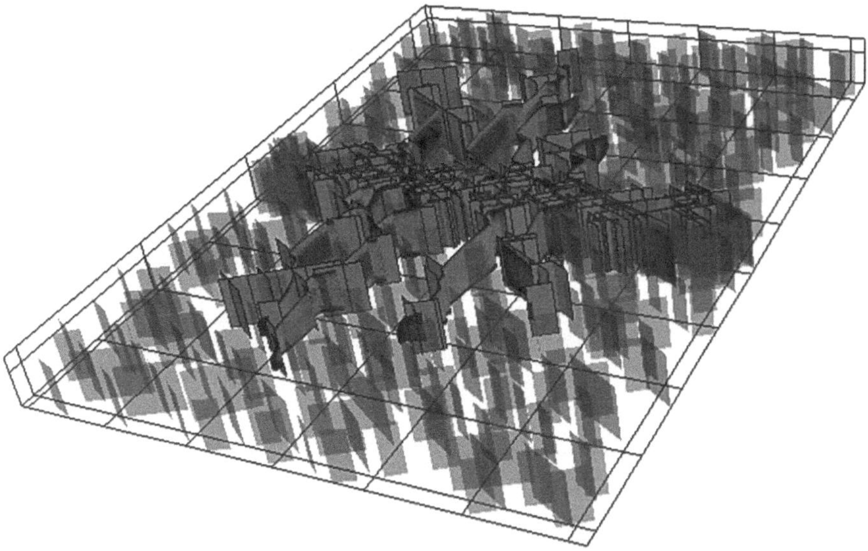

Figure 1.10 ***An example of complex hydraulic and activated natural fracture geometry, which was generated by a complex fracture propagation model (Wu and Olson, 2015).***

1.4 Effect of fractures on fluid flow

Complex hydraulic and natural fractures present challenges for reservoir modeling and production forecasting, which increases the difficulty of making appropriate plans to improve the recovery. Fractures may play different roles in the reservoir. Generally, because of the high permeability of fractures compared with the matrix, the fractures serve as main flow channels, and the fluid may preferably flow through fractures. Fractures can also create connectivity in the matrix, which enhances the effective reservoir permeability (Oda, 1985). In contrast, high-conductivity fractures also increase the permeability anisotropy. Furthermore, the recovery processes involved in fractured reservoirs may also be complicated. During water flooding, the injected fluid tends to flow through the fracture network, which limits the viscous displacement of oil from the matrix and leads to an early water breakthrough and an inefficient sweep (Bourbiaux, 2010; Lemonnier and Bourbiaux, 2010; Fernø, 2012). The counter-current spontaneous imbibition becomes an important mechanism in such processes, which depends on the capillary pressure curve and rock wettability (Fernø, 2012). The influence of fractures on other recovery processes such as gas drive, miscible gas flooding, and enhanced oil recovery has also been discussed in the literature (Firoozabadi, 2000; Manrique et al., 2007; Lemonnier and Bourbiaux, 2010). The uncertainty related to the fractures also increases the difficulty of production forecasting. For example, the connectivity between fractures has a great impact on recovery, but it is not easy to directly measure the fracture network connectivity (Lee et al., 1993).

To help evaluate production strategies in fractured reservoirs, reliable numerical models are needed for the representation of hydraulic and natural fractures. However, owing to the complexity of fracture geometry and complexity of recovery processes related to fractured reservoirs, it is challenging to effectively simulate fractured reservoirs in numerical simulation. For very densely fractured reservoirs with well-connected fracture networks, a homogenization procedure can be used to include the impact of fractures in the effective matrix permeability (Oda, 1985; Lee et al., 2001). For the modeling of other fractured reservoirs, two classes of models are often used, that is, dual-continuum models and discrete fracture models (DFMs). In simulators, dual porosity and dual permeability approaches have been widely applied, and they present a high computational efficiency. However, they are not adequate for modeling large-scale fractures or capturing the influence of fracture connectivity (Karimi-Fard et al., 2004; Bourbiaux, 2010; Moinfar et al., 2011). DFMs, using the finite-element method or finite-volume method, have been developed to tackle these challenges. Unstructured grids are used in most DFMs (Noorishad and Mehran, 1982; Karimi-Fard and Firoozabadi, 2003; Matthäi et al., 2005; Hoteit and Firoozabadi, 2006; Marcondes et al., 2010; Sandve et al., 2012; Hui et al., 2013; Karimi-Fard and Durlofsky, 2016) to explicitly represent each fracture. In these methods, conforming meshes are used, where the edges or faces of the matrix gridblocks around fractures match with the fracture surfaces. The usage of unstructured grids provides high flexibility in representing fracture geometries, and the result is more accurate compared with dual-continuum models. However, in real field studies, the application of unstructured-gridding-based DFMs is still limited because of the high complexity in gridding, especially around fracture intersections, and high computational cost (Du et al., 2017; Fumagalli et al., 2017).

1.5 Embedded discrete fracture model

In this book, we will introduce a new method, the EDFM, which is a combination of dual-continuum models and DFMs. In this method, the matrix is gridded regardless of the fracture geometry, and fractures are

Figure 1.11 ***Two examples of the application of the EDFM method for easily and efficiently modeling complex fracture patterns using structured grids (Yu et al., 2019).*** (A) A complex EDFM fracture pattern without natural fractures. (B) Pressure distribution after production simulation without natural fractures. (C) A complex EDFM fracture pattern with complex natural fractures. (D) Pressure distribution after production simulation with complex natural fractures.

discretized using the matrix gridblock boundaries, as shown in Fig. 1.11. Various types of fluid flows related to fractures are simulated by introducing special connections among matrix, fractures, and wellbores. Because in this method, the gridding of matrix is not affected by the locations or geometries of fractures, the common challenges involved in matrix gridding around fractures are effectively avoided, especially for cases with a large number of fractures or complex fracture geometries. It also offers the flexibility to conveniently simulate fractures with various types of computational grids. Furthermore, this approach is suitable for quick analysis of long-term reservoir performance under the influence of fracture uncertainties as no matrix re-gridding needs to be conducted after changing the fracture configuration. The EDFM has been successfully applied within several in-house and commercial reservoir simulators that have non-neighboring connection functionality. Recent studies have shown the convenience and efficiency of the EDFM approach compared with other methods using grid refinement or unstructured gridding (Moinfar et al., 2014; Panfili et al., 2015; Du et al., 2017). Various types of realistic fracture geometries can be effectively handled with this method, as will be discussed in later chapters.

The discretization of the matrix is of great importance for the implementation of the EDFM, as fracture discretization in the EDFM depends on the location of matrix block boundaries. Hence, the formulation and approach to implementation of the EDFM in different types of matrix gridding will be presented in this book as well. Overall, the primary objective of this book is to introduce the methodology to simulate complex fractures and related complex recovery processes using various types of computational grids with the EDFM.

1.6 Brief description of chapters

In Chapter 2, we introduce the complexities of naturally and hydraulically fractured reservoirs and the challenges they bring to reservoir simulation. Chapter 3 reviews current numerical technologies for modeling complex fractures in reservoir simulation, including dual-continuum models and DFMs. In Chapter 4, we present the basic methodology and formulations of the EDFM and its application in Cartesian Grids. The handling of dynamic fracture behaviors using the EDFM is discussed and demonstrated in Chapter 5. In Chapter 6, we present several field applications of the EDFM, including validation, history matching, sensitivity studies, and production forecasting. Chapter 7 discusses the application of the EDFM in geologically complex reservoirs represented by corner-point grids. The handling of geometrical calculation challenges is discussed in detail. In Chapter 8, we describe the methodology to apply the EDFM in two-dimensional and three-dimensional unstructured grids using an element-based finite-volume method. This demonstrates the flexibility of the EDFM in different types of computational grids. Finally, Chapter 9 concludes the book and includes recommendations based on our experience.

References

Barree, R.D., Fisher, M.K., Woodroof, R.A., 2002. A practical guide to hydraulic fracture diagnostic technologies. Paper SPE 77442. SPE Annual Technical Conference and Exhibition. San Antonio, Texas.

Bourbiaux, B., 2010. Fractured reservoir simulation: a challenging and rewarding issue. Oil Gas Sci. Technol. - Rev. IFP 65 (2), 227–238.

Bratton, T., Canh, D.V., Que, N.V., Duc, N.V., Gillespie, P., Hunt, D., Li, B., Marcinew, R., Ray, S., Montaron, B., Nelson, R., Schoderbek, D., Sonneland, L., 2006. The nature of naturally fractured reservoirs. Oilfield Rev. 18 (2), 4–23.

Du, S., Liang, B., Yuanbo, L., 2017. Field study: embedded discrete fracture modeling with artificial intelligence in Permian Basin for shale formation. Paper SPE 187202. SPE Annual Technical Conference and Exhibition. San Antonio, Texas.

Fernø M.A., Enhanced Oil Recovery in Fractured Reservoirs, *Introduction to Enhanced Oil Recovery (EOR) Processes and Bioremediation of Oil-Contaminated Sites*, Rijeka, Croatia. 89.110 (2012): 89-110.

Firoozabadi, A., 2000. Recovery mechanisms in fractured reservoirs and field performance. J. Can. Pet. Technol. 39 (11), 13–17.

Fisher, M.K., Heinze, J.R., Harris, C.D., Davidson, B.M., Wright, C.A., Dunn, K.P., 2004. Optimizing horizontal completion techniques in the Barnett Shale using microseismic fracture mapping. Paper SPE 90051. SPE Annual Technical Conference and Exhibition. Houston, Texas.

Fumagalli, A., Zonca, S., Formaggia, L., 2017. Advances in computation of local problems for flow-based upscaling in fractured reservoirs. Math. Comput. Simul. 137, 299–324.

Gale, J.F.W., Reed, R.M., Holder, J., 2007. Natural fractures in the Barnett Shale and their importance for hydraulic fracture treatments. AAPG Bull. 91 (4), 603–622.

Geiger, S., Matthäi, S.K., Niessner, J., Helmig, R., 2009. Black-oil simulations for three-component, three-phase flow in fractured porous media. SPE J. 14 (2), 338–354.

Hoteit, H., Firoozabadi, A., 2006. Compositional modeling of discrete-fractured media without transfer functions by the discontinuous Galerkin and mixed methods. SPE J. 11 (3), 341–352.

Hui, M.-H., Mallison, B.T., Fyrozjaee, M.H., Narr, W., 2013. The upscaling of discrete fracture models for faster, coarse-scale simulations of IOR and EOR processes for fractured reservoirs. Paper SPE 166075. SPE Annual Technical Conference and Exhibition New Orleans. Louisiana.

Karimi-Fard, M., Durlofsky, L.J., 2016. A general gridding, discretization, and coarsening methodology for modeling flow in porous formations with discrete geological features. Adv. Water Resour. 96, 354–372.

Karimi-Fard, M., Durlofsky, L.J., Aziz, K., 2004. An efficient discrete-fracture model applicable for general-purpose reservoir simulators. SPE J. 9 (2), 227–236.

Karimi-Fard, M., Firoozabadi, A., 2003. Numerical simulation of water injection in fractured media using the discrete-fracture model and the Galerkin method. SPE Reservoir Eval. Eng. 6 (2), 117–126.

Lee, C.H., Yu, J.L., Hwung, H.H., 1993. Fluid flow and connectivity in fractured rock. Water Resour. Manage. 7 (2), 169–184.

Lee, S.H., Lough, M.F., Jensen, C.L., 2001. Hierarchical modeling of flow in naturally fractured formations with multiple length scales. Water Resour. Res. 37 (3), 443–455.

Lemonnier, P., Bourbiaux, B., 2010. Simulation of naturally fractured reservoirs. State of the art. Oil Gas Sci. Technol. 65 (2), 239–262.

Manrique, E.J., Muci, V.E., Gurfinkel, M.E., 2007. EOR field experiences in carbonate reservoirs in the United States. SPE Reservoir Eval. Eng. 10 (6), 667–686.

Marcondes, F., Varavei, A., Sepehrnoori, K., 2010. An element-based finite-volume method approach for naturally fractured compositional reservoir simulation. 13th Brazilian Congress of Thermal Sciences and Engineering-ENCIT. Uberlândia, MG, Brazil.

Matthäi, S.K., Mezentsev, A., Belayneh, M., 2005. Control-volume finite-element two-phase flow experiments with fractured rock represented by unstructured 3D hybrid meshes. Paper SPE 93341. SPE Reservoir Simulation Symposium. The Woodlands, Texas.

Maxwell, S.C., Urbancic, T.I., Steinsberger, N., Zinno, R., 2002. Microseismic imaging of hydraulic fracture complexity in the Barnett Shale. Paper SPE 77440. SPE Annual Technical Conference and Exhibition. San Antonio, Texas.

Moinfar, A., Narr, W., Hui, M.-H., Mallison, B.T., Lee, S.H., 2011. Comparison of discrete-fracture and dual-permeability models for multiphase flow in naturally fractured reservoirs. Paper SPE 142295. SPE Reservoir Simulation Symposium. The Woodlands, Texas.
Moinfar, A., Varavei, A., Sepehrnoori, K., Johns, R.T., 2014. Development of an efficient embedded discrete fracture model for 3D compositional reservoir simulation in fractured reservoirs. SPE J. 19 (2), 289–303.
Nelson, R.A., 2001. Geologic Analysis of Naturally Fractured Reservoirs. Gulf Professional Publishing, Boston.
Noorishad, J., Mehran, M., 1982. An upstream finite element method for solution of transient transport equation in fractured porous media. Water Resour. Res. 18 (3), 588–596.
Oda, M., 1985. Permeability tensor for discontinuous rock masses. Geotechnique 35 (4), 483–495.
Olson, J.E., 2008. Multi-fracture propagation modeling: applications to hydraulic fracturing in shales and tight gas sands. Paper ARMA 08327. 42nd U.S. Rock Mechanics Symposium (USRMS). San Francisco, California.
Panfili, P., Colin, R., Cominelli, A., Giamminonni, D., Guerra, L., 2015. Efficient and effective field scale simulation of hydraulic fractured wells: methodology and application. Paper SPE 175542, SPE Reservoir Characterisation and Simulation Conference and Exhibition. Abu Dhabi, UAE.
Sandve, T.H., Berre, I., Nordbotten, J.M., 2012. An efficient multi-point flux approximation method for discrete fracture-matrix simulations. J. Comput. Phys. 231 (9), 3784–3800.
Settari, A., Cleary, M.P., 1986. Development and testing of a pseudo-three-dimensional model of hydraulic fracture geometry. SPE Prod. Eng. 1 (6), 449–466.
Sneddon, I.N., Elliot, H.A., 1946. The opening of a Griffith crack under internal pressure. Q. Appl. Math. 4 (3), 262–267.
Total, 2017. Carbonate geology know-how, https://www.ep.total.com/sites/g/files/wompnd1021/f/atoms/files/06_carbonate-reservoirs_knowhow.pdf.
U.S. Energy Information Administration (EIA), 2015. World shale resource assessments, https://www.eia.gov/analysis/studies/worldshalegas.
U.S. Energy Information Administration (EIA), 2016a. Shale gas and oil plays, lower 48 states (6/30/2016), https://www.eia.gov/maps/maps.htm.
U.S. Energy Information Administration (EIA), 2016b. Future U.S. tight oil and shale gas production depends on resources, technology, markets, https://www.eia.gov/todayinenergy/detail.php?id=27612.
U.S. Energy Information Administration (EIA), 2017. International energy outlook 2017, https://www.eia.gov/outlooks/ieo/pdf/0484(2017).pdf.
U.S. Energy Information Administration (EIA), 2018. Hydraulically fractured horizontal wells account for most new oil and natural gas wells, https://www.eia.gov/todayinenergy/detail.php?id=37815.
U.S. Energy Information Administration (EIA), 2019a. Tight oil development will continue to drive future U.S. crude oil production, https://www.eia.gov/todayinenergy/detail.php?id=38852.
U.S. Energy Information Administration (EIA), 2019b. EIA adds new play production data to shale gas and tight oil reports, https://www.eia.gov/todayinenergy/detail.php?id=38372.
Warpinski, N.R., Mayerhofer, M.J., Vincent, M.C., Cipolla, C.L., Lolon, E.P., 2009. Stimulating unconventional reservoirs: maximizing network growth while optimizing fracture conductivity. J. Can. Pet. Tech. 48 (10), 39–51.
Weng, X., Kresse, O., Cohen, C.E., Wu, R., Gu, H., 2011. Modeling of hydraulic-fracture-network propagation in a naturally fractured formation. SPE Prod. Oper. 26 (4), 368–380.

Wu, K., Olson, J.E., 2014. Mechanics analysis of interaction between hydraulic and natural fractures in shale reservoirs. Paper SPE 1922946. SPE/AAPG/SEG Unconventional Resources Technology Conference. Denver, Colorado.
Wu, K., Olson, J.E., 2015. Simultaneous multifracture treatments: fully coupled fluid flow and fracture mechanics for horizontal wells. SPE J. 20 (2), 337–346.
Yu, W., Wu, K., Zuo, L., Miao, J., Sepehrnoori, K., 2019. Embedded discrete fracture model assisted study of gas transport mechanisms and drainage area for fractured shale gas reservoirs. Paper URTeC 552. Unconventional Resources Technology Conference. Denver, Colorado.

CHAPTER TWO

Naturally and hydraulically fractured reservoirs

2.1 Complex fracture networks in shale

By combining multi-stage hydraulic fracturing and horizontal drilling, it is possible to effectively and efficiently develop shale/tight oil and gas reservoirs with natural fractures and low/ultralow permeability. During hydraulic fracturing treatments, new hydraulic fractures are created and pre-existing natural fractures may be activated, resulting in complex fracture networks that help the extraction of oil and gas from tight rocks, as shown in Fig. 2.1. When compared with sandstones in previous studies, the shales are more mechanically anisotropic and form complex fractures much easier (Raterman et al., 2017). Knowledge of the geometry and extent of underground hydraulic and natural fractures is very limited (Gale et al., 2018). Hydraulic fractures might be numerous, widespread, closely spaced, steeply dipping, or branching (Raterman et al., 2017). In addition, the proppant transport and proppant settling in hydraulic fractures are heterogeneous and a majority of hydraulic fractures may be void of proppant (Raterman et al., 2017).

2.2 Monitoring and observation of fracture complexity

Indirect methods such as observing microseismic activity are often applied to monitor fracturing behavior, predict stimulated reservoir volume, and forecast complex fracture geometry (Vermylen and Zoback, 2011; Fisher and Warpinski, 2012; Maity, 2018), as shown in Fig. 2.2.

Although understanding of complex fracture growth and the interaction between hydraulic and natural fractures has been improved by the microseismic data (Maity, 2018), verification of microseismic interpretation results is still required through the direct observation of fracture geometry by taking core intervals near the hydraulically fractured well (Gale et al., 2018).

During the hydraulic fracture test site (HFTS) pad in the Permian Basin, Maity et al. (2018) collected 600 ft of fracture cores in the well 6TW (see Fig. 2.3) covering both the upper and the middle Wolfcamp formations

Embedded Discrete Fracture Modeling and Application in Reservoir Simulation.
http://dx.doi.org/10.1016/B978-0-12-821872-3.00002-2

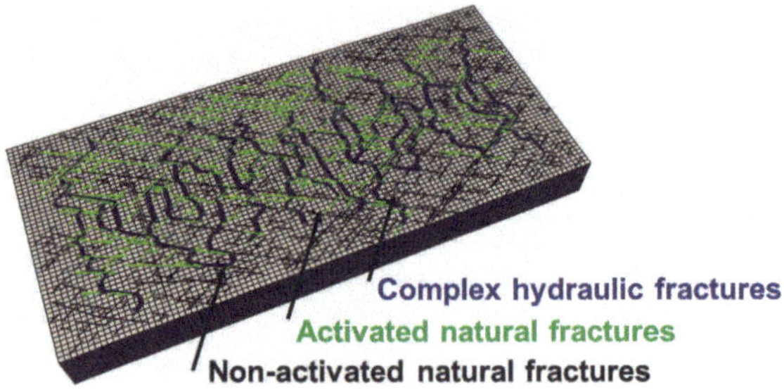

Figure 2.1 ***Generation of complex fracture networks during the hydraulic fracturing treatment (Yu et al., 2018).***

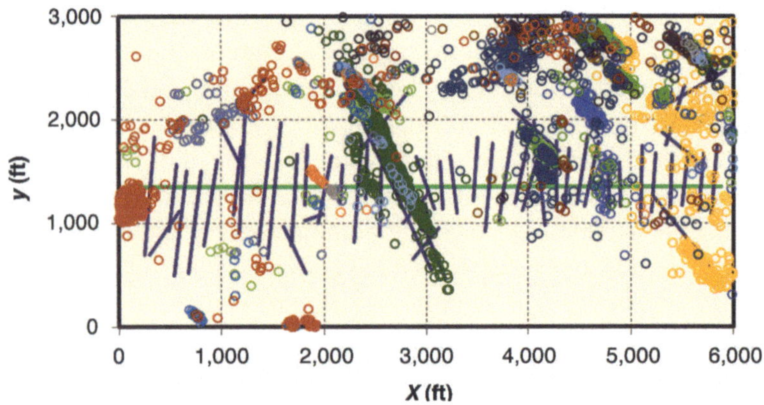

Figure 2.2 ***Microseismic data with possible complex hydraulic fractures (Miao et al., 2018).***

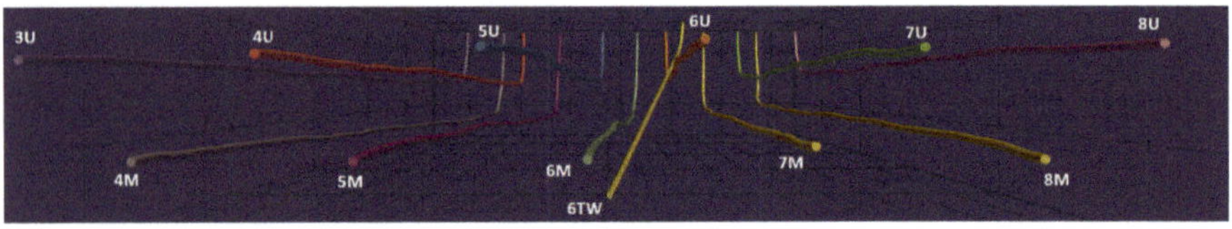

Figure 2.3 ***Well locations in the HFTS pad.*** *(Wood et al., 2018, reprinted by permission of the AAPG whose permission is required for further use).*

(~400 ft in the upper Wolfcamp and 200 ft in the middle Wolfcamp). Then they performed a comprehensive fracture characterization and analysis and found that the spatial proppant distribution is very complex and sporadic with sudden jumps and drops. The zones close to carbonate bed boundaries have a higher proppant concentration. This is because these boundaries experience a significant change of stress and brittleness, which creates

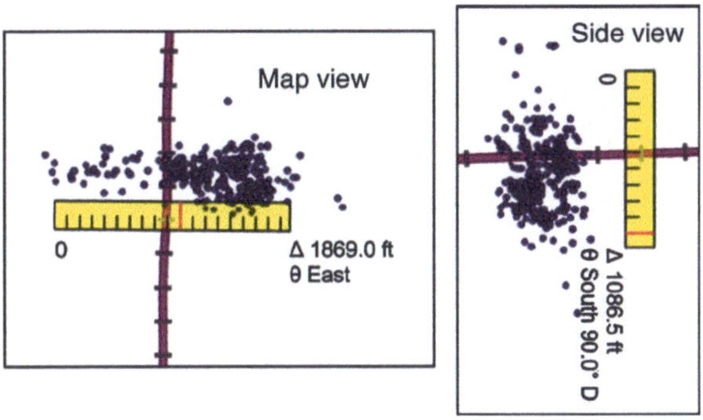

Figure 2.4 ***Interpretation of fracture length and height based on one stage of microseismic data.*** *(Stegent and Candler, 2018, reprinted by permission of the AAPG whose permission is required for further use).*

localized "proppant hot spots." More importantly, the propped fracture height in the upper Wolfcamp formation was observed as only about 30 ft high when close to treatment well at a distance of about 85 ft, which is less than 10% of the fracture height (~1000 ft) predicted based on microseismic events spreading in the vertical direction, as shown in Fig. 2.4. The upper Wolfcamp formation has a larger natural fracture density than the middle Wolfcamp formation, resulting in more interaction between hydraulic fractures and natural fractures and a higher risk of local proppant screen out. Accurately capturing the complexities of realistic proppant distribution and the interaction of hydraulic and pre-existing natural fractures in reservoir simulation is instrumental for the optimization of cluster spacing, stage spacing, and well spacing.

Wood et al. (2018) used fracturing fluid, oil, and proppant tracer analysis to examine the near-wellbore completion effectiveness and the inter-well communication between horizontal wells on the HFTS pad. They found that a high degree (>80%) of cluster efficiency was achieved for three wells with varying cluster density per stage based on the proppant tracer results. Two of the wells had three clusters per stage with cluster spacing of 90 ft, and one well had three or five clusters per stage with cluster spacing of 50 ft. However, the scheme with cluster spacing of 90 ft had a larger percentage of unstimulated reservoir volume than did the scheme with cluster spacing of 50 ft. In addition, inter-well communication occurred for all wells, although the degree of communication between wells in the middle Wolfcamp formation was larger than that in the upper Wolfcamp formation. In addition, there was greater vertical communication from the wells in the

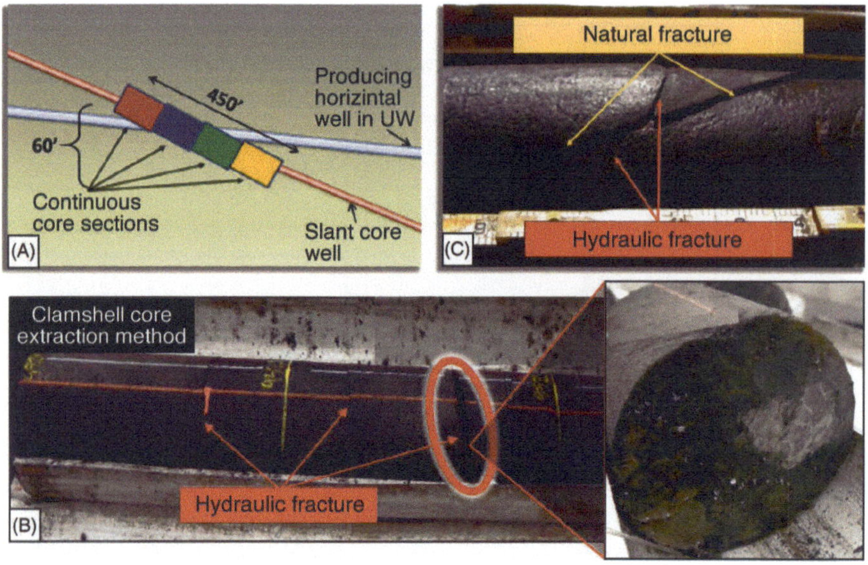

Figure 2.5 ***Observations of hydraulic and natural fractures from the cores.*** *(Ciezobka et al., 2018, reprinted by permission of the AAPG whose permission is required for further use).*

middle Wolfcamp formation to the wells in the upper Wolfcamp formation than the reverse. The well interference due to complex fracture interaction between wells bring challenges in reservoir simulation, necesssitating an accurate and convenient way of representing arbitrary fracture interactions.

The HFTS pad has provided key insights into understanding fracture complexity such as vertical and horizontal fracture geometries, as shown in Fig. 2.5 (Ciezobka et al., 2018). It has been observed that hydraulic fractures in the cores do not have a simple bi-wing geometry. Instead, they have highly complex geometries and interactions with varying densities. Current shale industry fracture simulation technology cannot easily and accurately handle such complexities (Ciezobka et al., 2018).

In addition, the reactivation of natural fractures adds more complexity to the hydraulic fracture geometry (Gale et al., 2018). Hydraulic fractures occur heterogeneously in cores: some of them have smooth fracture surfaces and some have a stepped surface geometry, as shown in Fig. 2.6 (Gale et al., 2018). Also, based on the core analysis of the pilot test of the Eagle Ford shale by ConocoPhillips (Raterman et al., 2017), the surface of drilling-induced fractures (Fig. 2.7A) is quite different from those of

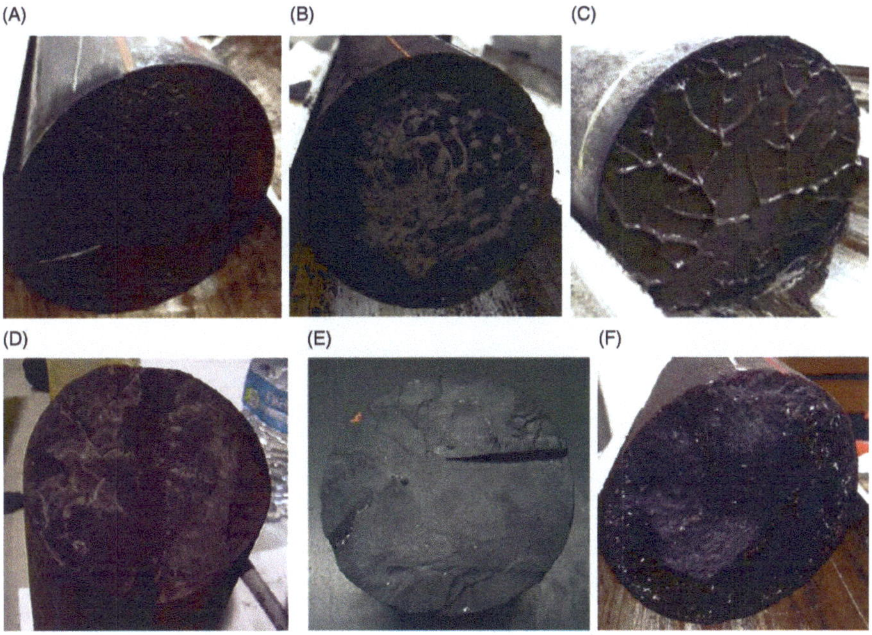

Figure 2.6 ***Varying hydraulic fracture surface features at the HFTS pad in the Permian basin.*** *(Gale et al., 2018, reprinted by permission of the AAPG whose permission is required for further use).*

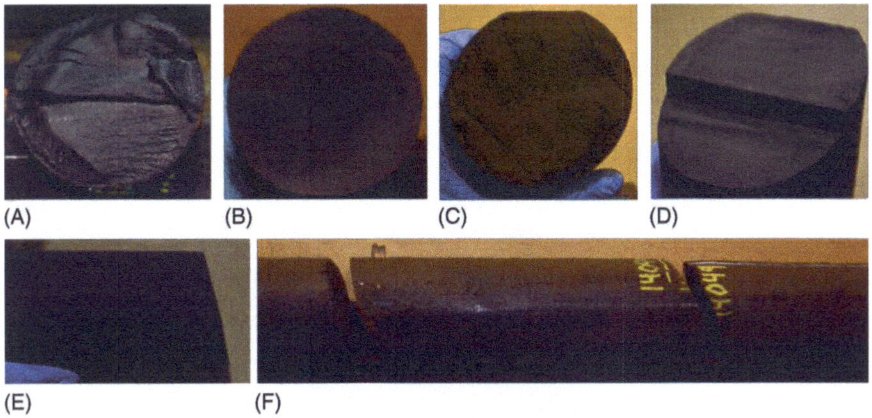

Figure 2.7 ***Fracture surface in the core from the Eagle Ford shale.*** *(Raterman et al., 2017, reprinted by permission of the AAPG whose permission is required for further use).*

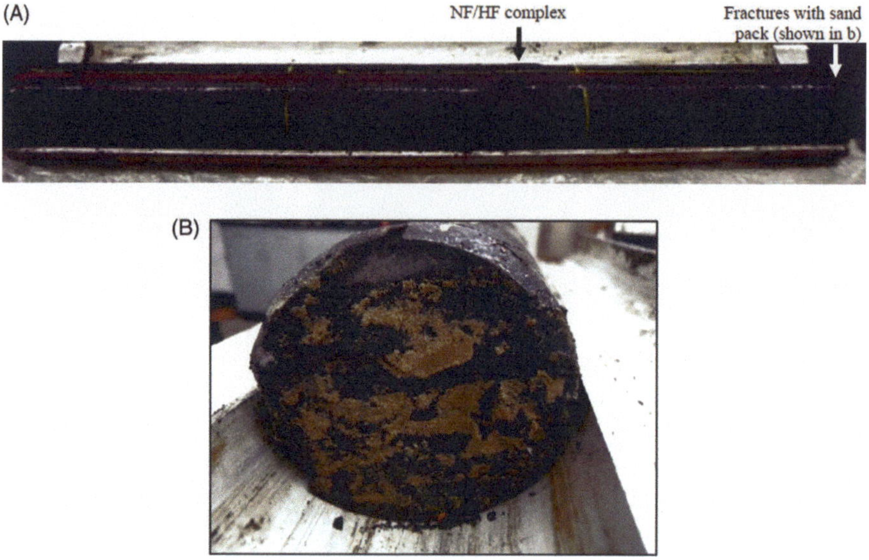

Figure 2.8 ***An observed hydraulic fracture doublet illustrating complex interactions between hydraulic and natural fractures.*** *(Gale et al., 2018, reprinted by permission of the AAPG whose permission is required for further use).*

hydraulic fractures (Fig. 2.7B–D). The surface of hydraulic fractures might be smooth, ridged, or occasionally stepped. Furthermore, a southeasterly dip in the fractures (Fig. 2.7E–F) was observed, which creates a significant challenge for current fracture and reservoir modeling approaches.

On the basis of the fracture analysis for the HFTS cores (Figs. 2.8 and 2.9), Gale et al. (2018) identified over 700 fractures including hydraulic fractures, calcite-filled natural fractures, drilling-induced fractures, and core-handling breaks. They found that the thickness of proppant in hydraulic fractures

Figure 2.9 ***A large propped hydraulic fracture with about 1 cm wide proppant pack.*** *(Gale et al., 2018, reprinted by permission of the AAPG whose permission is required for further use).*

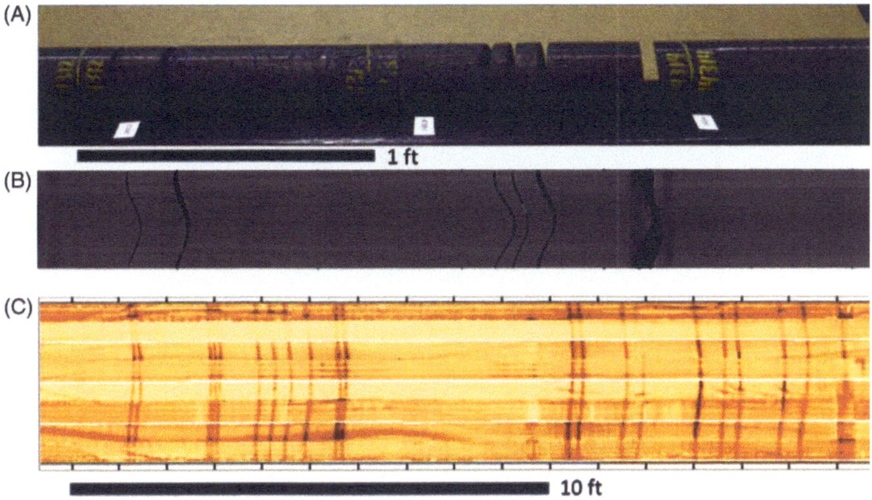

Figure 2.10 ***Observed hydraulic fracture doublet and triplet based on core analysis of the Eagle Ford shale formation.*** *(Raterman et al., 2017, reprinted by permission of the AAPG whose permission is required for further use).*

could be around 1 cm, which provides key insights into understanding the growth of complex hydraulic-fracture networks. In addition, a hydraulic fracture doublet where two hydraulic fractures are in close proximity was observed in one core, as shown in Fig. 2.8. Similar hydraulic fracture patterns such as doublets and triplets (which are often called hydraulic fractures swarms) were also observed in the cores of the Eagle Ford shale formation, as shown in Fig. 2.10. This indicates that two or three fractures may grow very closely within a few centimeters, and the occurrence of these hydraulic fracture swarms may be due to differences in mechanical stratigraphy and natural heterogeneities in the formation (Raterman et al., 2017). The spatial distribution of hydraulic fracture intensity is non-uniform, and the intensities decrease more quickly when distance increases by height rather than laterally, which is consistent with observations of the microseismic event cloud (Raterman et al., 2017).

The fracture geometries observed in the HFTS cores are likely to be crumbled (Rassenfoss, 2018), as shown in Fig. 2.11. Asymmetric fracture geometries along the horizontal wellbore may also be generated due to the reservoir depletion effect of parent wells, which can be evaluated based on the microseismic data (Stegent and Candler, 2018).

Fracture height is very difficult to determine exactly in unconventional oil and gas reservoirs. One method involves applying bottomhole pressure

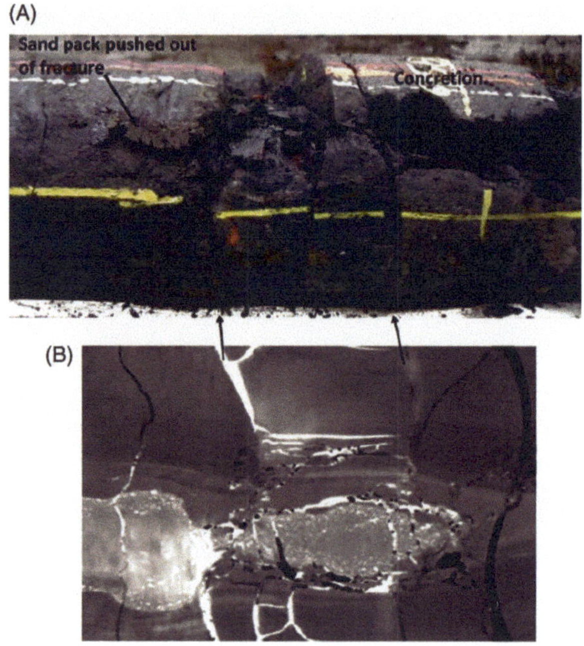

Figure 2.11 ***Fractured shale samples showing the fractures are like to be crumbled.*** (A) Core observation. (B) Computed tomographic scan. *(Gale et al., 2018, reprinted by permission of the AAPG whose permission is required for further use).*

gauges in a vertical well near the horizontal well to monitor the spatial depletion progress; this can provide key insights into understanding the effective fracture height, as shown in Fig. 2.12 (Raterman et al., 2017).

2.3 Distributed acoustic sensing and distributed temperature sensing

Distributed acoustic sensing (DAS) and distributed temperature sensing (DTS) are often applied to monitor the injected fluid distribution at the cluster level, as shown in Figs. 2.13 and 2.14, respectively. Fig. 2.13B displays acoustic energy against pumping time (Fig. 2.13C) for a stage with five clusters, where the red/yellow colors represent higher acoustic intensity, indicating higher injection rate and slurry volume. The interpretation of DAS data can quantitatively predict the distribution of injected fluid into each cluster, as shown in Fig. 2.13A. An uneven fluid distribution was observed for different clusters varying from 33% to 142% of the designed injected volume. The temperature against the injection time for different

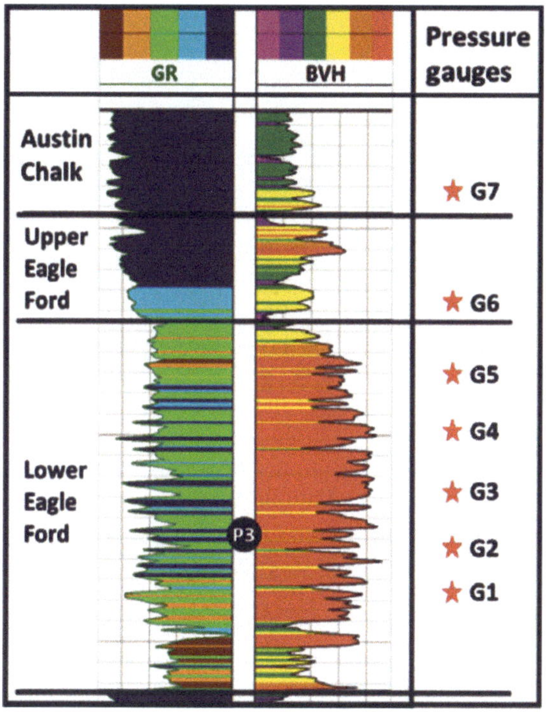

Figure 2.12 ***Fracture height monitoring based on pressure gauges in Eagle Ford*** *(Raterman et al., 2017, reprinted by permission of the AAPG whose permission is required for further use).*

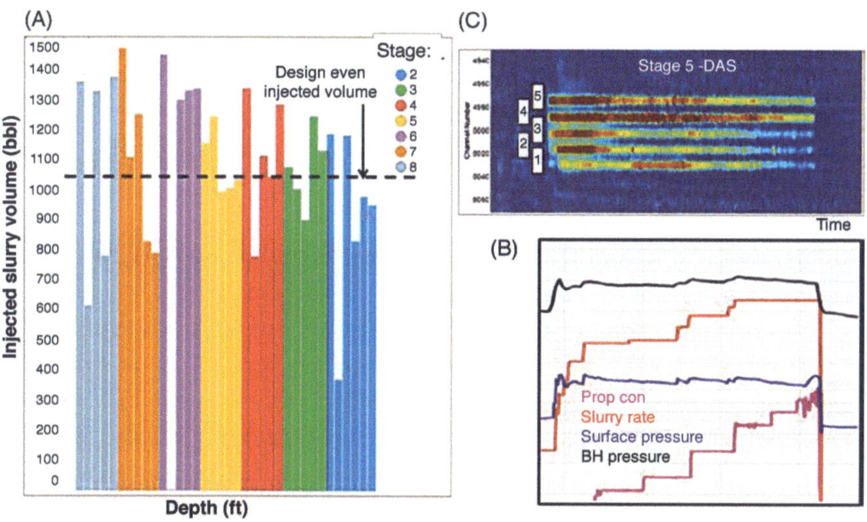

Figure 2.13 ***Distributed acoustic sensing (DAS) monitoring in Eagle Ford.*** *(Raterman et al., 2017, reprinted by permission of the AAPG whose permission is required for further use).*

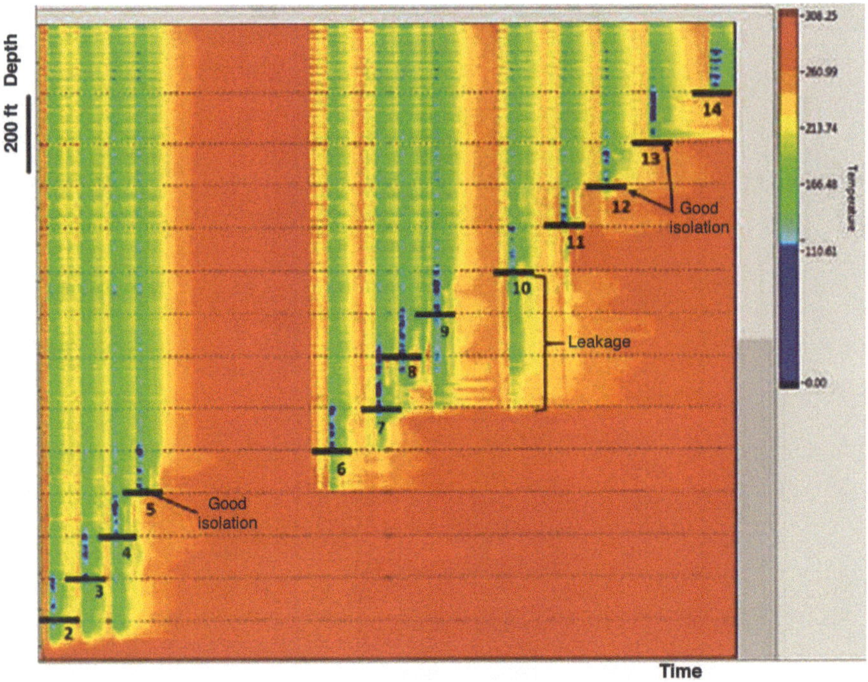

Figure 2.14 ***Distributed temperature sensing (DTS) monitoring in Eagle Ford.*** *(Raterman et al., 2017, reprinted by permission of the AAPG whose permission is required for further use).*

stages was also monitored, as shown in Fig. 2.14. The red color represents higher temperature, and the blue color indicates lower temperature. On the basis of the DTS data analysis for Fig. 2.14, three main observations were made: first, we can locate the fracture initiation points; second, we can qualitatively analyze the intake fracturing fluid for each cluster; lastly, we can monitor the quality of stage isolation. In addition, with the DTS analysis occurring during the production stage, the flow profile along the wellbore can be evaluated. Using the numerical models, the fracture geometry and property may be characterized during the production stage.

The DAS fiber of the monitoring well can also be used to calculate the strain rate during the stimulation of the fracturing well, as shown in Fig. 2.15. As illustrated, the fiber will produce an extended red signal if a hydraulic fracture is approaching the fiber. Once the fracture hits the fiber, the fiber will be compressed and produce a blue signal (Raterman et al., 2017). During the injection time of the fracturing well, new fracture opening (in red) in the monitoring well will be observed, which is

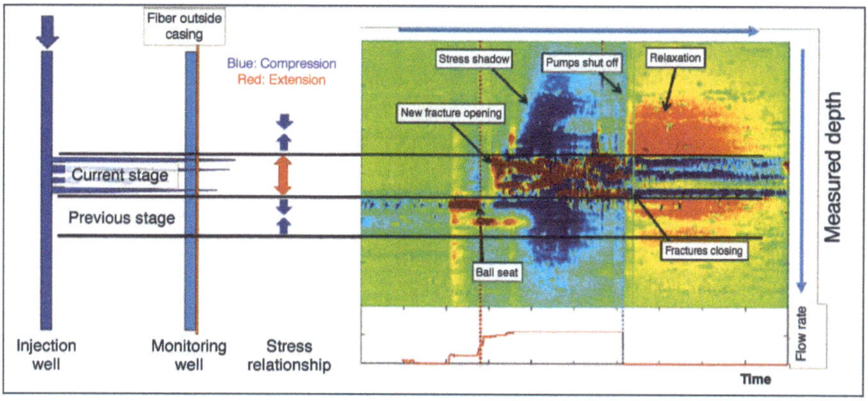

Figure 2.15 ***Strain rate of the DAS fiber in Eagle Ford.*** *(Raterman et al., 2017, reprinted by permission of the AAPG whose permission is required for further use).*

surrounded by the compression signal (in blue) from the formation due to stress shadowing effects. When the pumping stops, fracture closure (in blue) will be observed, which is surrounded by a relaxation of rock deformation. The information about the fracture location and number of fracture hits can be effectively determined. It has been reported that some fractures can extend for 1500 ft in Eagle Ford based on the strain analysis using the DAS fiber (Raterman et al., 2017).

2.4 Complex fracture hits

Fracture hits between parent and child wells result in higher fracture complexity, which has a large impact on the optimization of well spacing and gas huff-n-puff for enhanced oil recovery. The interpretation of ultrasonic wellbore image logs from the child wells can be used to identify the fracture hits from the parent wells (Fiallos et al., 2019, Fiallos Torres et al., 2019). Fig. 2.16 shows an example in Eagle Ford. As can be seen, wells W7H, W8H, and W9H were clearly crossed by a single hydraulic fracture at similar reservoir depths (8,445 ft, 8,457 ft, and 8,458 ft). Accordingly, the number of fracture hits between parent and child wells can be effectively identified. Fig. 2.17 presents a field example of strong well interference due to complex fracture hits in Eagle Ford, which were clearly examined using the interpretation of ultrasonic wellbore image logs. A total number of 28 fracture hits with varying fracture lengths from 750 to 3750 ft were determined and represented in the model, as shown in Fig. 2.18.

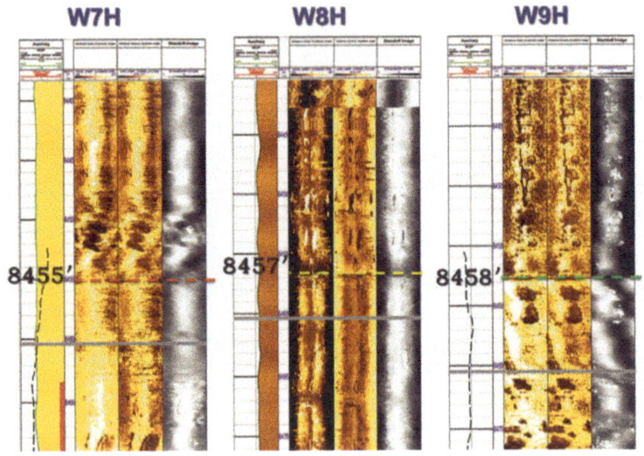

Figure 2.16 ***An example of the interpretation of ultrasonic wellbore image logs to identify fracture hits in Eagle Ford (Fiallos et al., 2019).***

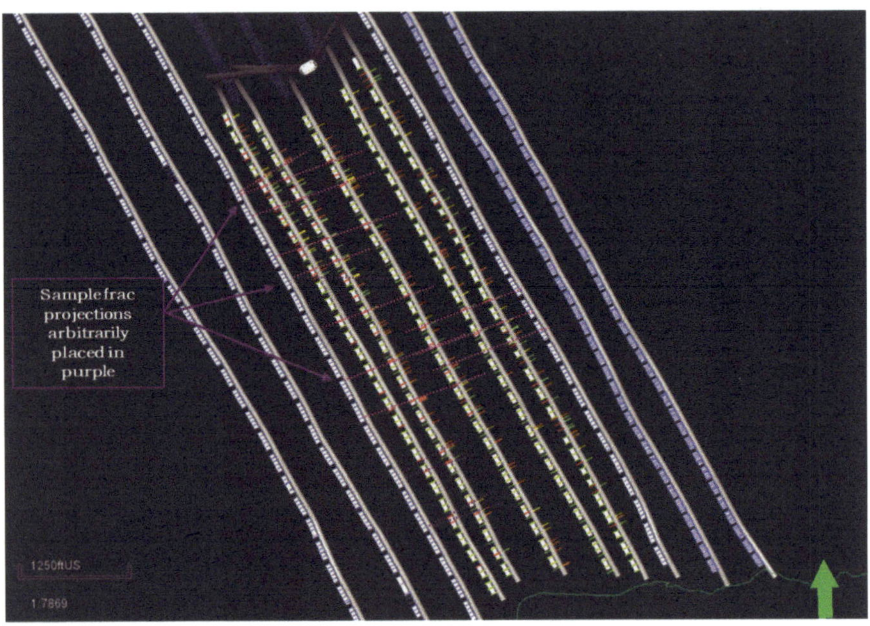

Figure 2.17 ***Fracture hits between outer parent wells and inner child wells in Eagle Ford (Fiallos et al., 2019).***

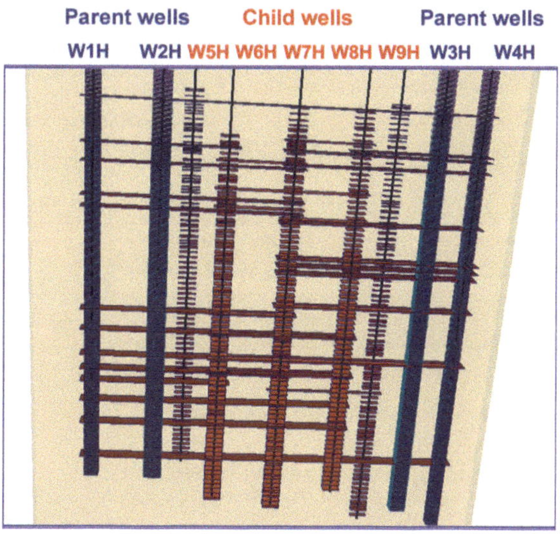

Figure 2.18 ***Twenty-eight fracture hits identified between parent and child wells in Eagle Ford.***

2.5 Three fracture classes

On the basis of the research of natural fracture effects on well productivity in Eagle Ford, Grossi (2015) defined three fracture classes based on the proportion of three fracture types: hydraulic-induced fractures, stratigraphic bound/confined natural fractures (NF_{EF}), and unconfined, non-stratigraphic bound or out of zone natural fractures (NF_{OZ}), as shown in Fig. 2.19. Fracture Class 1 has a dominant fracture type of hydraulically induced fractures with little-or-no effect from natural fractures. Fracture Class 2 has a strong intersection between hydraulic fractures and natural fractures, resulting in the formation of a complex fracture network and a significant increase in fracture surface area. This class yields the best production. Fracture Class 3 has some non-stratigraphic bound fractures within the stimulated reservoir volume. This transfers the stimulation energy through the fault displacement and diverts energy out of the intended zone, resulting in a large amount of unstimulated reservoir volume and yielding poor well production.

Transferring the aforementioned fracture complexity into the reservoir model plays an important role in adequately understanding the non-uniform distribution of the spatial drainage area and long-term production

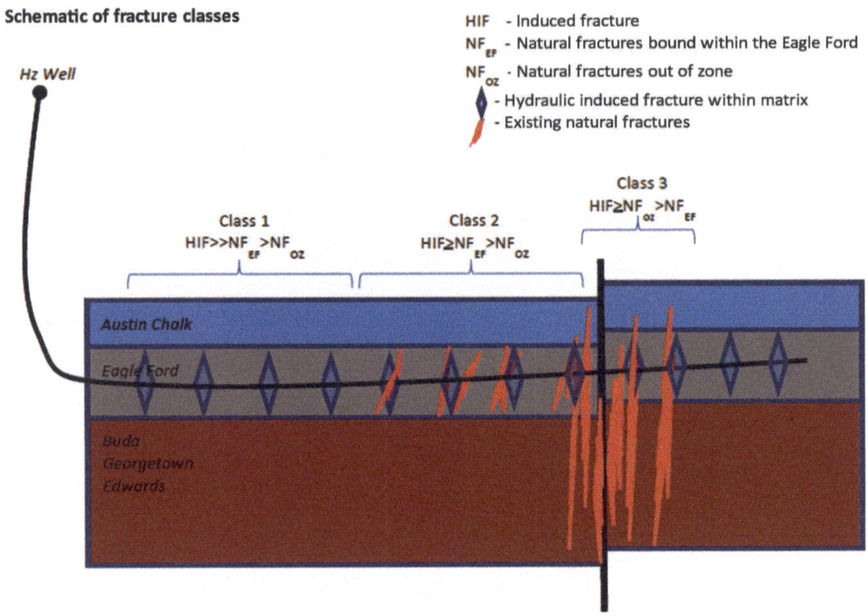

Figure 2.19 ***Schematic of three fracture classes in Eagle Ford.*** *(Grossi, 2015, reprinted by permission of the AAPG whose permission is required for further use).*

forecasting (Raterman et al., 2017). The simplified and widely used uniform planar fracture geometries make it difficult to accurately and efficiently capture the flow behavior from shale matrix to complex fractures and finally to the wellbore. In many cases, it is necessary to explicitly handle the fracture complexity through discrete fracture models in the production simulation of shale reservoirs, which will be covered in this book.

References

Ciezobka, J., Courtier, J., Wicker, J., 2018. Hydraulic Fracturing Test Site (HFTS) - project overview and summary of results. Paper URTeC 2937168, Pages 9, SPE/AAPG/SEG Unconventional Resources Technology Conference. Houston, Texas, http://archives.datapages.com/data/urtec/2018/2937168.html.

Fiallos, M.X., Yu, W., Ganjdanesh, R., Kerr, E., Sepehrnoori, K., Miao, J., Ambrose, R., 2019. Modeling interwell interference due to complex fracture hits in Eagle Ford using EDFM. Paper IPTC 19468, International Petroleum Technology Conference. Beijing, China.

Fiallos Torres, M.X., Yu, W., Ganjdanesh, R., Kerr, E., Sepehrnoori, K., Miao, J., Ambrose, R., 2019. Modeling interwell fracture interference and Huff-N-Puff pressure containment in Eagle Ford using EDFM. Paper SPE195240, SPE Oklahoma City Oil and Gas Symposium. Oklahoma City, Oklahoma.

Fisher, M.K., Warpinski, N.R., 2012. Hydraulic-fracture-height growth: real data. SPE Prod. Oper. 27 (1), 8–19.

Gale, J.F.W., Elliott, S.J., Laubach, S.E., 2018. Hydraulic fractures in core from stimulated reservoirs: core fracture description of HFTS slant core, Midland Basin, West Texas. Paper URTeC 2902624, Pages 18, SPE/AAPG/SEG Unconventional Resources Technology Conference. Houston, Texas, http://archives.datapages.com/data/urtec/2018/2902624.html.

Grossi, P., 2015. Investigating natural fracture effects on well productivity: Eagle Ford, La Salle County, Texas. Paper URTeC 2148347, Pages 13, Unconventional Resources Technology Conference. San Antonio, Texas, http://archives.datapages.com/data/urtec/2015/2148347.html.

Maity, D., 2018. Microseismicity analysis for HFTS pad and correlation with completion parameters. Paper URTeC 2902355, SPE/AAPG/SEG Unconventional Resources Technology Conference. Houston, Texas.

Maity, D., Ciezobka, J., Eisenlord, S., 2018. Assessment of in-situ proppant placement in SRV using through-fracture core sampling at HFTS. Paper URTeC 2902364, SPE/AAPG/SEG Unconventional Resources Technology Conference. Houston, Texas.

Miao, J., Yu, W., Xia, Z., Zhao, W., Xu, Y., Sepehrnoori, K., 2018. An easy and fast EDFM method for production simulation in shale reservoirs with complex fracture geometries. Paper ARMR-DFNE 18-0926, 2nd International Discrete Fracture Network Engineering Conference. Seattle, Washington.

Rassenfoss, S., 2018. A look into what fractures really look like. J. Pet. Technol. 70 (11), 28–36.

Raterman, K.T., Farrell, H.E., Mora, O.S., Janssen, A.L., Gomez, G.A., Busetti, S., McEwen, J., Davidson, M., Friehauf, K., Rutherford, J., Reid, R., Jin, G., Roy, B., Warren, M., 2017. Sampling a stimulated rock volume: an Eagle Ford example. Paper URTeC 2670034, Pages 18, SPE/AAPG/SEG Unconventional Resources Technology Conference. Austin, Texas, http://archives.datapages.com/data/urtec/2017/2670034.html.

Stegent, N., Candler, C., 2018. Downhole microseismic mapping of more than 400 fracturing stages on a multiwell pad at the hydraulic fracturing test site (HFTS): discussion of operational challenges and analytic results. Paper URTeC 2902311, Pages 28, SPE/AAPG/SEG Unconventional Resources Technology Conference. Houston, Texas, http://archives.datapages.com/data/urtec/2018/2902311.html.

Vermylen, J., Zoback, M.D., 2011. Hydraulic fracturing, microseismic magnitudes, and stress evolution in the Barnett shale, Texas, USA. Paper SPE 140507, SPE Hydraulic Fracturing Technology Conference. The Woodlands, Texas.

Wood, T., Leonard, R., Senters, C., Squires, C., Perry, M., 2018. Interwell communication study of UWC and MWC wells in the HFTS. Paper URTeC 2902960, Pages 10, SPE/AAPG/SEG Unconventional Resources Technology Conference. Houston, Texas, http://archives.datapages.com/data/urtec/2018/2902960.html.

Yu, W., Wu, K., Liu, M., Sepehrnoori, K., Miao, J., 2018. Production forecasting for shale gas reservoirs with nanopores and complex fracture geometries using an innovative non-intrusive EDFM method. Paper SPE 191666, SPE Annual Technical Conference and Exhibition. Dallas, Texas.

CHAPTER THREE

Numerical approaches for modeling complex fractures

3.1 Dual-continuum models

The concept of dual-continuum models was developed by Barenblatt et al. (1960). Compared with single-porosity models, dual-continuum models treat the reservoir as two intercommunicating systems, one with high storage capacity, representing the matrix, and the other with high flow capacity, representing the fractures. Warren and Root (1963) applied this idea to well test studies of naturally fractured reservoirs. Their "dual-porosity model" ignored the flow between matrix blocks. A "sugar-cube" idealization of the porous media was made, as shown in Fig. 3.1. In this idealization, the matrix system was represented as an array of identical, rectangular parallelepipeds, and the fractures were represented as an orthogonal system, where each fracture is perpendicular to one of the principal axes of permeability. Two sets of properties were defined at each point, and the equations describing pseudo-steady-state, single-phase flow in this system are

$$\frac{k_{fx}}{\mu}\frac{\partial^2 p_f}{\partial x^2}+\frac{k_{fy}}{\mu}\frac{\partial^2 p_f}{\partial y^2}-\phi_m c_m\frac{\partial p_m}{\partial t}=\phi_f c_f\frac{\partial p_f}{\partial t}, \tag{3.1}$$

$$\phi_m c_m\frac{\partial p_m}{\partial t}=\frac{\sigma k_m}{\mu}(p_f-p_m), \tag{3.2}$$

where k_{fx} and k_{fy} are the fracture permeability in the x- and y-directions, respectively, μ is the fluid viscosity, p_f is the pressure in fracture, ϕ_m is the matrix porosity, c_m is the matrix compressibility, p_m is the pressure in the matrix, ϕ_f is the fracture porosity, c_f is the fracture compressibility, t is time, k_m is the matrix permeability, and σ is called shape factor. σ describes the communication between the two continua. The expression given by Warren and Root (1963) is

$$\sigma=\frac{4N(N+2)}{L^2}, \tag{3.3}$$

where N is the number of normal sets of fractures and L is the characteristic length of the block. For three-dimensional cases where $N = 3$,

Embedded Discrete Fracture Modeling and Application in Reservoir Simulation.
http://dx.doi.org/10.1016/B978-0-12-821872-3.00003-4

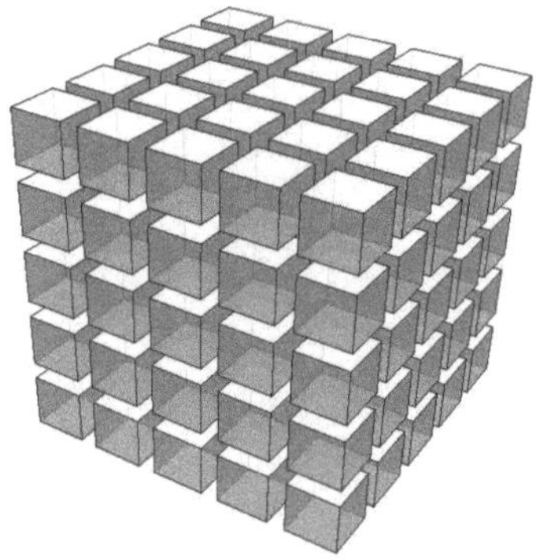

Figure 3.1 ***Illustration of the "sugar-cube" idealization in the dual-porosity model. Xu (2018).***

$$L = \frac{3L_x L_y L_z}{L_x L_y + L_y L_z + L_x L_z}, \tag{3.4}$$

where L_x, L_x, and L_x are the dimensions of each matrix block in the x-, y-, and z-directions.

Later, Kazemi et al. (1976) obtained an expression for shape factor in finite-difference simulations:

$$\sigma = 4\left(\frac{1}{L_x^2} + \frac{1}{L_y^2} + \frac{1}{L_z^2}\right), \tag{3.5}$$

where L_x, L_y and L_z have the same definitions as in Eq. (3.4). Eq. (3.5) was widely used in the later literature.

In Eqs. (3.1) and (3.2), only single-phase flow is considered, and several mechanisms impacting the fluid transport between the matrix and the fracture systems are not taken into account, including the gravitational effect, capillary effect, and diffusion. After Warren and Root (1963), the dual-porosity model was extended to two-phase and three-phase flows and implemented in different simulators to consider these effects (Kazemi et al., 1976; Rossen, 1977; Thomas et al., 1983; Rossen and Shen, 1989; Tarahhom, 2008).

The matrix–fracture fluid exchange in Eq. (3.2) ignores the gravity term. To perform simulations where the gravitational effect is of great importance, Gilman and Kazemi (1983) assigned different depths for the matrix and the fracture in the same location to add an explicit gravity term to the flow equation. However, they assumed a constant gravity head for the blocks. Later, Litvak (1986) introduced a time-dependent gravity term. Bossie-Codreanu et al. (1985) used the concept in the dual-porosity model and represented each elementary volume of the fractured reservoir as several cells, one for the matrix, and the others for fractures. As in Warren and Root's (1963) model, the fracture cells offered a continuous flow path, and the matrix cells were separated by the fracture cells. The gravitational and capillary effects could be directly simulated with this approach.

Another way to simulate the gravitational and capillary effects is to introduce pseudo-parameters. Thomas et al. (1983) used the concept of pseudo-relative permeability and pseudo-capillary pressure curves to account for capillary and gravitational forces. Dean and Lo (1988) used a pseudo-capillary pressure curve obtained from fine-grid simulation to account for the gravitational effect. Rossen and Shen (1989) used pseudo-capillary pressure curves for both the matrix and the fracture. The usage of pseudo-relative permeability and pseudo-capillary-pressure curves removes the requirement to include gravitational forces explicitly, but it involves the generation of pseudo-curves. As the pseudo-curves depend on many different factors, such as fluid composition, flow condition, and field history (Lemonnier and Bourbiaux, 2010), it is challenging to find a general way to generate such curves.

Improvements regarding the assumption of pseudo-steady-state flow were also made. Beckner et al. (1988) developed a dynamic transfer function where the matrix–fracture fluid transport was correlated with an "effective exposure time" of the matrix blocks to fracture water. The purpose was to simulate the transverse imbibition with advancing water level. Lim and Aziz (1995) combined analytical solutions of pressure diffusion equations for different flow geometries to derive the shape factor for single-phase flow, and the pseudo-steady-state assumption was removed. Sarma and Aziz (2006) derived an analytical expression to calculate the shape factor for transient pressure diffusion. They also proposed a technique to compute the shape factor for any arbitrary matrix-fracture geometry, and the technique accounted for both transient and pseudo-steady-state flow. However, for general cases, it is still difficult to derive a transfer function that is only dependent on time and valid for all different stages of field development (Lemonnier and Bourbiaux, 2010).

Subgridding techniques were also proposed to improve the accuracy of dual-porosity simulations (Saidi, 1983; Pruess and Narasimhan, 1985; Wu and Pruess, 1988; Aldejain, 1999). In the standard dual-porosity model, the property distribution within each matrix block is not considered. For cases with a large pressure drop rate or large block sizes, or when simulating certain processes such as gravity drainage and diffusion, not considering the property distribution may lead to large inaccuracies (Saidi, 1983). The purpose of using subgridding techniques is to improve the representation of pressure and saturation distribution in the matrix block. Saidi (1983) used gridded matrix blocks in his study and calculated the matrix–fracture fluid transfer using a transient concept. By using a discretization of the matrix blocks in both the radial and the vertical directions, he successfully matched 22 years' production history of the Haft Kel Field. Another subgridding technique called "multiple interacting continua" (MINC) (Pruess and Narasimhan, 1982; Pruess and Narasimhan, 1985; Wu and Pruess, 1988) was also proposed, where nested control volumes were created, as shown in Fig. 3.2. The assumption is that the property change in a matrix block is mainly affected by its distance from the nearest fractures. The benefit of this approach is that the matrix–fracture flow in a two-dimensional fractured media can be treated as one-dimensional flow, and the number of gridblocks introduced by MINC is lower than that using explicit discretization (as shown in Fig. 3.2C). The MINC method was widely applied in the later literature (Chen et al., 1987; Beckner et al., 1991; Naimi-Tajdar et al., 2007; Jiang and Younis, 2016).

The dual-porosity model was also extended to include the flow between matrix blocks (Blaskovich et al., 1983; Hill and Thomas, 1985) when such type of flow cannot be ignored in the reservoir. The extended model is called the dual-permeability model. This method is more computationally expensive than the dual-porosity model as it requires a linear system twice as large as that in the dual-porosity model (Dean and Lo, 1988). However, it is more general and can be used to simulate a wide range of fractured reservoirs. Currently, the dual-permeability model is implemented in most of commercial simulators.

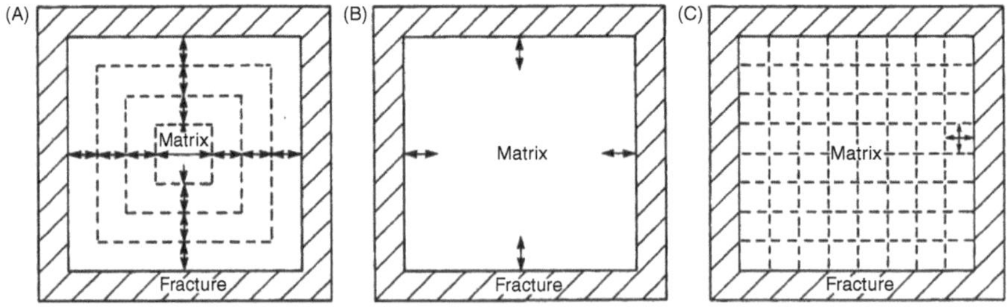

Figure 3.2 ***Matrix discretization using different methods (Wu and Pruess, 1988).*** (A) MINC. (B) Dual-porosity. (C) Explicit discretization.

Dual-porosity and dual-permeability models are effective for modeling naturally fractured reservoirs with high fracture density, where the locations or geometries of individual fractures are ignored, and a well-connected fracture system is assumed. They are extremely useful for simulating reservoirs where the flow response in the matrix is delayed to some extent compared with that in the fracture network (Lemonnier and Bourbiaux, 2010). However, the evaluation of the transfer function for different processes in the dual-porosity and dual-permeability models is generally challenging (Karimi-Fard et al., 2004; Sandve et al., 2012). Also, limitations exist for these models when simulating disconnected fractures, a small number of large-scale fractures, or highly localized anisotropy caused by natural fractures (Karimi-Fard and Firoozabadi, 2003; Karimi-Fard et al., 2004; Moinfar et al., 2011). In such cases, the contribution of individual fractures becomes significant, and it is desirable and sometimes necessary to use an explicit representation of each fracture in the simulation model for reliable simulation results.

3.2 Discrete fracture models

Discrete fracture models (DFMs) were developed to simulate the impact of individual fractures explicitly. Most DFMs use conforming meshes, where the edges or faces of the matrix blocks match with the surfaces of the fractures. When fracture width is ignored in the grid generation process, the fractures can be represented as elements with a lower dimension than the matrix elements.

Because of the usage of conforming meshes in most DFMs, the permeability values of fractures can be directly assigned to blocks at the fracture locations for an explicit representation of fractures. When the fracture plane is perpendicular to one of the axes of the matrix grid, a structured grid can be used as long as careful grid refinement around the fracture is conducted. Cipolla et al. (2011) summarized the gridding techniques in structured grids for the explicit representation of fractures, including the usage of nested local grid refinement (LGR), logarithmic LGR, and global tartan grid. The usage of these techniques is limited by their capability to represent fractures with arbitrary orientations. Unstructured gridding techniques were often used to simulate such fractures (Noorishad and Mehran, 1982; Matthäi et al., 2007; Marcondes et al., 2010; Karimi-Fard and Durlofsky, 2016; Ahmed et al., 2017; Hui et al., 2018; Fumagalli and Keilegavlen, 2019).

In unstructured-gridding-based DFMs, different numerical schemes have been used, such as the finite-element method (Noorishad and Mehran, 1982; Kim and Deo, 2000), the finite-volume method (Karimi-Fard et al., 2004; Ahmed et al., 2017), the mimetic finite-difference method (Antonietti et al., 2016; Formaggia et al., 2018), and the virtual finite-element method (Benedetto et al., 2016; Fumagalli and Keilegavlen, 2019).

Using the finite-element method, Noorishad and Mehran (1982) developed a DFM to simulate two-dimensional solute transport problems. The fractures were represented as two-nodal point elements. Similarly, Baca et al. (1984) developed a DFM based on the finite-element method to simulate single-phase flow in two-dimensional fractured media, and the fractures were treated as line elements. Juanes et al. (2002) provided a general finite-element formulation applicable to elements of any dimension. They used the formulation for the study of fractured porous media, where in three-dimensional media, the fractures are represented as two-dimensional surfaces. For the simulation of multiphase flow, Kim and Deo (2000) and Karimi-Fard and Firoozabadi (2003) developed DFMs based on the finite-element method, where triangular elements were used for the matrix, and line elements were used for the fractures (as shown in Fig. 3.3). The models were used to simulate water–oil two-phase flow in fractured media.

The mixed-finite-element (MFE) method was also used in DFMs. Martin et al. (2005) developed a DFM based on the MFE method to simulate

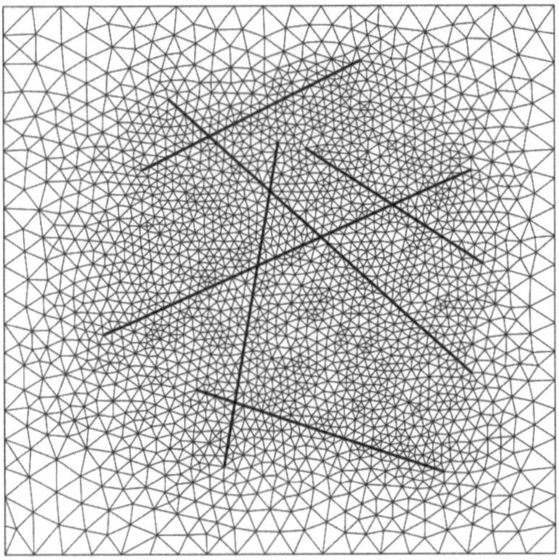

Figure 3.3 ***Discretization of the fractured porous media in the DFM (Karimi-Fard and Firoozabadi 2003).***

single-phase flow with a limited number of fractures. Hoteit and Firoozabadi (2006) developed a DFM where the discontinuous Galerkin (DG) method was combined with the MFE method. The MFE method was used to approximate the pressure and velocity along the fracture, and the DG method was used to approximate the species-balance equations. The model was used for two-dimensional, two-phase compositional simulations.

As many reservoir simulators are based on the finite-volume formulations, DFMs were also developed using the finite-volume scheme to maintain compatibility with existing simulators. It is also easier to maintain local mass conservation and get a clear upwind criterion with this method (Monteagudo and Firoozabadi, 2004). In some early work, Bastian et al. (2000) and Granet et al. (2001) used the finite-volume scheme to develop DFMs for two-dimensional, two-phase simulations. These are simulations where the fractures were represented by linear elements, and the matrix was represented by triangular elements. A similar approach was used by Marcondes et al. (2010). Monteagudo and Firoozabadi (2004) extended this approach to three-dimensional problems, where the matrix was discretized by tetrahedron elements, and the fractures were discretized by triangular elements. The effects of gravity and capillary pressure were also included in their two-phase simulations. Later, Fu et al. (2005), using an idea similar to Monteagudo and Firoozabadi's (2004), discretized the matrix and fractures and implemented the DFM in a simulator using the control-volume finite-element method. The simulator was used for three-dimensional, three-phase simulations with discrete fractures. A similar approach was used by Matthäi et al. (2005), but with an "indirect hybrid meshing," where the matrix was discretized as a combination of tetrahedra, hexahedra, pyramids, and prisms. The hybrid meshing method was also used by Matthäi et al. (2007) and Geiger et al. (2009).

As mentioned earlier, a great benefit of using the finite-volume method in DFMs is the compatibility with existing simulators. To make full use of this advantage, Karimi-Fard et al. (2004) developed a DFM compatible with any connectivity-list-based simulator. As nearly orthogonal grids were used in their model, the two-point flux approximation scheme could be applied with an isotropic permeability. For non-K-orthogonal grids, DFMs based on the multi-point flux approximation were developed (Sandve et al., 2012; Gläser et al., 2017; Ahmed et al., 2017).

For unstructured-gridding-based DFMs, the improvement in gridding techniques in recent years allowed for the representation of more realistic fracture geometries. Cipolla et al. (2011) presented simulation examples

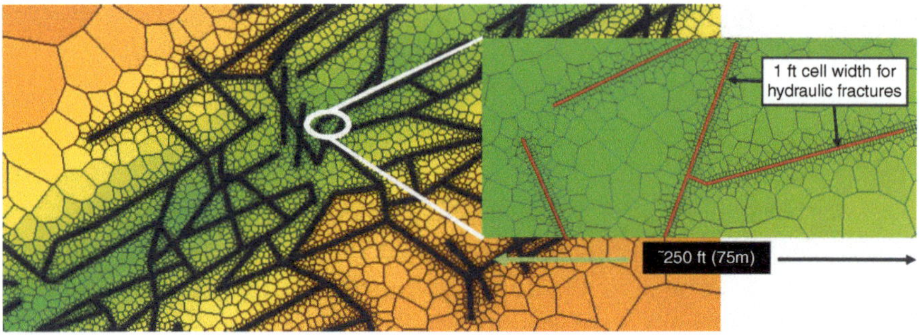

Figure 3.4 ***An unstructured grid used to simulate complex hydraulic fractures (Cipolla et al., 2011).***

with complex hydraulic fracture geometries interpreted from microseismic measurements, and the simulation grid used in their study is shown in Fig. 3.4. Karimi-Fard and Durlofsky (2016) developed a DFM based on Karimi-Fard et al.'s (2004) work to simulate fractures with irregular geometries, such as intersecting nonplanar fractures. Tetrahedron elements were used for the matrix, and triangular elements were used for fractures (as shown in Fig. 3.5). In these models, to maintain computational efficiency as the number of fractures and the complexity of fracture geometries increase, upscaling techniques could be used to create coarse-scale models with lower numbers of control volumes (Hui et al., 2013; Karimi-Fard and Durlofsky, 2016; Hui et al., 2018).

Figure 3.5 ***Triangular discretization of fractures (Karimi-Fard and Durlofsky, 2016).***

3.3 Limitations of unstructured gridding based DFMs

Although DFMs based on unstructured gridding are able to capture the detailed geometry of each fracture accurately, several disadvantages still exist in these models for solving large-scale, realistic problems. First, it is still complicated to generate a conforming mesh for a complex fracture network, especially when considering the intersection angle between fractures in general three-dimensional cases (Fumagalli et al., 2017). Second, the generation of mesh could introduce many small elements that may limit the time step and affect the quality of the numerical solution. Third, the number of matrix blocks is generally large due to the discretization around fractures. The resulting linear system is typically hard to solve despite the development of solvers and parallel computing in recent years, which leads to a low computational efficiency (Du et al., 2017; Fumagalli et al., 2017). Lastly, as many commercial simulators only support structured gridding and a large number of reservoir models have already been built with structured grids, the use of unstructured grids in these models requires significant effort in simulator development and model re-gridding. All of these aspects limit the field applications of unstructured-gridding-based DFMs. In recent years, DFMs using non-conforming meshes have been developed in an effort to tackle these challenges (Huang et al., 2011; D'Angelo and Scotti, 2012; Berrone et al., 2013; Moinfar et al., 2014; Schwenck et al., 2015). As one of these models, the embedded discrete fracture model has gained much attention in recent years and will be discussed in the next chapters.

References

Ahmed, R., Edwards, M.G., Lamine, S., Huisman, B.A.H., Pal, M., 2017. CVD-MPFA full pressure support, coupled unstructured discrete fracture-matrix Darcy-flux approximations. J. Comput. Phys. 349, 265–299.

Aldejain, A.A., 1999. Implementation of a dual porosity model in a chemical flooding simulator. PhD Dissertation. The University of Texas at Austin.

Antonietti, P.F., Formaggia, L., Scotti, A., Verani, M., Verzott, N., 2016. Mimetic finite difference approximation of flows in fractured porous media. ESAIM-Math. Model. Num. 50 (3), 809–832.

Baca, R.G., Arnett, R.C., Langford, D.W., 1984. Modelling fluid flow in fractured-porous rock masses by finite-element techniques. Int. J. Numer. Methods Fluids 4 (4), 337–348.

Barenblatt, G.I., Zheltov, I.P., Kochina, I.N., 1960. Basic concepts in the theory of seepage of homogeneous liquids in fissured rocks. J. Appl. Math. Mech. 24 (5), 1286–1303.

Bastian, P., Chen, Z., Ewing, R.E., Helmig, R., Jakobs, H., Reichenberger, V., 2000. Numerical simulation of multiphase flow in fractured porous media. Numerical Treatment of Multiphase Flows in Porous Media: 50-68. Springer, Berlin.

Beckner, B.L., Chan, H.M., McDonald, A.E., Wooten, S.O., Jones, T.A., 1991. Simulating naturally fractured reservoirs using a subdomain method. Paper SPE 21241, SPE Symposium on Reservoir Simulation. Anaheim, California.

Beckner, B.L., Firoozabadi, A., Aziz, K., 1988. Modeling transverse imbibition in double-porosity simulators. Paper SPE 17414, SPE California Regional Meeting. Long Beach. California.

Benedetto, M.F., Berrone, S., Borio, A., Pieraccini, S., Scialò, S., 2016. A hybrid mortar virtual element method for discrete fracture network simulations. J. Comput. Phys. 306, 148–166.

Berrone, S., Pieraccini, S., Scialo, S., 2013. On simulations of discrete fracture network flows with an optimization-based extended finite element method. SIAM J. Sci. Comput. 35 (2), A908–A935.

Blaskovich, F.T., Cain, G.M., Sonier, F., Waldren, D., Webb, S.J., 1983. A multicomponent isothermal system for efficient reservoir simulation. Paper SPE 11480, Middle East Oil Technical Conference and Exhibition. Manama, Bahrain.

Bossie-Codreanu, D., Bia, P.R., Sabathier, J.-C., 1985. The "checker model," an improvement in modeling naturally fractured reservoirs with a tridimensional, triphasic, black-oil numerical model. SPE J. 25 (5), 743–756.

Chen, W.H., Wasserman, M.L., Fitzmorris, R.E., 1987. A thermal simulator for naturally fractured reservoirs. Paper SPE 16008, SPE Symposium on Reservoir Simulation. San Antonio, Texas.

Cipolla, C.L., Fitzpatrick, T., Williams, M.J., Ganguly, U.K., 2011. Seismic-to-simulation for unconventional reservoir development. Paper SPE 146876, SPE Reservoir Characterisation and Simulation Conference and Exhibition. Abu Dhabi, UAE.

D'Angelo, C., Scotti, A., 2012. A mixed finite element method for Darcy flow in fractured porous media with non-matching grids. ESAIM-Math. Model. Num. 46 (2), 465–489.

Dean, R.H., Lo, L.L., 1988. Simulations of naturally fractured reservoirs. SPE Reservoir Eng. 3 (2), 638–648.

Du, S., Liang, B., Yuanbo, L., 2017. Field study: embedded discrete fracture modeling with artificial intelligence in Permian basin for shale formation. Paper SPE 187202, SPE Annual Technical Conference and Exhibition. San Antonio, Texas.

Formaggia, L., Scotti, A., Sottocasa, F., 2018. Analysis of a mimetic finite difference approximation of flows in fractured porous media. ESAIM-Math. Model. Num. 52 (2), 595–630.

Fu, Y., Yang, Y.-K., Deo, M., 2005. Three-dimensional, three-phase discrete-fracture reservoir simulator based on control volume finite element (CVFE) formulation. Paper SPE 93292, SPE Reservoir Simulation Symposium. The Woodlands, Texas.

Fumagalli, A., Keilegavlen, E., 2019. Dual virtual element methods for discrete fracture matrix models. Oil Gas Sci. Technol. 74, 41.

Fumagalli, A., Zonca, S., Formaggia, L., 2017. Advances in computation of local problems for a flow-based upscaling in fractured reservoirs. Math. Comput. Simul. 137, 299–324.

Geiger, S., Matthäi, S.K., Niessner, J., Helmig, R., 2009. Black-oil simulations for three-component, three-phase flow in fractured porous media. SPE J. 14 (2), 338–354.

Gilman, J.R., Kazemi, H., 1983. Improvements in simulation of naturally fractured reservoirs. SPE J. 23 (4), 695–707.

Gläser, D., Helmig, R., Flemisch, B., Class, H., 2017. A discrete fracture model for two-phase flow in fractured porous media. Adv. Water Resour. 110, 335–348.

Granet, S., Fabrie, P., Lemonnier, P., Quintard, M., 2001. A two-phase flow simulation of a fractured reservoir using a new fissure element method. J. Pet. Sci. Eng. 32 (1), 35–52.

Hill, A.C., Thomas, G.W., 1985. A new approach for simulating complex fractured reservoirs. Paper SPE 13537, Middle East Oil Technical Conference and Exhibition, Bahrain.

Hoteit, H., Firoozabadi, A., 2006. Compositional modeling of discrete-fractured media without transfer functions by the discontinuous Galerkin and mixed methods. SPE J. 11 (3), 341–352.

Huang, H., Long, T.A., Wan, J., Brown, W.P., 2011. On the use of enriched finite element method to model subsurface features in porous media flow problems. Comput. Geosci. 15 (4), 721–736.

Hui, M.-H., Karimi-Fard, M., Mallison, B., Durlofsky, L.J., 2018. A general modeling framework for simulating complex recovery processes in fractured reservoirs at different resolutions. SPE J. 23 (2), 598–613.
Hui, M.-H., Mallison, B.T., Fyrozjaee, M.H., Narr, W., 2013. The upscaling of discrete fracture models for faster, coarse-scale simulations of IOR and EOR processes for fractured reservoirs. Paper SPE 166075, SPE Annual Technical Conference and Exhibition. New Orleans, Louisiana.
Jiang, J., Younis, R.M., 2016. Hybrid coupled discrete-fracture/matrix and multicontinuum models for unconventional-reservoir simulation. SPE J. 21 (3), 1009–1027.
Juanes, R., Samper, J., Molinero, J., 2002. A general and efficient formulation of fractures and boundary conditions in the finite element method. Int. J. Numer. Methods Eng. 54 (12), 1751–1774.
Karimi-Fard, M., Durlofsky, L.J., 2016. A general gridding, discretization, and coarsening methodology for modeling flow in porous formations with discrete geological features. Adv. Water Resour. 96, 354–372.
Karimi-Fard, M., Durlofsky, L.J., Aziz, K., 2004. An efficient discrete-fracture model applicable for general-purpose reservoir simulators. SPE J. 9 (2), 227–236.
Karimi-Fard, M., Firoozabadi, A., 2003. Numerical simulation of water injection in fractured media using the discrete-fracture model and the Galerkin method. SPE Reservoir Eval. Eng. 6 (2), 117–126.
Kazemi, H., Merrill, Jr., L.S., Porterfield, K.L., Zeman, P.R., 1976. Numerical simulation of water-oil flow in naturally fractured reservoirs. SPE J. 16 (6), 317–326.
Kim, J.-G., Deo, M.D., 2000. Finite element, discrete-fracture model for multiphase flow in porous media. AIChE J. 46 (6), 1120–1130.
Lemonnier, P., Bourbiaux, B., 2010. Simulation of naturally fractured reservoirs. State of the art. Oil Gas Sci. Technol. 65 (2), 239–262.
Lim, K.T., Aziz, K., 1995. Matrix-fracture transfer shape factors for dual-porosity simulators. J. Pet. Sci. Eng. 13 (3–4), 169–178.
Litvak, B.L., 1986. Simulation and characterization of naturally fractured reservoirs. Reservoir Characterization, 561-584. Academic Press, Orlando.
Marcondes, F., Varavei, A., Sepehrnoori, K., 2010. An element-based finite-volume method approach for naturally fractured compositional reservoir simulation. 13th Brazilian Congress of Thermal Sciences and Engineering-ENCIT. Uberlândia, MG, Brazil.
Martin, V., Jaffré, J., Roberts, J.E., 2005. Modeling fractures and barriers as interfaces for flow in porous media. SIAM J. Sci. Comput. 26 (5), 1667–1691.
Matthäi, S.K., Geiger, S., Roberts, S.G., Paluszny, A., Belayneh, M., Burri, A., Mezentsev, A., Lu, H., Coumou, D., Driesner, T., Heinrich, C.A., 2007. Numerical simulation of multiphase fluid flow in structurally complex reservoirs. Geol. Soc., London, Spec. Publ. 292 (1), 405–429.
Matthäi, S., Menzentsev, A., Belayneh, M., 2005. Control-volume finite-element two-phase flow experiments with fractured rock represented by unstructured 3D hybrid meshes. Paper SPE 93341, SPE Reservoir Simulation Symposium. The Woodlands, Texas.
Moinfar, A., Narr, W., Hui, M.-H., Mallison, B.T., Lee, S.H., 2011. Comparison of discrete-fracture and dual-permeability models for multiphase flow in naturally fractured reservoirs. Paper SPE 142295, SPE Reservoir Simulation Symposium. The Woodlands, Texas.
Moinfar, A., Varavei, A., Sepehrnoori, K., Johns, R.T., 2014. Development of an efficient embedded discrete fracture model for 3D compositional reservoir simulation in fractured reservoirs. SPE J. 19 (2), 289–303.
Monteagudo, J.E.P., Firoozabadi, A., 2004. Control-volume method for numerical simulation of two-phase immiscible flow in two- and three-dimensional discrete-fractured media. Water Resour. Res. 40 (7), 1–20.

Naimi-Tajdar, R., Sepehrnoori, K., Delshad, M., 2007. Matrix subgridding and its effects in dual porosity simulators. Paper IPTC 11195, International Petroleum Technology Conference. Dubai, UAE.

Noorishad, J., Mehran, M., 1982. An upstream finite element method for solution of transient transport equation in fractured porous media. Water Resour. Res. 18 (3), 588–596.

Pruess, K., Narasimhan, T.N., 1982. On fluid reserves and the production of superheated steam from fractured, vapor-dominated geothermal reservoirs. J. Geophys. Res. 87 (B11), 9329–9339.

Pruess, K., Narasimhan, T.N., 1985. A practical method for modeling fluid and heat flow in fractured porous media. SPE J. 25 (1), 14–26.

Rossen, R.H., 1977. Simulation of naturally fractured reservoirs with semi-implicit source terms. SPE J. 17 (3), 201–210.

Rossen, R.H., Shen, E.I.C., 1989. Simulation of gas/oil drainage and water/oil imbibition in naturally fractured reservoirs. SPE Reservoir Eval. Eng. 4 (4), 464–470.

Saidi, A.M., 1983. Simulation of naturally fractured reservoirs. Paper SPE 12270, SPE Reservoir Simulation Symposium. San Francisco, California.

Sandve, T.H., Berre, I., Nordbotten, J.M., 2012. An efficient multi-point flux approximation method for discrete fracture–matrix simulations. J. Comput. Phys. 231 (9), 3784–3800.

Sarma, P., Aziz, K., 2006. New transfer functions for simulation of naturally fractured reservoirs with dual porosity models. SPE J. 11 (3), 328–340.

Schwenck, N., Flemisch, B., Helmig, R., Wohlmuth, B.I., 2015. Dimensionally reduced flow models in fractured porous media: crossings and boundaries. Comput. Geosci. 19 (6), 1219–1230.

Tarahhom, F., 2008. Development of an implicit full-tensor dual porosity compositional reservoir simulator. PhD Dissertation. The University of Texas at Austin.

Thomas, L.K., Dixon, T.N., Pierson, R.G., 1983. Fractured reservoir simulation. SPE J. 23 (1), 42–54.

Warren, J.E., Root, P.J., 1963. The behavior of naturally fractured reservoirs. SPE J. 3 (3), 245–255.

Wu, Y.-S., Pruess, K., 1988. A multiple-porosity method for simulation of naturally fractured petroleum reservoirs. SPE Reservoir Eng. 3 (1), 327–336.

Xu, Y., 2018. Development and application of embedded discrete fracture model for conventional and unconventional reservoir simulation. PhD Dissertation, University of Texas, Austin.

CHAPTER FOUR

Basic EDFM approach using Cartesian grid

4.1 Overview of the embedded discrete fracture model

The embedded discrete fracture model (EDFM) was developed as a combination of dual-continuum models and discrete fracture models (DFMs). Different from most DFMs, the EDFM is based on non-conforming meshes. The basic idea of the EDFM includes several key points:

1. Generate the matrix grid without considering the locations of fractures, and discretize the fractures using matrix gridblock boundaries.
2. Separate the domain of fractures from the matrix.
3. Use special connections to represent the flow within the fracture domain and the flow between the matrix and the fracture domains.

By applying the above ideas, the EDFM avoids the requirement for conforming mesh, which makes it easy to generate the mesh for simulation. The initial idea of using a separate grid domain to represent fractures was described by Hearn et al. (1997) for the modeling of conductive faults. In their study, the fault was modeled using gridblocks outside the basic model area, and non-neighboring connections (NNCs) were used to link these gridblocks to the basic model. The authors indicated that this idea could avoid re-gridding if the fault locations needed to be changed during history matching. However, the details about the discretization of fractures and the calculation of NNC factors were not presented in their work.

Later, a similar idea was utilized by several other researchers as well (Lee et al., 2001; Philip et al., 2005; Hustedt et al., 2005; Vitel and Souche, 2007). Lee et al. (2001) developed a hierarchical approach to model fractures with different length scales. In their study, long fractures (fracture length larger than gridblock dimension) were treated similarly as wells. These fractures were discretized by the gridblock boundaries, and for each matrix gridblock that was intersected by any long fracture, a large transport index was used to correlate the matrix–fracture fluid transfer rate with the potential difference between the matrix gridblock and the part of fracture inside the gridblock. By assuming that the fracture length in the horizontal direction

Embedded Discrete Fracture Modeling and Application in Reservoir Simulation.
http://dx.doi.org/10.1016/B978-0-12-821872-3.00004-6

was much larger than fracture height, the authors were able to treat the flow in fractures as one-dimensional flow.

Philip et al. (2005) borrowed the idea from Hearn et al. (1997) to model the flow through fractures using NNCs. However, in their study, fractures were not explicitly represented as separate gridblocks. Instead, assuming the fractures were infinitely conductive, NNCs were added between each pair of gridblocks that were crossed by the same fracture. The authors applied the methodology to two-dimensional problems with fractures parallel to the coordinate axis.

Hustedt et al. (2005) developed a methodology to model dynamically growing fractures induced by water injection, in which a void block in the simulation grid was converted to a so-called "fracture-grid-block" (FGB). The FGB was connected to matrix gridblocks intersected by the fracture through special connections, and the area of intersection determined the extent of the connection. To model the dynamic fracture growth, the authors modified the volume of the FGB to reflect the growing or shrinking of the fracture. It should be noted that the authors did not mention the discretization of fractures. The conductivity of the fracture was assumed to be infinitely high.

Later, Vitel and Souche (2007) developed a pipe network model that integrated the discretization of fractures using a similar idea to that in Lee et al. (2001). The fractures could have a finite conductivity in their model, and each fracture was discretized into a series of control volumes by the matrix gridblock boundaries. The connections between segments of the same fracture, intersecting fractures, matrix and fracture, and neighboring matrix blocks were represented as "pipes" to construct a graph. The authors compared their model with an unstructured-gridding-based DFM and obtained a good match of simulation results.

Li and Lee (2008) further developed the method by Lee et al. (2001) for simulating long natural fractures. They proposed formulations for calculating matrix–fracture connection transmissibilities. Furthermore, connections between fractures and intersections between fractures and vertical wellbores were also included in their model. A series of two-dimensional case studies with vertical fractures were conducted to demonstrate the viability of several concepts underpinning the EDFM. Hajibeygi et al. (2011) combined the idea presented by Li and Lee (2008) with the multiscale finite-volume method for two-dimensional simulations with fractures.

Later, Moinfar et al. (2014) extended the idea of Li and Lee (2008) to three-dimensional studies which included inclined fractures. Four types of connections were proposed to simulate fluid flow between matrix, fractures, and wellbores (Moinfar et al., 2013b; Moinfar et al., 2014). For the first time, the name "embedded discrete fracture model" was used for this methodology. They systematically proposed three types of NNC in the EDFM and later added another type of connection between fractures and wellbores (Moinfar et al., 2013b). The EDFM was implemented in an in-house, fully implicit, compositional reservoir simulator [general purpose adaptive simulator (GPAS)] with Cartesian grids, and the EDFM was used to simulate both hydraulic and natural fractures.

In recent years, extensive studies have been performed to apply the EDFM to different types of simulators. The EDFM was applied in several in-house and commercial simulators using Cartesian grids (Shakiba, 2014; Panfili and Cominelli, 2014; Cavalcante Filho et al., 2015a; Xu, 2015; Tang et al., 2017). Shakiba (2014) implemented this model in two in-house, IMPEC-type reservoir simulators. Panfili and Cominelli (2014) applied the EDFM to a commercial simulator, and they performed simulations of miscible gas injection in carbonate reservoirs. Cavalcante Filho et al. (2015a) developed an EDFM preprocessor and applied the EDFM to an in-house, compositional simulator. Xu (2015) implemented the EDFM in an in-house, chemical-flooding reservoir simulator and two commercial simulators. These studies showed the compatibility of the EDFM with existing reservoir simulators.

The high accuracy and computational efficiency of the EDFM were demonstrated by different researchers through comparisons of the EDFM with grid refinement models and DFMs using unstructured grids (Moinfar et al., 2014; Xu, 2015; Panfili et al., 2015; Du et al., 2017; Flemisch et al., 2018). Moinfar et al. (2014) compared the computational efficiency of the EDFM with a fine model for simulating non-orthogonal fractures, where the fracture was represented through the stair-stepping method. A good agreement was obtained between the results of both models. Panfili et al. (2015) compared the EDFM with the local grid refinement (LGR) method and the equivalent fractured well method, and in their study, the EDFM obtained a high computational efficiency compared with the LGR method. Du et al. (2017) compared the EDFM with an unstructured gridding fracture model for full-field well interference studies, and the EDFM was found to have a significant advantage regarding computational efficiency while maintaining high accuracy. Flemisch et al. (2018) did a benchmark study comparing the EDFM with six DFMs based on either unstructured gridding or the extended finite element method (XFEM). The results of the EDFM obtained a good agreement with other models.

Owing to its advantages, the EDFM has been combined with various physical models and techniques to solve different types of problems. Moinfar et al. (2013a) combined the EDFM with the Barton–Bandis model to simulate the variation of fracture width due to changes in effective normal stress. Shakiba and Sepehrnoori (2015) used the EDFM to simulate complex fracture networks interpreted from microseismic monitoring data. Cavalcante Filho et al. (2015b) modified the formulations of the EDFM to simulate well branches created with a multi-lateral well stimulation technology. de Sousa Junior et al. (2016) used a simplified version of the EDFM to calculate equivalent transmissibility factors between matrix blocks to represent faults in corner-point grids. Ren et al. (2016) coupled the EDFM with XFEM to simulate two-dimensional flow in fractured tight reservoirs while incorporating geomechanical effects.

Besides regular Cartesian grids, the EDFM has been combined with other types of matrix gridding to combine the flexibility provided by these types of gridding with the advantages of the EDFM. Jiang and Younis (2016) incorporated the EDFM with the multiple interacting continua (MINC) method. Ding et al. (2018) developed an EDFM–MINC approach where they used a MINC proximity function to subdivide the matrix and calculate matrix–fracture fluid exchange. Hui et al. (2019) coupled the EDFM with the dual-porosity dual-permeability model. Panfili and Cominelli (2014) and Fumagalli et al. (2016) implemented the EDFM in matrix gridblocks with corner-point geometry. Panfili et al. (2015) and Yang et al. (2018) demonstrated the use of moderate LGR with the EDFM to improve its simulation accuracy. All these studies showed the generality and effectiveness of the EDFM in different types of grids.

As the EDFM has gained significant attention in recent years, methods have also been proposed to improve its accuracy and computational efficiency. Ding et al. (2014) developed a method to compute matrix–fracture transmissibility based on either an analytical steady-state solution or an upscaling procedure. Moinfar et al. (2013b) proposed a coupled dual-continuum model and EDFM approach where small natural fractures in a gridblock were upscaled using the transmissibility factors calculated by the EDFM to improve the computational efficiency. Fumagalli et al. (2016, 2017) developed a flow-based upscaling procedure based on transmissibility calculations in the EDFM. Tene et al. (2017) developed a projection-based EDFM (pEDFM) approach where the matrix–matrix and matrix–fracture transmissibility factors were modified based on the projection area of the fracture on gridblock faces. The purpose of developing this methodology was primarily to simulate fractures with a lower permeability compared with the matrix, which behave like flow barriers. The method reduces to regular DFM when the fractures are placed at the interface between gridblocks. Jiang and Younis (2017) further improved the pEDFM formulation and emphasized the improvement in accuracy for conductive fractures. Chai et al. (2018) presented another method called compartmental EDFM (cEDFM) to deal with flow barriers. In this approach, whenever a discrete fracture crosses a matrix gridblock, it splits the matrix gridblock, hence the flows from both parts of the split matrix gridblock to the fracture are treated separately. Similar to the pEDFM, the cEDFM also improved the simulation accuracy for fluid flow near faults.

4.2 Description of EDFM methodology in conventional reservoir simulators using Cartesian grids

The EDFM borrows the concept from the dual-continuum approach which creates fracture cells in contact with corresponding matrix cells to account for the mass transfer between continua. Once a fracture penetrates a matrix cell, an additional cell is created to represent the fracture segment in the physical domain. Each individual fracture may be discretized into several fracture segments by the matrix cell boundaries. To differentiate the newly added cells from the original matrix cells, these additional cells are called "fracture cells." For each fracture, only one fracture cell is created inside each matrix gridblock it penetrates, and the fracture segment is only in touch with one matrix gridblock. This makes it convenient to evaluate the transmissibility and set up connections. By creating fracture control volumes in such a manner, the EDFM avoids the use of conforming meshes around fractures, and the fluid flow related to fractures is simulated through the introduction of transport indices.

In Fig. 4.1, we illustrate the procedure to add fracture cells in the EDFM using a simple case with only three matrix blocks and two fractures. Fig. 4.1A and B shows the physical domain and the computational domain, respectively. Each matrix block and fracture segment in the physical domain is represented by a cell in the computational domain with the same color. Before adding the fractures, there are three matrix cells: Cells 1, 2, and 3. After adding the fractures, the total number of cells increases. Fracture 1

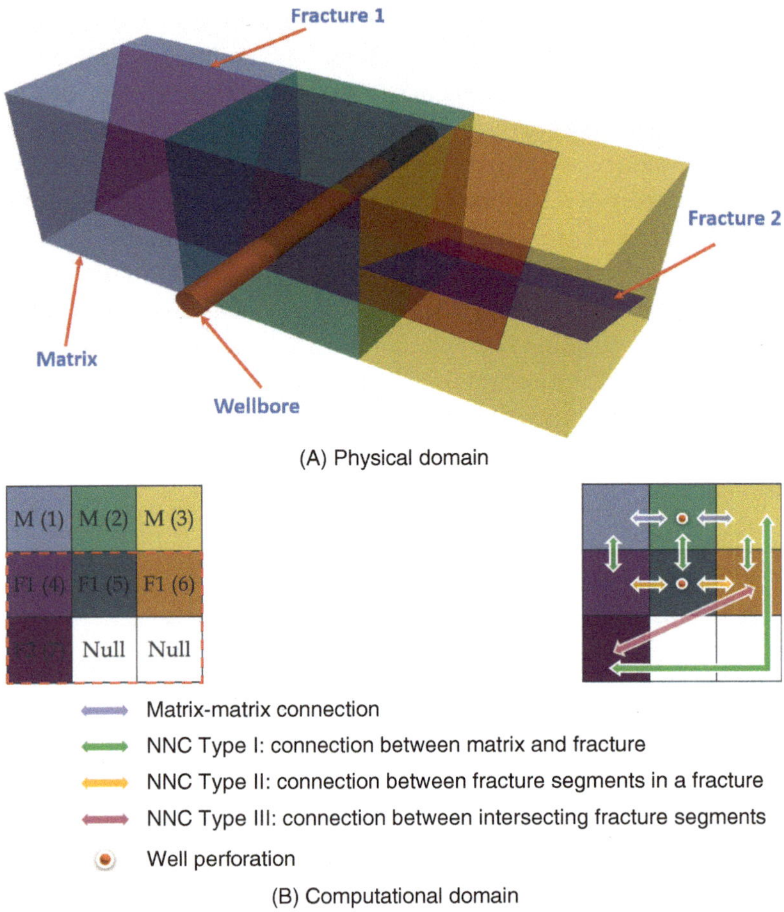

Figure 4.1 ***Explanation of the discretization of fractures and the construction of non-neighboring connections in the EDFM (Xu, 2018).*** (A) The physical domain with three matrix gridblocks, a long fracture, a short fracture, and a horizontal wellbore. (B) The computational domain. On the left-hand side, "M" represents "matrix," "F1" represents "Fracture 1," and "F2" represents "Fracture 2." On the right-hand side, the arrows show the connections in the model.

penetrates all three matrix blocks, hence three fracture cells (Cells 4, 5, and 6) are introduced into the computational domain to represent the corresponding fracture segments. Only one extra cell (Cell 7) is added for Fracture 2 as it only penetrates one matrix block. To keep the grid structured, every row should have the same number of cells, therefore two null cells (Cells 8 and 9) are also introduced. Finally, the total number of cells increases from three ($1 \times 3 = 3$) to nine ($3 \times 3 = 9$).

After adding the fracture cells, the properties of the fracture cells, such as depth, permeability, and pore volume, need to be modified to reflect the properties of the fractures. In Fig. 4.1, these properties of Cells 4, 5, 6, and 7 are modified. The depth of a fracture cell is the depth at the centroid of the corresponding fracture segment, and the permeability of a fracture cell is the permeability of the fracture segment, which is the same as the fracture permeability at that location. The volume of a fracture segment can be calculated as

$$V_f = A_{seg} w_f, \tag{4.1}$$

where A_{seg} is the area of the corresponding fracture segment, and w_f is the fracture width. In some problems, V_f may be used as the pore volume of the fracture segment, which means

$$PV_f = V_f, \tag{4.2}$$

where PV_f is the pore volume of the fracture segment. However, if an accurate fracture pore volume is critical for the simulation, a porosity other than 1.0 can be used for the fracture segment (e.g., considering the volume occupied by proppants), and

$$PV_f = \phi_f A_{seg} w_f. \tag{4.3}$$

PV_f is used as the pore volume of the fracture cell. In structured grids, the gridblock used to represent a fracture segment typically has larger dimensions than the actual fracture segment. Therefore, an effective fracture porosity can be defined as

$$\phi_{f,eff} = \frac{PV_f}{V_b}, \tag{4.4}$$

where V_b is the bulk volume of the gridblock that is used to represent the fracture segment. In Fig. 4.1, V_b is set to be the same as the bulk volume of the matrix gridblocks.

After adding cells to represent fracture segments, one needs to cancel the connections automatically generated for these cells by the simulator. In Fig. 4.1, these connections include those between Cells 1 and 4, 2 and 5, 3 and 6, 4 and 7, 4 and 5, and 5 and 6. NNCs are subsequently defined. The aim of introducing NNCs is to allow flow communication between cells that are physically connected but not neighboring in the computational domain. Three types of NNCs are defined as follows:

- NNC type I: connection between a fracture segment and the matrix block it penetrates.
- NNC type II: connection between fracture segments within the same fracture.
- NNC type III: connection between intersecting fracture segments.

All three types of NNCs are shown by arrows in Fig. 4.1B. The cells in each NNC pair are connected by a transmissibility factor.

In addition to these NNCs, the connections between fractures and wells are also introduced by the EDFM. When a fracture segment intersects the wellbore trajectory (as shown in Fig. 4.1A), the corresponding fracture cell needs to be treated as a wellblock by adding a well perforation to this cell, as shown in Fig. 4.1B (Cell 5).

The procedure discussed here can be implemented in traditional finite-difference or finite-volume reservoir simulators in a non-intrusive manner if the simulators have the NNC functionality. Many simulators allow the definition of NNCs by providing a connection list and transmissibility factors, which makes it possible and convenient to apply the EDFM in them. Also, the calculations of the connection factors, including the NNC transmissibility factors and the fracture well indices, need to be performed before running the reservoir simulator. Therefore, they can be treated as a part of the "preprocessing" work for reservoir simulation. An EDFM preprocessor needs to be developed to perform the geometrical calculations, compute the connection factors, and modify the input data of reservoir simulators. Detailed formulations used for the calculation will be discussed in the next section.

4.3 Calculation of NNC transmissibility factors and fracture well indices

This section presents a brief review of the formulations used to calculate connection factors in the EDFM. The assumptions and limitations are also discussed.

4.3.1 Basic flux formulations

In the EDFM, after the NNCs have been constructed, a two-point flux approximation is used to evaluate the fluid flow between two cells in an NNC pair. The volume flow rate of phase j between two cells in an NNC pair is

$$q_j = \lambda_{rj} T_{nnc} \Delta P_j, \tag{4.5}$$

where λ_{rj} is the relative mobility of phase j, T_{nnc} is the NNC transmissibility factor, and ΔP_j is the flow potential difference of phase j between the cells. If the units of q_j, λ_j, T_{nnc}, and ΔP_j are ft^3/day, cp^{-1}, md-ft, and psi, respectively, a unit conversion factor of 0.006328 is required on the right-hand side of the equation.

A general form for calculating T_{nnc} is (Moinfar et al., 2014)

$$T_{nnc} = \frac{k_{nnc} A_{nnc}}{d_{nnc}}, \tag{4.6}$$

where k_{nnc} is an average permeability of the cells in an NNC pair, A_{nnc} is the area open to flow, and d_{nnc} is a characteristic distance between the two gridblocks. Detailed definitions of k_{nnc}, A_{nnc}, and d_{nnc} for different types of connections in the EDFM will now be presented.

4.3.2 Matrix–fracture intersection

In the case of a fracture fully penetrating a matrix gridblock, the matrix gridblock is divided into two parts, as shown in different colors in Fig. 4.2. If a uniform pressure gradient that is normal to the fracture plane on each side of the fracture (i.e., a linear pressure distribution) is assumed, as shown in Fig. 4.2, the transmissibility factor for the matrix–fracture intersection can be calculated as

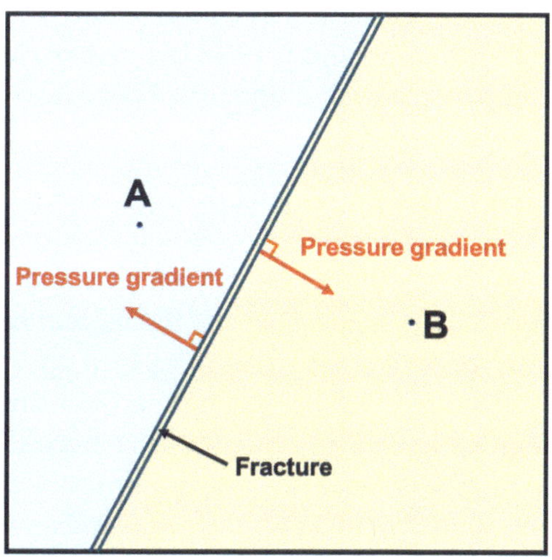

Figure 4.2 ***Illustration of matrix–fracture intersection (modified from Xu et al., 2017).*** The two parts of the matrix block on both sides of the fracture are shown in different colors.

$$T_{f-m} = \frac{2A_f \vec{n} \cdot (\overline{\overline{K}} \cdot \vec{n})}{d_{f-m}}, \tag{4.7}$$

where $T_{f\text{-}m}$ is the transmissibility factor, A_f is the area of fracture segment on one side, $\overline{\overline{K}}$ is the matrix permeability tensor, $\vec{n}$ is the unit normal vector of the fracture plane, and $d_{f\text{-}m}$ is the average normal distance between the fracture plane and the matrix block. $\vec{n} \cdot (\overline{\overline{K}} \cdot \vec{n})$ is the same as $\vec{n}^T \cdot (\overline{\overline{K}} \cdot \vec{n})$, and it is sometimes also written as $(\overline{\overline{K}} \cdot \vec{n}) \cdot \vec{n}$. Owing to the small value of fracture width and high value of fracture permeability, the influence of fracture properties on matrix–fracture flow is ignored. A derivation of Eq. (4.7) can be found in Section 4.3.7.

In Eq. (4.7), both A_f and $d_{f\text{-}m}$ require accurate geometrical calculations, whose complexity heavily depends on block geometries. For example, $d_{f\text{-}m}$ can be obtained through an integration:

$$d_{f-m} = \frac{\int_V x_n \, dV}{V}, \tag{4.8}$$

where V is the matrix block volume, and x_n is the distance from a volume element in V to the fracture plane. The integration should be performed on both sides of the fracture, and it can be calculated analytically or numerically. For example, for 2D cases, it can be easily calculated by computing the distances from the centroids of both parts of the matrix block ("A" and "B" in Fig. 4.2) to the fracture plane and using the following equation:

$$d_{f-m} = \frac{V_A d_{f-A} + V_B d_{f-B}}{V_A + V_B}, \tag{4.9}$$

where V_A and V_B are the volumes of the left part and the right part in Fig. 4.2, respectively, and $d_{f\text{-}A}$ and $d_{f\text{-}B}$ are the distances from points A and B to the fracture plane, respectively.

In the case of a fracture not fully penetrating the matrix gridblock, the pressure distribution in the matrix block deviates from the assumption of linear distribution, and the calculation of the transmissibility factor becomes complicated. As an approach to keep the "non-intrusive" nature of the EDFM, the method by Li and Lee (2008) can be used, which assumes that the value of the transmissibility factor is proportional to the area of fracture segment, and the actual area of the fracture segment is used in Eq. (4.7).

However, it is recommended that when possible, the matrix grid can be adjusted to reduce the simulation error introduced by this assumption.

4.3.3 Connection between fracture segments within an individual fracture

In the EDFM, a fracture is discretized into a number of fracture segments. In Cartesian grids, when the fracture fully penetrates a matrix block, the resulting fracture segments can have three to six vertices, as shown in Fig. 4.3. The fluid flow through the fracture is represented by the connections between the fracture segments, as also shown in Fig. 4.3. The transmissibility factor between two neighboring fracture segments in a fracture can be calculated as

$$T_{seg} = \frac{T_1 T_2}{T_1 + T_2}$$
$$T_1 = \frac{k_f A_c}{d_{seg1}}, T_2 = \frac{k_f A_c}{d_{seg2}}, \tag{4.10}$$

which can also be written as

$$T_{seg} = \frac{k_f A_c}{d_{seg1} + d_{seg2}}, \tag{4.11}$$

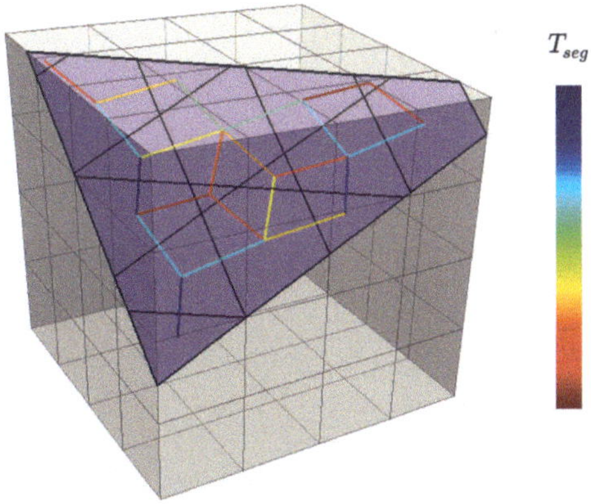

Figure 4.3 ***Illustration of connections between fracture segments within an individual fracture (Xu, 2018).*** The black lines split the fracture into fracture segments. The lines in different colors show the connections between fracture segments, and the color represents the value of the transmissibility factor.

where k_f is the fracture permeability, A_c is the common flow area between the fracture segments, which can be calculated as the product of the fracture width and the length of the common line between the two fracture segments, and d_{seg1} and d_{seg2} are the average distances from the two fracture segments to the common face, respectively.

It should be noted that the two-point flux approximation used in Eq. (4.11) is a simplified approach to ease handling by traditional reservoir simulators, and it is a source of error when the fracture segments are not forming an orthogonal grid. This may occur in three-dimensional cases with inclined fractures.

4.3.4 Fracture intersection

Accurate and efficient modeling of fracture intersection is a challenging problem for discrete fracture modeling. The complexity of flow behaviors at fracture intersections makes it difficult to model accurately. Following the work of Moinfar et al. (2014), this problem can be simplified by assigning a transmissibility factor between intersecting fracture segments to approximate the mass transfer at the fracture intersection. The transmissibility factor is calculated as

$$T_{int} = \frac{T_1 T_2}{T_1 + T_2}$$
$$T_1 = \frac{k_{f1} w_{f1} L_{int}}{d_{f1}}, T_2 = \frac{k_{f2} w_{f2} L_{int}}{d_{f2}}, \tag{4.12}$$

where k_{f1} and k_{f2} are the permeabilities of fracture segments, w_{f1} and w_{f2} are the widths of fracture segments, L_{int} is the length of the intersection line, and d_{f1} and d_{f2} are the average normal distances from the fracture segments to the intersection line. It should be noted that d_{f1} and d_{f2} should be calculated by considering the two parts of the fracture segment on both sides of the intersection line. In Fig. 4.4,

$$d_{f1} = \frac{A_1 d_{c1} + A_2 d_{c2}}{A_1 + A_2}$$
$$d_{f2} = \frac{A_3 d_{c3} + A_4 d_{c4}}{A_3 + A_4}, \tag{4.13}$$

where $d_{c1}, d_{c2}, d_{c3}, d_{c4}$ are the distances from the centroids of Parts 1, 2, 3, and 4 to the intersection line, respectively, and A_1, A_2, A_3, and A_4 are areas of Parts 1, 2, 3, and 4, respectively.

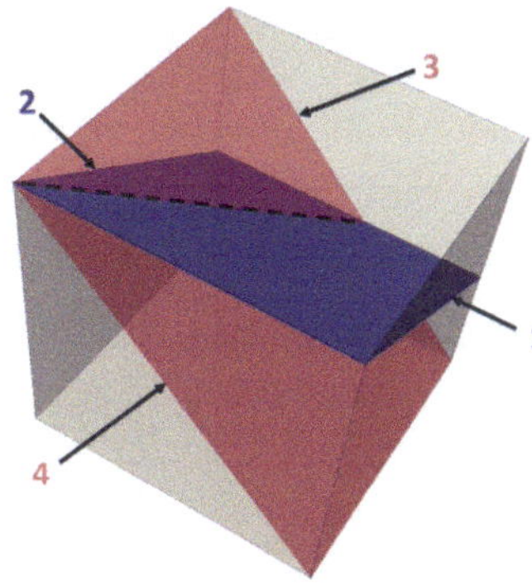

Figure 4.4 ***An example of fracture intersection (modified from Xu, 2018).*** The two fracture segments are shown in blue and pink, respectively. Each fracture segment is split into two parts by the intersection line. "1" and "2" represent the two parts of the first fracture segment, and "3" and "4" represent the two parts of the second fracture segment.

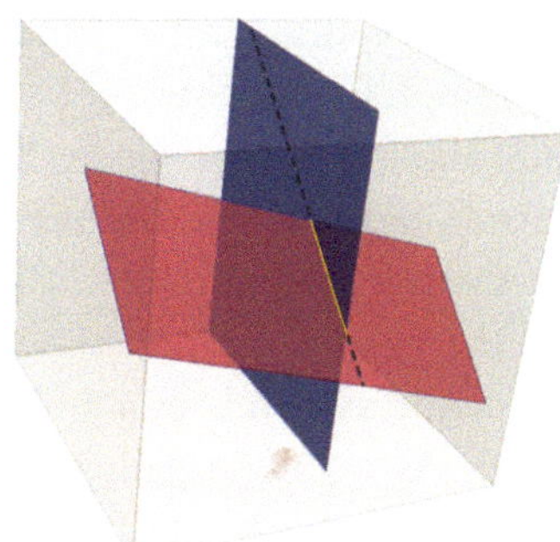

Figure 4.5 ***An example of partially intersecting fracture segments (Xu and Sepehrnoori, 2019).*** The two fracture segments are shown in pink and blue, respectively, and they partially intersect. The yellow solid line represents the true intersection line, and the black line shows the intersection line used to help calculate the average normal distances.

If two fracture segments partially intersect with each other, as shown in Fig. 4.5, the intersection line should first be extended to assist in calculating the normal distances. After that, the true length of the intersection line (shown in yellow in Fig. 4.5) is used in Eq. (4.12) to calculate the transmissibility factor.

The limitation of this fracture intersection model is that it cannot consider the complex flow mechanisms at the fracture intersection such as stream tube routing (Berkowitz et al., 1994) as the flow direction in each subsegment is unknown. The impact of the intersection angle is not considered either. Therefore, for multi-component flow with a high local Peclet number ($Pe > 10^{-2}$) (Berkowitz et al., 1994), Eq. (4.12) should be used with caution.

4.3.5 Fracture–well intersection

In the EDFM, if a fracture segment is intersected by the trajectory of a wellbore, the control volume representing the fracture segment should be treated as a well block, and an effective well index will be calculated. Modified from Peaceman (1983)'s model, the formulation to calculate the effective well index for a fractured horizontal well is (Moinfar et al., 2013b)

$$WI_f = \frac{2\pi k_f w_f}{ln(\frac{r_e}{r_w})}, r_e = 0.14\sqrt{L_{seg}^2 + H_{seg}^2}, \tag{4.14}$$

where k_f is the fracture permeability, w_f is the fracture width, r_e is the effective radius, k_w is the wellbore radius, and L_{seg} and H_{seg} are the length and height of the fracture segment, respectively.

Eq. (4.14) applies when the angle θ between the wellbore axis and the fracture plane is 90 degrees. When the fracture plane is non-orthogonal or inclined, or in the case of slanted well, θ may not be 90 degrees, as shown in Fig. 4.6A. In such cases, the cross-section of the wellbore is an ellipse instead of a circle. The semi-major axis and semi-minor axis of the ellipse are $r_w / sin(\theta)$ and r_w, respectively. As w_f is typically several orders of magnitude smaller than L_s and H_s, the flow from the wellbore to the fracture segment can be treated as two-dimensional flow. In addition, as rw is typically much smaller than L_s and H_s, at some distance from the wellbore, the flow is

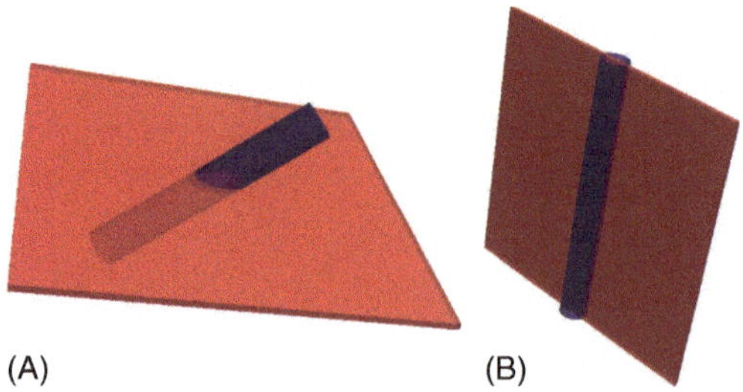

Figure 4.6 ***Examples of fracture–well intersections (Xu and Sepehrnoori, 2019).*** (A) An example of fracture–well intersection where the angle between the wellbore axis and the fracture plane is not 90 degrees. (B) An example where a vertical well is connected to a vertical fracture. The fracture segments are shown in pink, and the wellbores are shown in purple.

almost radial. Batchelor (1970) showed that the flow with an elliptical well cross-section is almost identical to that with a circular well cross-section, if the wellbore radius is replaced by $\frac{1}{2}(a+b)$, where a and b are the semi-major axis and semi-minor axis of the ellipse. Therefore, an effective well radius can be calculated as

$$rw_{\theta} = \frac{1}{2}\left(\frac{rw}{sin(\theta)} + rw\right) = \left(\frac{1}{2} + \frac{1}{2sin(\theta)}\right)rw. \tag{4.15}$$

Substitution into Eq. (4.14) gives

$$WI_{f\theta} = \frac{2\pi k_f w_f}{ln(\frac{re}{rw_{\theta}})} = \frac{2\pi k_f w_f}{ln(\frac{2sin(\theta)}{1+sin(\theta)}\frac{re}{rw})}. \tag{4.16}$$

Eq. (4.16) shows that if the influence of θ is ignored, the value of the effective well index will be underestimated.

Another situation where the effective well index needs to be calculated is when a vertical well is fractured, as shown in Fig. 4.6B. In this case, the flow is almost linear in the fracture segment, and the well index can be calculated as

$$WI_f = \frac{k_f w_f H_w}{d_{s-w}}, \tag{4.17}$$

where H_w is the height of the well in the fracture segment and d_{s-w} is the average normal distance from the fracture segment to the wellbore axis, which can be easily calculated through numerical integration or some geometrical method.

4.3.6 Fracture pore-volume cutoff

It can be observed from Eqs. (4.11), (4.12), (4.16), and (4.17) that when computing the transport indices for fracture–fracture or fracture–well flow, the product of fracture width and fracture permeability always appears in the equations. This leads to the definition of fracture conductivity

$$C_f = w_f k_f. \tag{4.18}$$

If w_f and k_f are varied while $w_f k_f$ is kept constant, the transport indices calculated in the EDFM will remain unchanged. Therefore, when the simulation results are not very sensitive to the fracture pore volume (e.g., in some simulations of primary production from hydraulically fractured reservoirs), the following equations can be used to assign new values to w_f and k_f and increase the pore volume of fractures:

$$w_f^{new} = w_f \times \frac{PV_{cutoff}}{PV_f}, \tag{4.19}$$

$$k_f^{new} = k_f \times \frac{PV_f}{PV_{cutoff}}, \tag{4.20}$$

$$PV_f^{new} = PV_{cutoff}, \tag{4.21}$$

where PV_{cutoff} is a predefined value of the smallest fracture pore volume that can be accepted in the simulation. The cutoff value could also be defined for $\phi_{f,eff}$. As the fracture cells typically have the smallest pore volume in the simulation, using Eqs. (4.19)–(4.21) can greatly improve the computational performance of EDFM simulations.

4.3.7 Derivation of matrix–fracture transmissibility factor

When considering the matrix–fracture flow, the matrix gridblock is divided into two parts, A and B, as shown in Fig. 4.2. The volumes of part A and part B are denoted as V_A and V_B, respectively. The average pressure in the whole gridblock is

$$\begin{aligned} p_m &= (\phi V_A p_A + \phi V_B p_B) / (\phi V_A + \phi V_B) \\ &= (V_A p_A + V_B p_B) / (V_A + V_B) \end{aligned}, \tag{4.22}$$

where ϕ is the porosity, and p_A and p_B are the average pressure in part A and part B, respectively. Using two-point flux approximation, the flow rate of phase j from the fracture to part A is

$$q_{f-A} = T_{f-A} \lambda_j (p_f - p_A), \tag{4.23}$$

where p_f is the average pressure in the fracture segment, $T_{f\text{-}A}$ is the phase independent part of transmissibility between the fracture and part A, and λ_j is the relative mobility of phase j. Assuming a uniform pressure gradient that is normal to the fracture plane, as shown in Fig. 4.2, $T_{f\text{-}A}$ can be calculated as

$$T_{f-A} = A_f \vec{n} \cdot (\overline{\overline{K}} \cdot \vec{n}) / d_{f-A}, \tag{4.24}$$

where A_f is the area of the fracture segment on one side, $\overrightarrow{\overrightarrow{K}}$ is the matrix permeability tensor, $\vec{n}$ is the normal vector of the fracture plane, d_{f-A} is the average normal distance from part A to the fracture, which can be calculated by

$$d_{f-A} = \frac{\int_{V_A} x_n dV_A}{V_A}. \tag{4.25}$$

$(p_f - p_A)\vec{n} / d_{f-A}$ is the pressure gradient. In the case of anisotropic matrix permeability, the flow direction may be different from the direction of the pressure gradient. Therefore, the other $\vec{n}$ in the equation projects the flow velocity onto the normal direction of the fracture plane.

Similarly, the flow rate of phase j from the fracture to part B is

$$q_{f-B} = T_{f-B}\lambda_j(p_f - p_B), \tag{4.26}$$

and

$$T_{f-B} = A_f\, \vec{n} \cdot (\overrightarrow{\overrightarrow{K}} \cdot \vec{n}) / d_{f-B}, \tag{4.27}$$

$$d_{f-B} = \frac{\int_{V_B} x_n dV_B}{V_B}. \tag{4.28}$$

The total flow from the fracture to the matrix is

$$q_{f-m} = q_{f-A} + q_{f-B}. \tag{4.29}$$

By the definition of T_{f-m},

$$q_{f-m} = T_{f-m}\lambda_j(p_f - p_m). \tag{4.30}$$

Assuming the same magnitude of pressure gradients on both sides of the fracture,

$$\frac{p_f - p_A}{p_f - p_B} = \frac{d_{f-A}}{d_{f-B}}. \tag{4.31}$$

Combining all these equations,

$$T_{f-m} = \frac{2A_f \vec{n} \cdot (\overrightarrow{\overrightarrow{K}} \cdot \vec{n})}{(V_A d_{f-A} + V_B d_{f-B}) / (V_A + V_B)}. \tag{4.32}$$

Note that the denominator of Eq. (4.32) is the same as the definition of $d_{f\text{-}m}$ in Eq. (4.7).

4.4 Modeling complex fracture geometries

Experimental studies have shown that the shape of a single hydraulic fracture can be quite complex, for example, nonplanar fracture with varying width and height (El Rabaa, 1989; Weijers and de Pater, 1992; Abass et al., 1996; Bahorich et al., 2012). Recent developments in fracture propagation modeling also indicate that irregular hydraulic fracture shapes originate due to the "stress shadow" effect, proppant transport, or the influence of natural fracture networks (Olson, 2008; Weng et al., 2011; Wu and Olson, 2015). As the early behaviors of a hydraulically fractured reservoir are dominated by hydraulic fracture geometry, it is important to adequately consider the influence of fracture geometry in production forecasting. This section discusses the methodology to simulate some types of complex fracture geometries using the EDFM.

4.4.1 Nonplanar fracture geometry

To account for the complexity in fracture shape, the EDFM is extended to handle nonplanar fracture shape by discretizing a nonplanar fracture into several interconnected planar fracture segments. The connections between these planar fracture segments are treated as fracture intersections.

For two intersecting fracture segments, if two sub-segments have small areas (in Fig. 4.4, $A_2 \rightarrow 0$, $A_4 \rightarrow 0$), the transmissibility factor between the fracture segments is mainly determined by the sub-segments with larger areas. In Eq. (4.13),

$$d_{f1} \approx d_{c1}, d_{f2} \approx d_{c3}. \tag{4.33}$$

The formulation for the transmissibility factor calculation (T_{int}) has the same form as the one used for two fracture segments within an individual fracture (T_{seg}) if the width and permeability of the two intersecting fracture segments are the same. Therefore, some connections between the fracture segments within a fracture can be represented as fracture intersections, and this makes it possible to represent a nonplanar fracture as a series of intersecting fractures. Fig. 4.7 illustrates how a nonplanar fracture is dealt with using the EDFM. The fracture in the figure is discretized into six interconnected planar fractures. For each intersection, the ratios of $L1/L3$ and

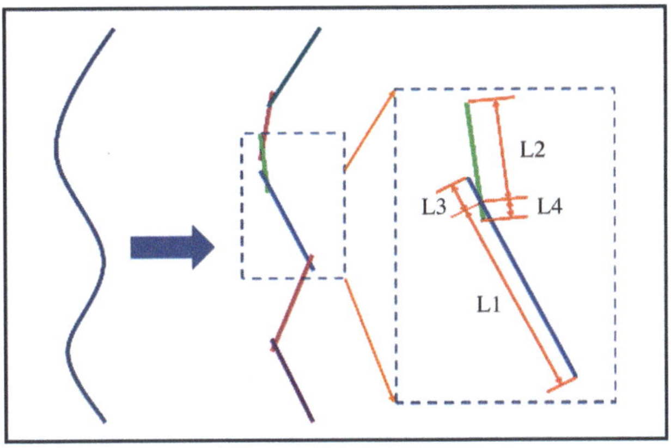

Figure 4.7 *Illustration of the modeling of nonplanar fracture (Xu et al., 2017).*

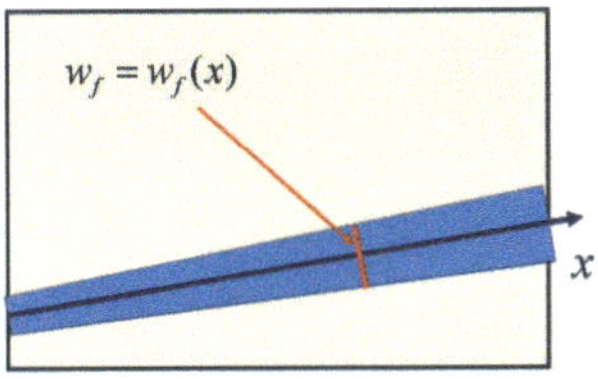

Figure 4.8 *A 2D example showing a fracture segment with varying width (Xu et al., 2017).*

$L2/L4$ should be high enough to eliminate the influence of the small subsegments. For example, the ratios can be set as 100.

4.4.2 Varying fracture width

A fracture with varying width can be modeled with the EDFM by discretizing it into interconnected segments and assigning each segment an "average width" $(\overline{w_f})$ and "effective permeability" $(k_{f,eff})$. Attention needs to be paid to compute both parameters.

Fig. 4.8 shows a 2D case where a fracture segment with a length of L_s has varying width, and s is the distance from a cross-section to one end of the fracture segment. The width is a function of x. The total volume of the segment is

$$V_{seg} = H \int_0^{L_s} w_f(x)dx, \tag{4.34}$$

where H is the height of the fracture segment. The average width to calculate the fracture segment volume should be

$$\overline{w_f} = V_{seg} / HL_s = \frac{1}{L_s} \int_0^{L_s} w_f(x)\,dx. \tag{4.35}$$

For the calculation of transmissibility, if the cubic law for fracture conductivity applies,

$$C_f(x) = k_f(x) w_f(x) = \lambda w_f^3(x), \tag{4.36}$$

where λ is 1/12 for smooth fracture surfaces and $\lambda < 1/12$ for coarse fracture surfaces.

For the fluid flow in fractures, based on Darcy's law,

$$q_j = HC_f(x)\lambda_j \frac{dP}{dx}, \tag{4.37}$$

where q_j is flow rate of phase j and λ_j is the relative mobility of phase j. For each fracture segment, assuming a constant q_j, the pressure drop along the fracture segment is

$$\Delta p = \int_0^{L_s} \frac{q_j}{HC_f(x)} dx. \tag{4.38}$$

To keep the pressure-drop constant between both ends of the segments, an effective fracture conductivity can be defined which satisfies the following equation:

$$\int_0^{L_s} \frac{q_j}{HC_f(x)} dx = \int_0^{L_s} \frac{q_j}{HC_f^{eff}} dx, \tag{4.39}$$

which gives

$$C_f^{eff} = \frac{L_s}{\int_0^{L_s} \frac{dx}{C_f(x)}} = \frac{L_s}{\int_0^{L_s} \frac{dx}{\lambda w_f^3(x)}}. \tag{4.40}$$

As fracture conductivity is the product of fracture width and fracture permeability, if $\overline{w_f}$ is used for the whole fracture segment, the effective fracture permeability $k_{f,eff}$ is

$$k_{f,eff} = C_f^{eff} / \overline{w_f} = \frac{L_s^2}{\left(\int_0^{L_s} w_f(x)\,dx\right)\left(\int_0^{L_s} \frac{dx}{\lambda w_f^3(x)}\right)}. \tag{4.41}$$

Similarly, if the fracture has a constant fracture permeability but varying width, the effective fracture permeability should be

$$k_{f,eff} = \frac{L_s^2}{\left(\int_0^{L_s} w_f(x)\,dx\right)\left(\int_0^{L_s} \frac{dx}{k_f w_f(x)}\right)}. \tag{4.42}$$

4.4.3 Special handling of small fracture segments

In the EDFM, the discretization of fractures by matrix block boundaries may generate some fracture segments with extremely small volumes. This happens more frequently when modeling complex fracture geometries, where many small fractures are used to represent the nonplanar fracture shape and variation in fracture width. These small control volumes may cause problems in preconditioning, and they may limit the simulation time step to an unreasonable value. Simply eliminating these segments may cause the loss of connectivity, as shown in Fig. 4.9. The star-delta transformation (Karimi-Fard et al., 2004) can be used as an alternative to the fracture pore volume cutoff method to prevent such problems and improve the stability of EDFM simulations. For a fracture segment with extra small volume, if N denotes the number of NNCs related to this fracture segment, C_1, C_2, ..., and C_N denote the cells connected to this segment by NNCs, and T_1, T_2, ..., T_N denote the NNC transmissibility factors for these connections, the small fracture segment can be eliminated following the steps given below:

1. Remove the cell for this fracture segment in the computational domain. Then, eliminate all NNCs related to this cell.

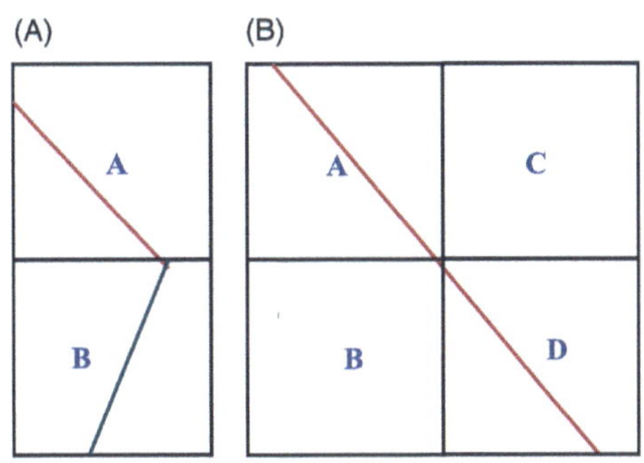

Figure 4.9 ***Two-dimensional examples showing the importance of small fracture segments for connectivity (Xu et al., 2017).*** The red and blue lines represent fractures. In both (A) and (B), removing the small fracture segment in Cell B will lead to loss of connectivity.

2. Add $N(N-1)/2$ connections for all pairs of cells in C_1, C_2, …, and C_N, and the transmissibility factor between C_i and C_j is

$$T_{i,j} = \frac{T_i T_j}{\sum_{k=1}^{N} T_k}. \tag{4.43}$$

The above method eliminates the small control volumes while keeping the appropriate connectivity. However, as only the phase-independent part of the transmissibility is considered in this transformation, it is only an approximation for multiphase flow. Furthermore, as this method ignores the volume of the small fracture segments, it is recommended only when there is a very high pore volume contrast (e.g., 1,000) between fracture segments. It is also worth noticing that this method should not be applied to fracture segments with well intersections as it will cause loss of fracture–well connections.

4.5 Model verifications

In the following simulation studies, five cases are presented to demonstrate the applicability of the EDFM for various types of fracture geometries. As stated before, the EDFM is a non-intrusive method that may be used in any finite-difference or finite-volume reservoir simulator capable of handling NNCs. Therefore, the range of problems the EDFM can handle and the physics involved depend mainly on the capability of the simulator. The EDFM is applicable to both 2D and 3D cases, for both black-oil and compositional formulations. In this study, two commercial reservoir simulators (referred to as RS1 and RS2) with both the black-oil and compositional formulations were used to perform the case studies to show the generality of the EDFM approach. The results obtained with RS1 are presented for all cases, and the results obtained with RS2 are compared with those of RS1. The examples consider shale gas (methane) production from a multi-fractured horizontal well. Water is at residual saturation and the only flowing phase is gas. The gas relative permeability at residual water saturation is assumed to be 1.0 and no capillary effect is considered. Two-dimensional reservoir models were created for verification purposes, and we used the black-oil formulations in the simulators. Most of the reservoir properties and operation parameters were kept the same for all five cases as shown in Table 4.1. The Langmuir gas desorption effect was considered following the method of Seidle and Arri (1990). In their method, the gas adsorbed to the

Table 4.1 Basic reservoir and operation parameters in simulations for Case 1 (Xu et al., 2017).

Parameter	Value	Unit
Reservoir permeability	0.0005	md
Reservoir porosity	12%	–
Initial reservoir pressure	4,500	psi
Reservoir thickness	80	ft
Rock compressibility	1×10^{-6}	psi^{-1}
Initial water saturation	10%	–
Initial gas saturation	90%	–
Depth of top layer	1,000	ft
Langmuir pressure	535	psi
Langmuir volume	196.4	scf/ton
Rock density	164.2	lb/ft^3
Bottomhole pressure	1,000	psi
Wellbore radius	0.25	ft
Simulation time	1,000	day

rock surface can be mathematically treated as gas dissolved in an immobile oil (which could be added artificially). By assuming a small oil saturation, the Langmuir gas desorption can be modeled using solution gas–oil ratio

$$R_s = \frac{0.17525 v_L \rho_B B_o p}{\phi_m S_{om}(p_L + p)}, \tag{4.44}$$

where v_L = Langmuir isotherm constant, scf/ton; ρ_B = rock density, g/cm^3; B_o = oil formation volume factor, assumed to be constant; p_L = Langmuir pressure, psi; p = pressure, psi; ϕ_m = reservoir porosity in this equivalent model; $\phi_m \approx \phi$ for small S_{om}; S_{om} = assumed oil saturation; and R_s = solution gas–oil ratio, scf/STB. The revised PVT table assuming B_o = 0.001 and S_{om} = 0.001 is provided in Table 4.2.

We demonstrate the accuracy of the EDFM by comparing the gas flow rates and pressure profiles of the EDFM against LGR models or a semi-analytical solution, which is briefly introduced here. The semi-analytical solution was used to verify the EDFM for nonplanar fracture geometry with varying fracture width and fracture permeability along the fracture length. Similar to the EDFM, it discretizes the nonplanar fracture geometry into a number of small fracture segments. Its formulation mainly consists of two parts for simulating shale gas production. First, an analytical solution is used to solve the diffusivity equation for gas transport in shale. Second, a numerical solution is used to solve gas flow in each fracture segment. Below

Table 4.2 PVT table for the simulation (Xu et al., 2017).

P	R_s	B_o	B_g	μ_g (cp)
14.7	201.6	0.001	0.2189	0.0129
400	3,224.5	0.001	0.0078	0.0132
800	4,516.7	0.001	0.0038	0.0136
1,200	5,213.1	0.001	0.0025	0.0141
1,600	5,648.5	0.001	0.0018	0.0149
2,000	5,946.5	0.001	0.0015	0.0158
2,400	6,163.3	0.001	0.0012	0.0168
2,800	6,328.1	0.001	0.0010	0.0180
3,200	6,457.6	0.001	0.0009	0.0192
3,600	6,562.0	0.001	0.0008	0.0206
4,000	6,648.0	0.001	0.0007	0.0219
9,000	7,114.3	0.001	0.0004	0.0376

we briefly introduced the fundamental equations for gas transport in shale and gas flow in fracture. More details about the model development and verification can be found in the work by Yu (2015) and Yu et al. (2016).

The diffusivity equation of conventional gas reservoirs is modified by considering the gas desorption effect for modeling gas transport in shale as follows:

$$\left[\frac{\partial}{\partial x}\left(\frac{\rho_g k}{\mu_g}\frac{\partial p}{\partial x}\right)+\frac{\partial}{\partial y}\left(\frac{\rho_g k}{\mu_g}\frac{\partial p}{\partial y}\right)+\frac{\partial}{\partial z}\left(\frac{\rho_g k}{\mu_g}\frac{\partial p}{\partial z}\right)\right]=\left[S_g\phi+(1-\phi)K_a\right]c_g\rho_g\frac{\partial p}{\partial t}, \tag{4.45}$$

where p_g is the gas density, ϕ is the rock porosity, k is the reservoir permeability, c_g is the isothermal gas compressibility factor, p is the pressure, μ_g is the gas viscosity, K_a is the differential equilibrium portioning coefficient of gas at a constant temperature, which is a function of pressure and temperature and defined as (Patzek et al., 2013):

$$K_a=\left(\frac{\partial\rho_a}{\partial\rho_g}\right)_T, \tag{4.46}$$

where ρ_a is the adsorbed gas mass per unit shale volume (kilograms of adsorbed gas per cubic meter of solid). In this study, we used Langmuir isotherm to model the gas desorption effect (Langmuir, 1918), which is given below

$$v=\frac{pv_L}{p+p_L}, \tag{4.47}$$

where v_L is Langmuir volume and p_L is Langmuir pressure. According to the form of Langmuir isotherm, the differential equilibrium partitioning coefficient of gas can be expressed as

$$K_a = \frac{\rho_g(p_{ST}, T_{ST})\rho_b}{c_g\rho_g(1-\phi)} \frac{v_L p_L}{(p_L + p)^2}, \tag{4.48}$$

where ρ_b is the bulk density of shale and $\rho_g(p_{ST}, T_{ST})$ is the stock tank gas density. The analytical solution for the above-modified diffusivity equation can be found in the work by Yu (2015) and Yu et al. (2016).

For gas flow in a fracture, the pressure drop along the j-th fracture segment is proportional to the fluid velocity and can be calculated below:

$$p_j - p_{j+1} = \int_{y_j}^{y_{j+1}} \left(\mu_g \middle/ \rho_g k_f w_f h_f \right)_j \left[q_j + q_{fj}(y - y_j) \right] dy, \tag{4.49}$$

where k_f is fracture permeability, w_f is fracture width, h_f is fracture height, q_j is gas flow rate at node j of the j-th fracture segment, and q_{fj} is the flux of the j-th fracture segment. In addition, the mass balance is satisfied for each intersection point of the fracture segments.

In the LGR models, only the cells containing fractures are refined (into three sub-cells in the direction perpendicular to the fracture plane) to speed up the simulations, as shown in Fig. 4.10. The coarse grids in both the EDFM and the LGR models are kept the same in each case to make a fair comparison in central processing unit (CPU) time. The widths of the fractures are modeled as the real sizes, for example, 0.015 ft.

4.5.1 Case 1: Bi-wing fractures

We first consider an ideal case with four transverse planar fractures perpendicular to the wellbore. The reason why we choose this case is that most simulators can handle this case accurately through LGR. The reservoir is rectangular with dimensions of 3,100 × 1,675 × 80 ft, and it is discretized by 200 × 99 gridblocks horizontally. The fracture half-length is 348 ft, and the distance between neighboring fractures is 75 ft (shown in Fig. 4.11). By changing the fracture width along the fracture length, two sub-cases were created. In the first sub-case (Case 1a), we assume the fractures have a constant width of 0.015 ft and constant permeability of 10,000 md, which gives a high fracture conductivity of 150 md-ft. In this sub-case, the fractures behave as infinite-conductivity fractures. In the second sub-case (Case 2b),

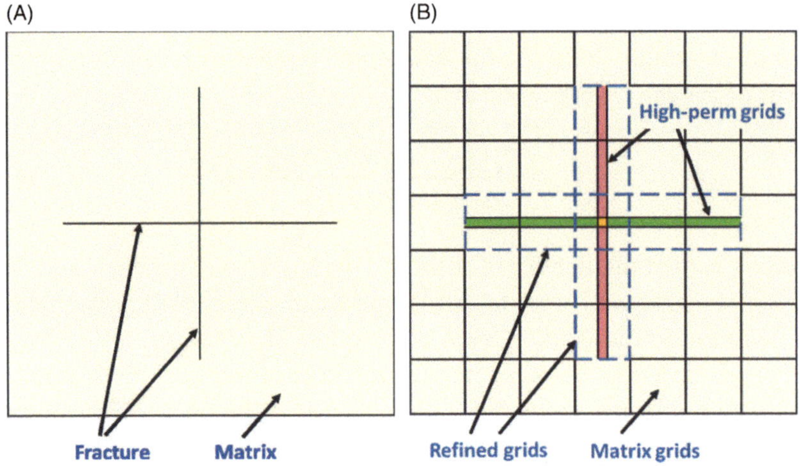

Figure 4.10 ***An example to illustrate the local grid refinement model (Xu et al., 2017).*** The two intersecting fractures in (A) are represented by high-perm sub-grids in (B). In (B), the purple and green cells are assigned high permeability (the same as the corresponding fractures) and high porosity (1.0). The orange block represents the fracture intersection, and this block has a very small volume.

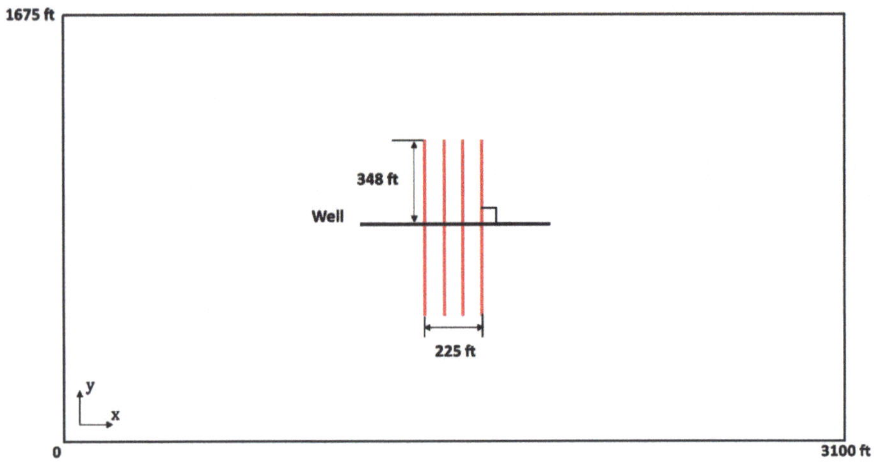

Figure 4.11 ***Reservoir dimensions and transverse hydraulic fractures in Case 1 (Xu et al., 2017).*** The purple lines represent the fractures.

we assume the fracture width has a linear relationship with the distance x from the wellbore, and the fracture permeability is proportional to the square of fracture width:

$$w_f(x) = w_{f,max} \times (1 - \frac{x}{x_f}), \tag{4.50}$$

$$k_f(x) = k_{f,max} \times \frac{w_f^2(x)}{w_{f,max}^2}, \tag{4.51}$$

where x_f is fracture half-length, $w_{f,max}$ is the maximum fracture width and $k_{f,max}$ is the maximum fracture permeability. In Case 2b, $w_{f,max}$ and $k_{f,max}$ are set to be 0.015 ft and 100 md, respectively. An LGR model is created to approximately simulate the fractures in Case 2b and each fracture is represented by 49 segments with different fracture widths and permeabilities. While in the EDFM, each fracture is discretized into seven segments, as shown in Fig. 4.12. The average width and effective permeability of each segment are calculated using Eqs. (4.35) and (4.41), respectively.

Figs. 4.13 and 4.14 show the pressure profiles after one year of production for Cases 1a and 1b, respectively. Both the results predicted by the LGR model and the EDFM are shown in the figures. As can be seen, the results are quite similar with a negligible difference. Fig. 4.15 shows a good agreement of the gas flow rate between both models. This case study tests the Types I and II NNCs and shows that the EDFM gives consistent results with the LGR model for bi-wing fracture geometry.

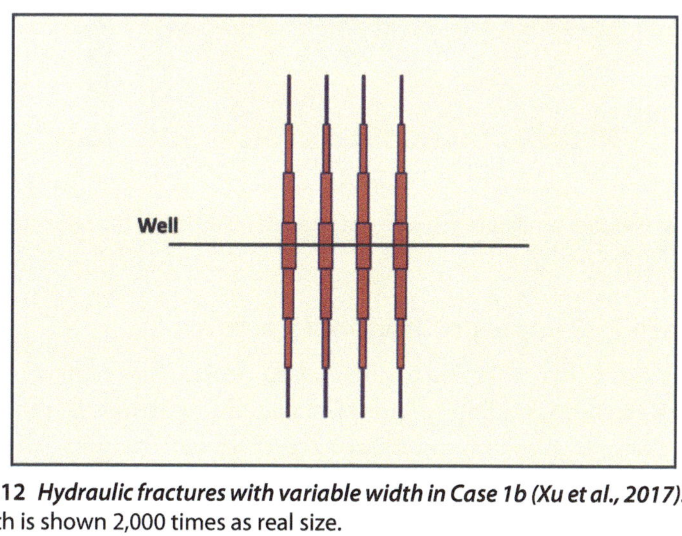

Figure 4.12 ***Hydraulic fractures with variable width in Case 1b (Xu et al., 2017).*** The fracture width is shown 2,000 times as real size.

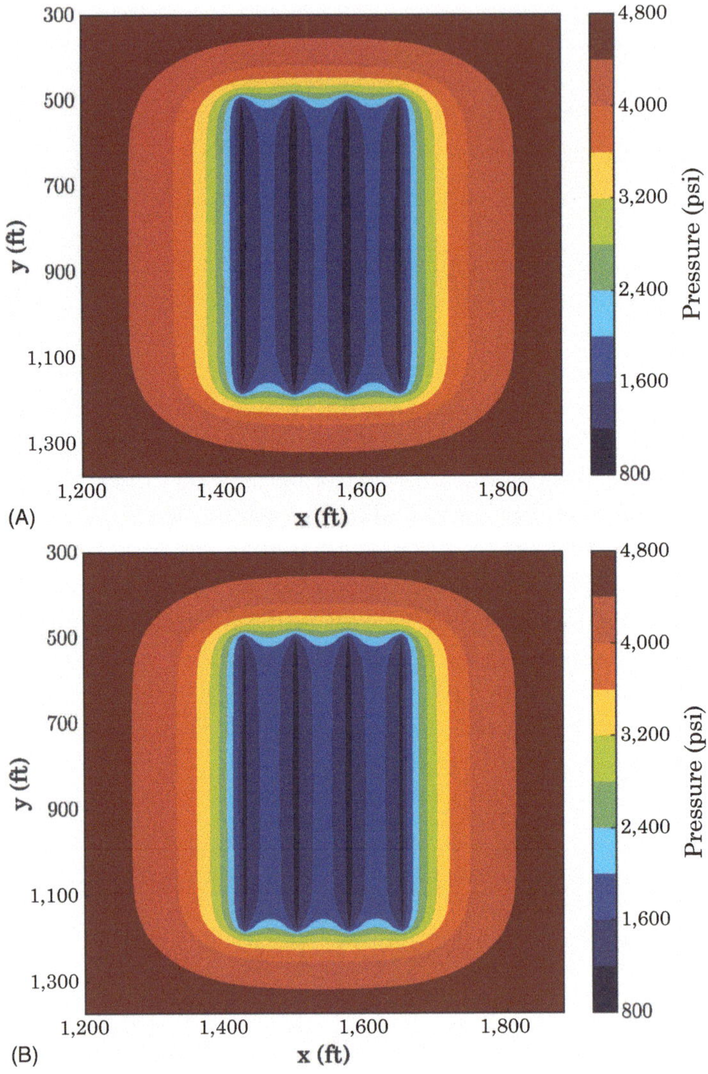

Figure 4.13 Pressure profiles (1 year) for Case 1a predicted by (A) the LGR model and (B) the EDFM *(Xu et al., 2017)*.

4.5.2 Case 2: Complex orthogonal fractures

The operation of hydraulic fracturing may cause the reopening of pre-existing natural fractures, leading to complex fracture networks. In this case, we show an example involving intersecting orthogonal fractures. All fractures are aligned with one of the axes of the matrix grid; hence, the LGR model can give an accurate solution. The reservoir dimensions and the locations

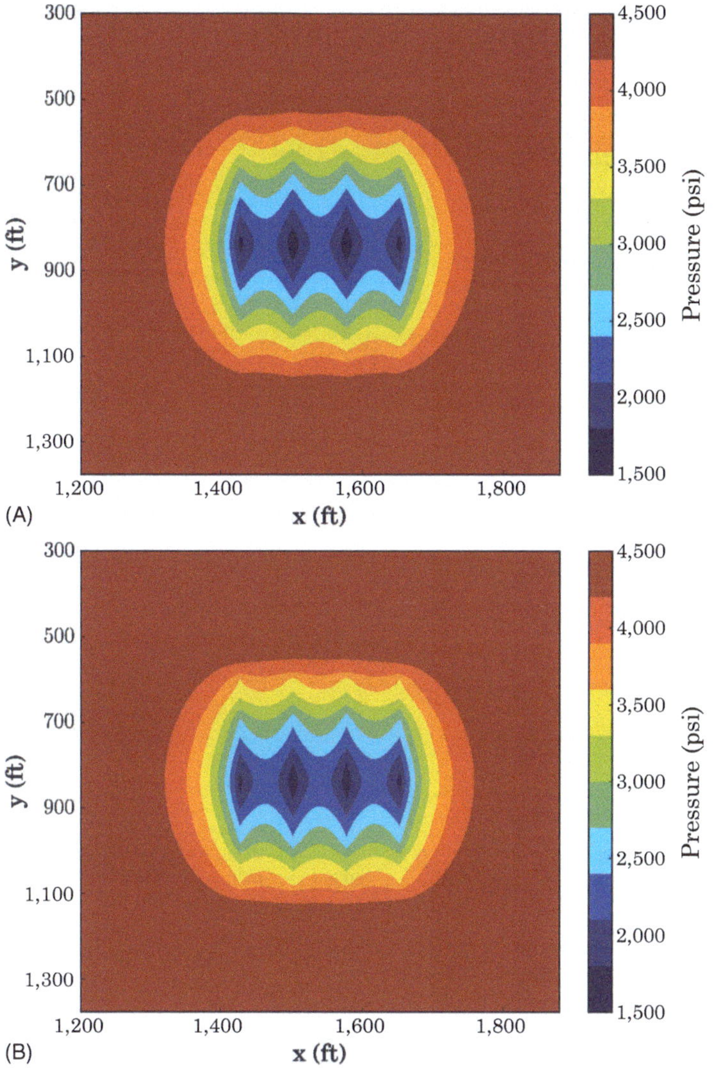

Figure 4.14 Pressure profiles (1 year) for Case 1b predicted by (A) the LGR model and (B) the EDFM *(Xu et al., 2017)*.

of hydraulic fractures are kept the same as in Case 1. Twelve natural fractures are added into the reservoir to form a fracture network, as shown in Fig. 4.16. By changing the conductivity of natural fractures, two sub-cases were investigated. In the first sub-case (Case 2a), the natural fractures have the same width and permeability as hydraulic fractures. In the second sub-case (Case 2b), the permeability of all natural fractures is reduced to 5 md

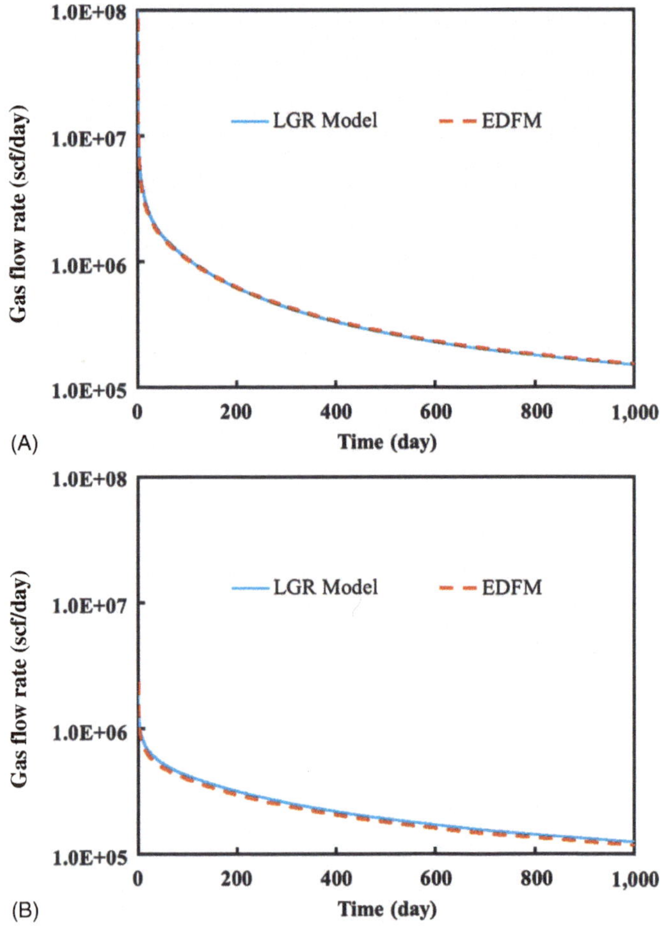

Figure 4.15 Gas flow rate for (A) Case 1a and (B) Case 1b predicted by the LGR model and the EDFM (*Xu et al., 2017*).

so that they have a finite conductivity. To capture the fracture intersections in the LGR model, the gridblocks with fracture intersections are refined in both the x- and y-directions (3×3). This introduces a small control-volume at each fracture intersection.

Fig. 4.17 shows the similar pressure profiles predicted by the LGR model and the EDFM after 240 days of production for Case 2a. The slight pressure difference near the fractures is caused by simplification of transmissibility calculation in the EDFM. Fig. 4.18 shows a good agreement between the gas flow rate simulated by both models for Cases 2a and 2b. The EDFM accurately handled the fracture intersections with finite fracture conductivity, illustrating that the EDFM has the capability of modeling fracture networks.

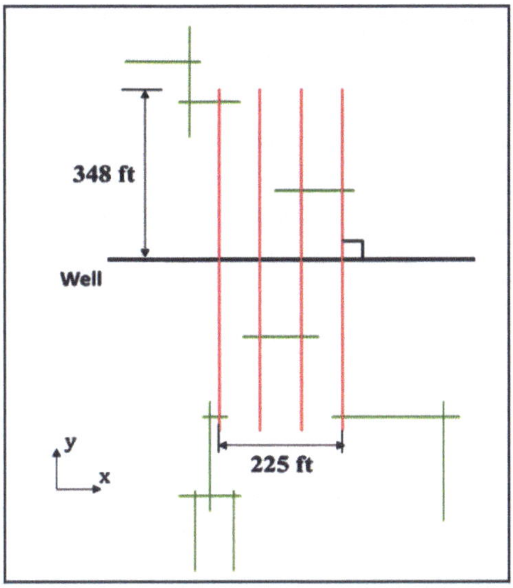

Figure 4.16 ***Fracture network of Case 2 (Xu et al., 2017).*** The purple lines represent the hydraulic fractures and the green lines represent the natural fractures.

4.5.3 Case 3: Non-orthogonal fractures

During hydraulic fracturing, the fractures tend to grow in the direction perpendicular to the minimum horizontal stress. In some cases, when the well axis deviates from the direction of the minimum horizontal stress, the preferred direction of fracture propagation will not be perpendicular to the wellbore. This often leads to the propagation of non-orthogonal fractures. One of the advantages of the EDFM is its flexibility in modeling non-orthogonal fractures in any orientation.

In Case 3, we present an example where the fractures have an angle of 63 degrees with the wellbore, as shown in Fig. 4.19. The fracture width and permeability are the same as in Case 1. A "zigzag" LGR model with a gridblock size of 1.5 × 1.5 ft was created in an attempt to approximately represent the non-orthogonal fractures as shown in Fig. 4.20, as it is always difficult to "exactly" model a non-orthogonal fracture by grid refinement. The reservoir dimensions were decreased (523.5 × 600 × 80 ft) compared with previous cases to reduce the simulation time.

The pressure profiles simulated by both the LGR model and the EDFM are shown in Fig. 4.21, and the gas flow rate curves are presented in Fig. 4.22, where the close agreement demonstrates the accuracy of the EDFM in modeling non-orthogonal fractures. By using the EDFM, one can avoid using unstructured grids and empower a commercial simulator to model

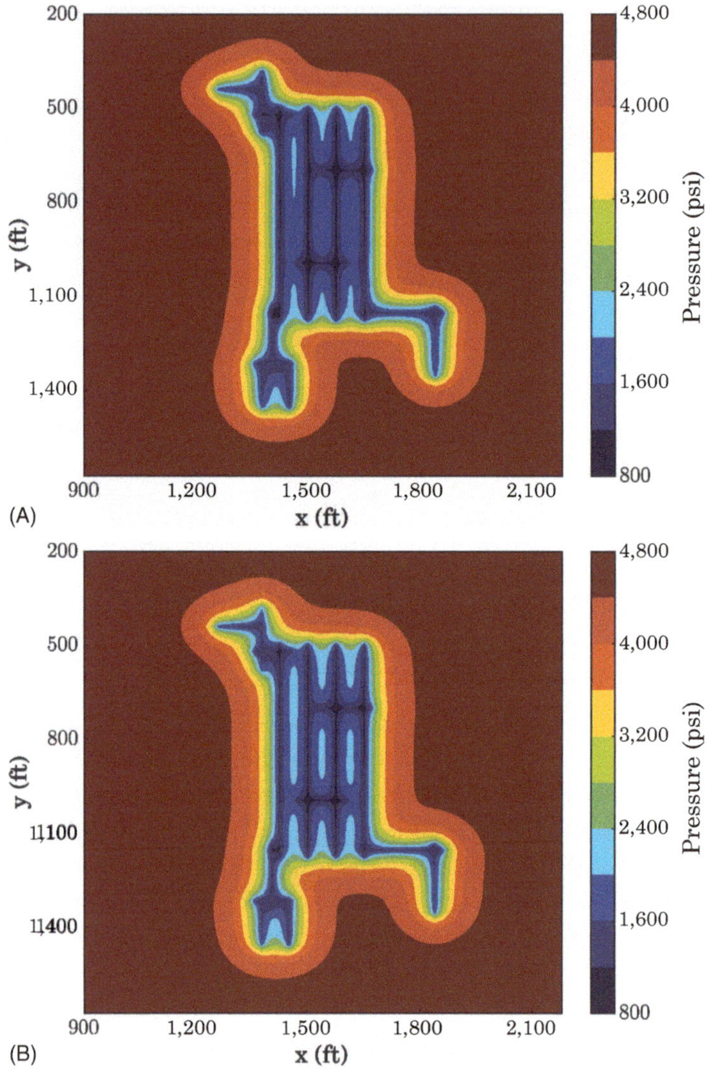

Figure 4.17 Pressure profiles (240 days) for Case 2a predicted by (A) the LGR model and (B) the EDFM *(Xu et al., 2017)*.

non-orthogonal fractures accurately and effectively without having access to the source code of the simulator.

4.5.4 Case 4: Nonplanar fractures

The propagation of hydraulic fractures is controlled by several mechanisms. One of them is the stress interaction between fractures, which is commonly

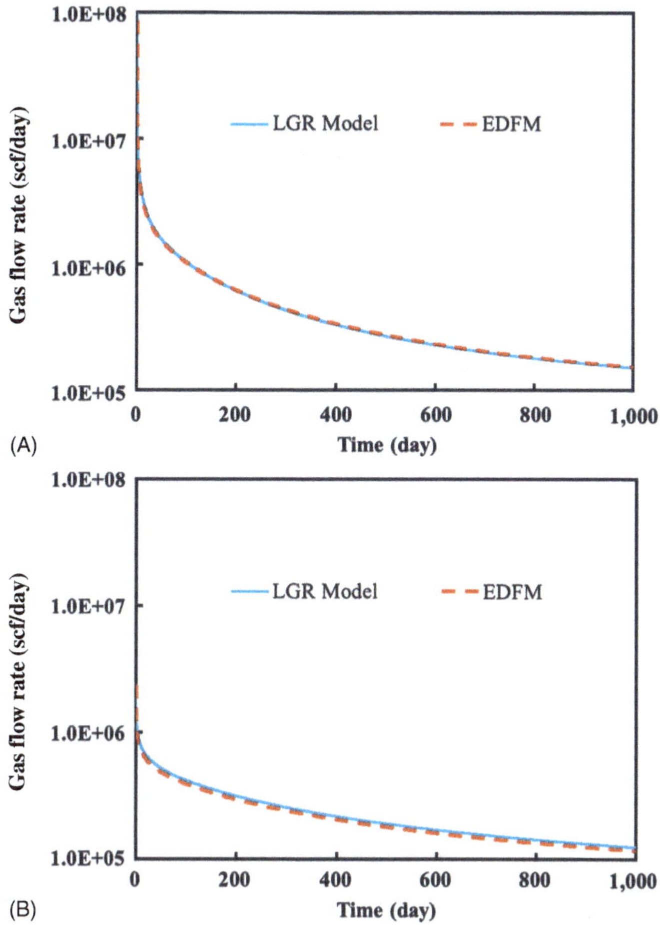

(A) (B)

Figure 4.18 Gas flow rate for (A) Case 2a and (B) Case 2b predicted by the LGR model and the EDFM *(Xu et al., 2017)*.

referred to as the "stress shadow" effect. Because of this effect, the fractures in a single fracturing stage may have different lengths and geometries, and some fractures will have a nonplanar fracture shape. Here, we present an example where the fractures are nonplanar and the fracture length differs in a single fracturing stage. An LGR model was built through the "zigzag" approximation, as shown in Fig. 4.20. Each nonplanar fracture was discretized into two non-orthogonal planar fractures. These two connecting fractures behave as a single nonplanar fracture, as illustrated in previous section. Several small fracture segments were removed with Eq. (4.43) due to their extra-small volume.

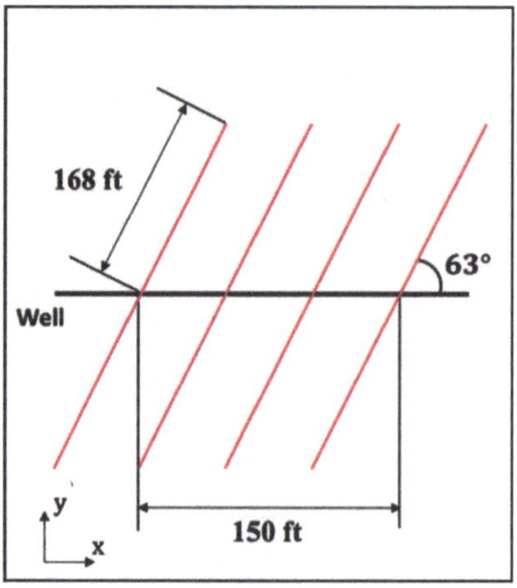

Figure 4.19 ***Non-orthogonal fractures in Case 3 (Xu et al., 2017).*** The purple lines represent the hydraulic fractures.

For both the LGR model and the EDFM, the predicted pressure profiles are shown in Fig. 4.23, and the gas flow rate curves are presented in Fig. 4.24. The good agreement between both models demonstrates the accuracy of the EDFM in modeling non-orthogonal fractures. The EDFM allows a general simulator to model nonplanar fractures accurately and conveniently.

4.5.5 Case 5: Complex nonplanar fractures with variable width

The EDFM has shown its applicability and accuracy compared with the LGR models in previous case studies. The LGR model is not able to handle cases where the fractures are highly nonplanar and have a variable width along the length. However, the "embedded" nature of the EDFM gives it the flexibility to model such fractures.

Here, we present a case where the hydraulic fracture geometries are generated by a fracture propagation model (Wu and Olson, 2015), as shown in Fig. 4.25. The model provided the fracture width at the end of each segment, and we calculated the average width and effective permeability using Eqs. (4.35) and (4.41) for each segment, respectively. We modeled each nonplanar fracture as ten interconnected planar fracture segments. The fracture permeability was calculated by the cubic law assuming a smooth surface. The reservoir dimensions and grid size are the same as those in Case 1 (3,100 × 1,675 × 80 ft, discretized by 200 × 99 gridblocks horizontally).

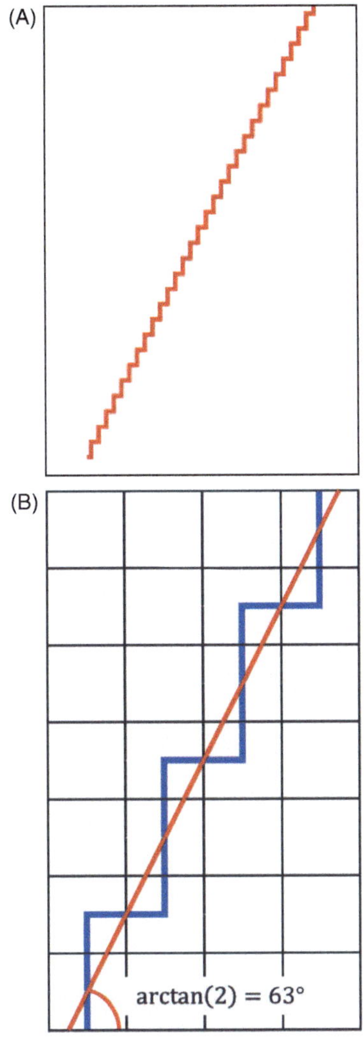

Figure 4.20 ***Explanation of "zigzag" fracture representation in Case 3 (Xu et al., 2017).*** The red lines in (A) represent the high-permeability gridblocks for fractures. A closer view is given in (B), where the red line shows the actual fracture shape, and the blue lines show the "zigzag" representation of the fracture shape.

The semi-analytical solution by Yu (2015) and Yu et al. (2016) is used as a reference solution.

Fig. 4.26 shows the gas flow rate for both models, and Fig. 4.27 gives the pressure profiles at 120 days. As expected, the EDFM successfully modeled the shape of the complex nonplanar fractures and reached a good agreement with the semi-analytical solution in the gas flow rate.

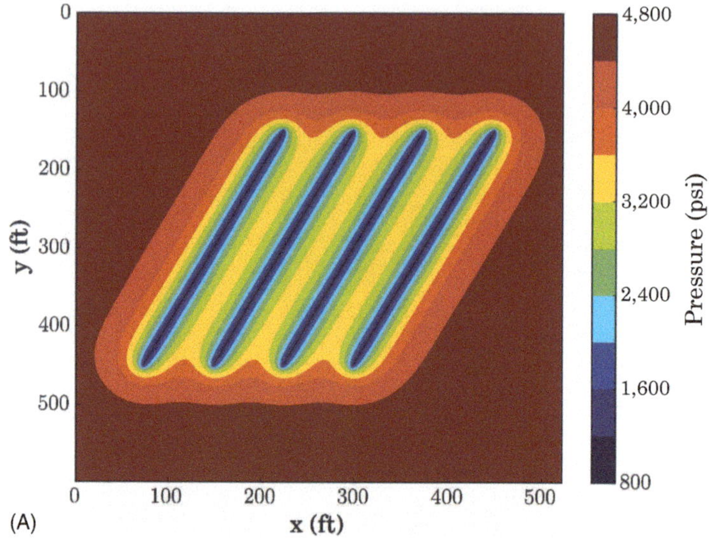

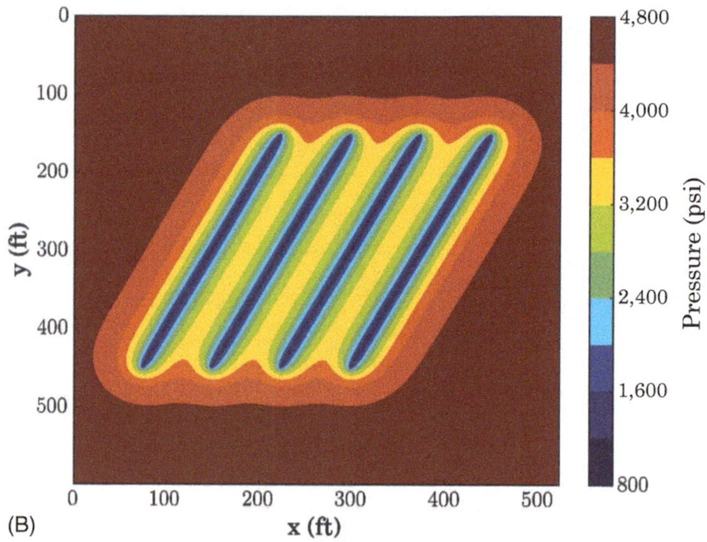

Figure 4.21 Pressure profiles (60 days) for Case 3 predicted by (A) the LGR model and (B) the EDFM *(Xu et al., 2017)*.

4.5.6 Computational efficiency comparison

Computational efficiency is an important factor in determining the applicability of a model. Table 4.3 summarizes the comparison in CPU time between the LGR models and the EDFM using RS1.

In Table 4.3, T_{total} is the CPU time for a case with the LGR model or the EDFM, T_{nofrac} is the CPU time for the same case without fractures,

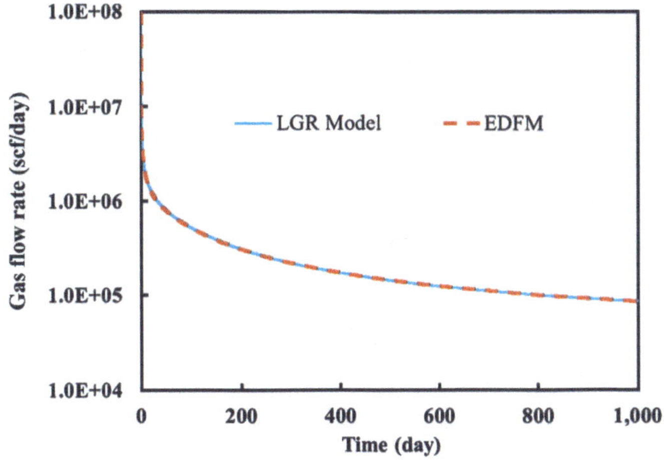

Figure 4.22 ***Gas flow rate for Case 3 predicted by the LGR model and the EDFM (Xu et al., 2017).***

T_{EDFM}/T_{LGR} is the CPU time ratio of the EDFM to the LGR model, N_m is the total number of matrix cells, N_{extra} is the extra number of cells introduced by the LGR or the EDFM, and the last column shows the total number of Newton cycles.

It can be observed that although N_{extra} is much smaller than N_m, there is always a large increase in CPU time after introducing these extra cells. Therefore, these extra cells play a critical role in limiting the simulation efficiency. For orthogonal cases (Cases 1 and 2), the fractures can be precisely modeled by LGR, but the number of extra cells in the LGR model is higher than that in the EDFM. In Cases 1a and 1b, the EDFM has no advantage over the LGR model. However, as the number of extra cells increases, the EDFM has a better computational performance than the LGR models in Cases 2a and 2b.

For non-orthogonal or nonplanar fracture cases (Cases 3 and 4), to approximate the inclination angle, the LGR models require many more extra cells due to grid refinement. In addition, for LGR models, the small control volumes at fracture intersections undermine the simulation time step. As a result, the EDFM has a great advantage over the LGR models regarding CPU performance. Furthermore, the EDFM is flexible in modeling irregular fracture shapes, which is very challenging for the LGR model.

The comparison clearly illustrates the difference between the LGR model and the EDFM. For the LGR model, a large number of extra cells are required around the fractures to model the complex fracture shape, leading

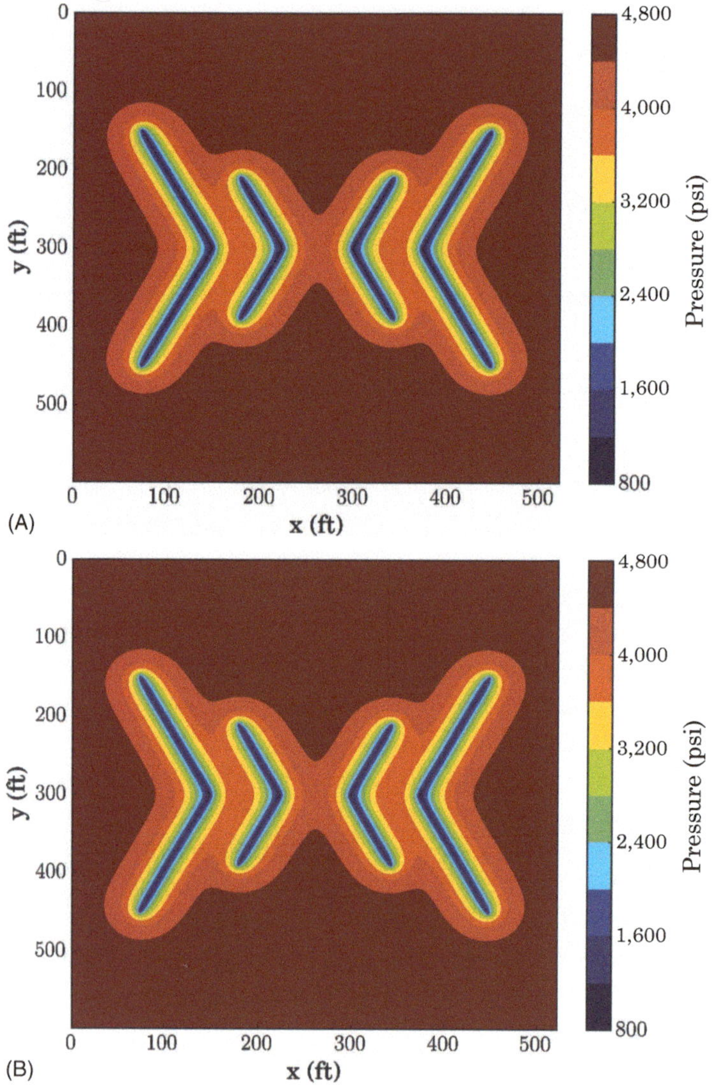

Figure 4.23 Pressure profiles (30 days) for Case 4 predicted by (A) the LGR model and (B) the EDFM *(Xu et al., 2017)*.

to a high computational cost. However, the EDFM avoids this problem by handling the connections mathematically without losing accuracy.

As we performed all the case studies using both RS1 and RS2, the CPU time of the EDFM for all cases using RS1 and RS2 are summarized in Table 4.4. In both simulators, the EDFM shows high computational performance. Table 4.5 shows the difference in cumulative gas production at

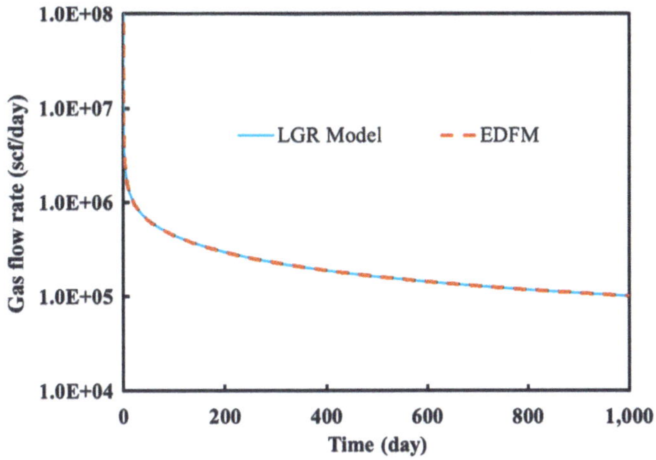

Figure 4.24 ***Gas flow rate for Case 4 predicted by the LGR model and the EDFM (Xu et al., 2017).***

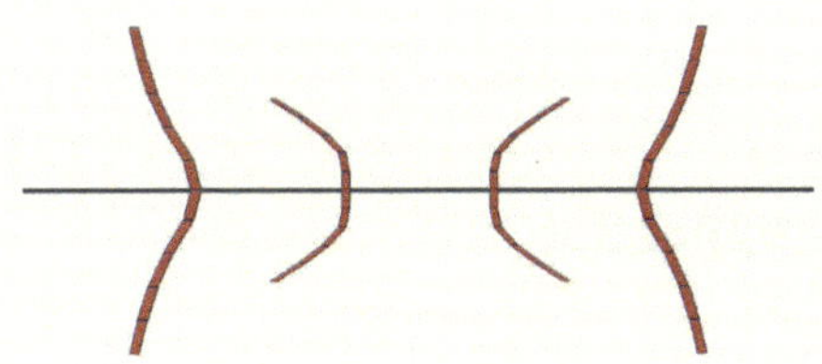

Figure 4.25 ***Hydraulic fracture geometries for Case 5 (Xu et al., 2017).*** The length in the *y*-direction is shown 0.2 times as the real size to clearly show the nonplanar shape. The fracture width is shown 1,000 times the real size.

1,000 days between RS1 and RS2. It can be observed that high consistency was obtained between both simulators.

It should be noted that the EDFM introduces off-diagonal terms in the discretized pressure matrix, and this may lead to more solver iterations in the computations. Therefore, the computational speed of the EDFM can be affected by the performance of the linear solver used in the simulators. For example, In RS2, we observed a great difference in CPU performance for different solvers. For cases with non-orthogonal fractures, the constrained pressure residual preconditioner has a better performance than the nested factorization preconditioner. We recommend this preconditioner for EDFM simulations due to its efficient reordering of the operating matrix.

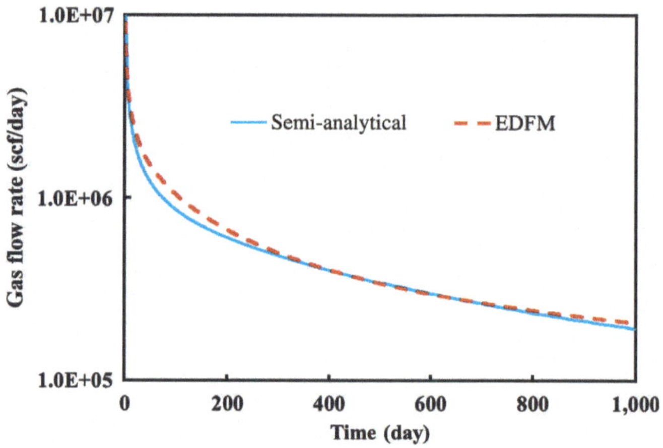

Figure 4.26 ***Gas flow rate for Case 5 predicted by the EDFM and the semi-analytical solution (Xu et al., 2017).***

The case studies show the high computational efficiency of the EDFM in the two reservoir simulators used in this work. Although the performance of the EDFM may vary in different simulators, we expect the EDFM to be faster than the LGR model for complex fracture geometries.

4.6 Model application

In the previous sections, the EDFM was shown to be a powerful extension of conventional reservoir simulators, able to handle shale reservoirs with complex fracture. It is applicable to naturally fractured reservoirs as well as hydraulically fractured reservoirs (Moinfar et al., 2014; Cavalcante Filho et al., 2015a, 2015b). In this section, we present two case studies to demonstrate the application of the EDFM in shale reservoirs with nonplanar hydraulic fractures and the inclusion of many natural fractures.

4.6.1 Case 6: Shale gas production in a naturally fractured reservoir

The same matrix grid and reservoir parameters as in Case 1 are used in this study (3,100 × 1,675 × 80 ft, discretized by 200 × 99 gridblocks horizontally). The hydraulic fractures and natural fractures are shown in Fig. 4.28. The pre-existing natural fractures have a dominant direction as shown in the figure. Each nonplanar hydraulic fracture is discretized into 75 planar fractures with different widths, and each natural fracture is represented by a single fracture plane. The natural fracture conductivity is assumed to be

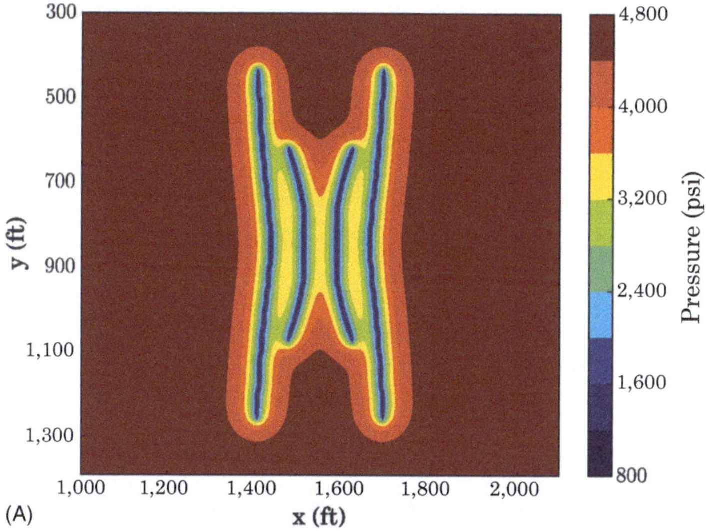

(A)

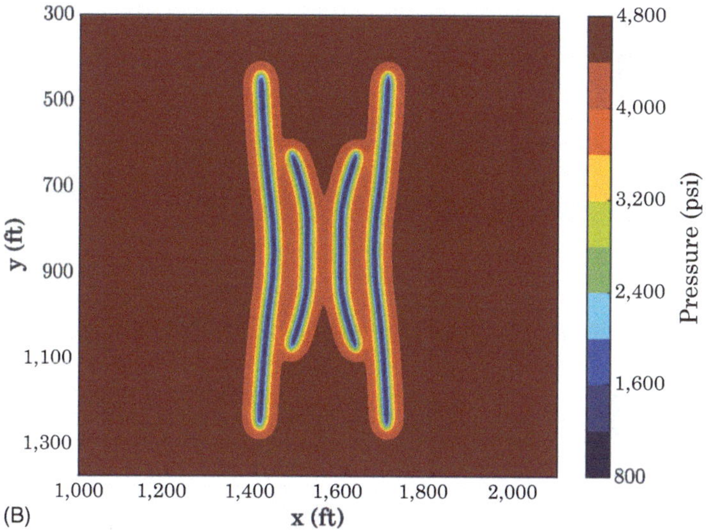

(B)

Figure 4.27 Pressure profiles (120 days) for Case 5 predicted by (A) the EDFM and (B) the semi-analytical solution *(Xu et al., 2017)*.

1 md-ft. Some natural fractures near the wellbore impacted the propagation of hydraulic fractures by diverting their growth path. All natural fractures are divided into two groups, as shown in different colors in Fig. 4.28. The blue lines represent the natural fractures that intersect the hydraulic fractures. They become part of the fracture network and increase the contact area between hydraulic fractures and the reservoir. The green lines represent

Table 4.3 CPU time comparison between the LGR models and the EDFM for different cases using RS1 (Xu et al., 2017).

Case	Model	T_{total} (s)	T_{nofrac} (s)	T_{EDFM}/T_{LGR}	N_m	N_{extra}	Number of Newton cycles
Case 1a	LGR	18	7	1.28	19,800	392	141
	EDFM	23				196	210
Case 1b	LGR	17		1.09		392	138
	EDFM	19				196	137
Case 2a	LGR	52		0.87		802	328
	EDFM	45				373	320
Case 2b	LGR	34		0.91		802	269
	EDFM	31				373	218
Case 3	LGR	7,676	46	0.05	139,600	9,696	2,253
	EDFM	361				1,204	332
Case 4	LGR	8,814		0.05		7,776	2,032
	EDFM	466				984	383

Table 4.4 CPU time of the EDFM simulations using both RS1 and RS2 (Xu et al., 2017).

Case	Case 1a	Case 1b	Case 2a	Case 2b	Case 3	Case 4	Case 5	Case 6a	Case 6d	Case 7
T_{total} of RS1 (sec)	23	19	45	31	361	466	41	60	102	3995
T_{total} of RS2 (sec)	7	4	5	4	64	40	4	18	20	2921

Table 4.5 Difference in cumulative gas production at 1,000 days between RS1 (EDFM) and RS2 (EDFM) (Xu et al., 2017).

Case	Case 1a	Case 1b	Case 2a	Case 2b	Case 3	Case 4	Case 5	Case 6a	Case 6d
Error	0.30%	0.19%	0.84%	1.61%	0.39%	0.86%	1.75%	0.31%	1.01%

other natural fractures, which are disconnected and relatively far from the well. The intersections between the wellbore and natural fractures are not considered in this study.

Four comparative simulation runs were performed to investigate how these two groups of fractures affect the well performance separately. In Case 6a, only the hydraulic fractures are considered. In Cases 6b and 6c, in addition to the hydraulic fractures, the natural fractures, "green" and "blue," are added into the reservoir, respectively. In Case 6d, all the hydraulic and natural fractures are considered in the simulation.

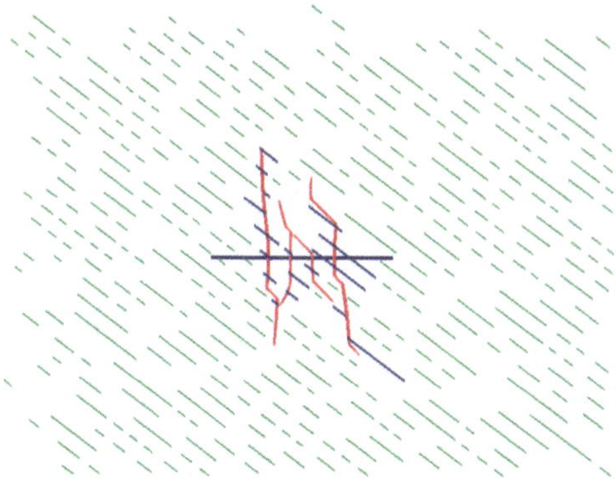

Figure 4.28 ***Hydraulic fractures (purple) and natural fractures (green and blue) for Case 6 (Xu et al., 2017).*** The width of hydraulic fractures is shown 1,000 times the real size. The blue lines represent the natural fractures that intersect the hydraulic fractures, and the green lines represent the "disconnected" natural fractures. The number of "blue" fractures is 17, and the number of "green" fractures is 360.

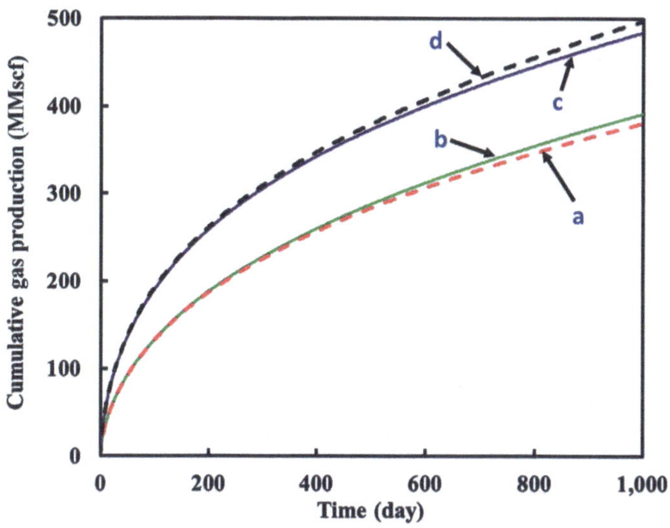

Figure 4.29 ***Cumulative gas production for Case 6 (Xu et al., 2017).***

The influence of different types of natural fractures on production is shown in Fig. 4.29. Gas production in Cases 6c and 6d is much higher than that of Cases 6a and 6b, indicating the sizable influence of the connected "blue" fractures. In addition, early in the time observed (before 200 days), the cumulative production curves of Cases 6a and 6b almost overlap, and the gas production of 6c is almost the same as that of Case 6d. This indicates the "green" fractures are not evidently influencing the early time production.

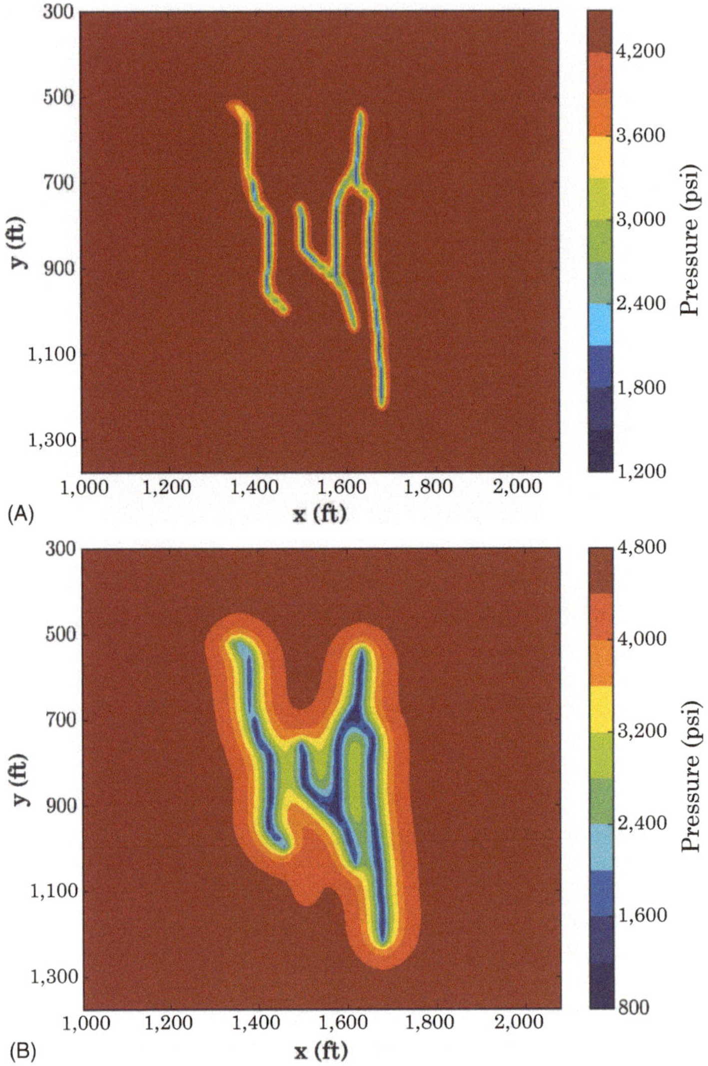

Figure 4.30 Pressure profiles for Case 6a (no natural fracture) (A) after 10 days and (B) after 105 days *(Xu et al., 2017)*.

Later in the time observed (after 200 days), the "green" fractures began to affect the production, indicated by the gas rate of 6b being evidently higher than Case 6a. The "green" fractures enhance the flow in the region far from the wellbore and decrease the time for gas to reach the fracture network.

The pressure profiles shown in Figs. 4.30 and 4.31 further illustrate the impact of both types of natural fractures. The influence of the "blue"

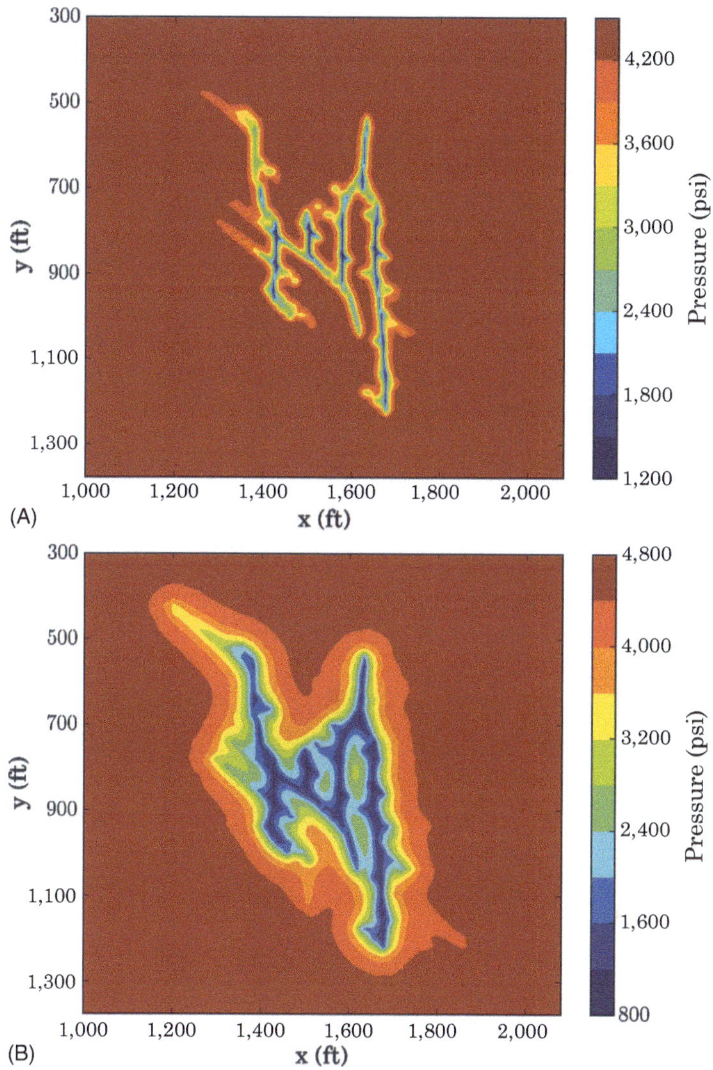

Figure 4.31 Pressure profiles for Case 6d (with all natural fractures) (A) after 10 days and (B) after 105 days *(Xu et al., 2017)*.

natural fractures can be easily observed in Fig. 4.31. Although they are not as highly conductive as hydraulic fractures, their influence on the pressure distribution around the fractures cannot be ignored. In contrast, the "green" natural fractures are not obviously impacting the pressure field around the hydraulic fractures, but they are implicitly enhancing the overall matrix permeability.

This example shows the influences of different types of natural fractures on production. The connectivity between natural fractures and hydraulic fractures can greatly impact both the early time and late time production, and the existence of a large number of natural fractures affects the late time production significantly. When predicting only the early-time production, it is possible to ignore the "disconnected" natural fractures.

It is also shown in this case that the EDFM can effectively model both fracture networks and disconnected natural fractures. It provides a powerful tool in rapid production analysis for highly fractured reservoirs.

4.6.2 Case 7: Three-dimensional, three-phase, multi-component simulation

As mentioned before, the EDFM can be used as a general procedure in both 2D and 3D cases, and for both compositional and black-oil formulations. In real field applications, compositional simulations are often used. In addition, a reservoir may have multiple layers, and the height of hydraulic fractures may be smaller than the reservoir height. Therefore, we present a 3D, compositional simulation example to show the application of the EDFM in a typical field study.

The reservoir dimensions are 6,000 × 1,500 × 80 ft and a uniform 300 × 75 × 5 matrix grid is used in this study. The reservoir parameters are summarized in Table 4.6. The fluid properties of the Bakken formation are used in the simulation by assuming seven pseudo components: CO_2,

Table 4.6 Basic reservoir and operation parameters in simulation for Case 7 (Xu et al., 2017).

Parameter	Value	Unit
Reservoir permeability	0.002	md
Reservoir porosity	7%	–
Initial reservoir pressure	2,000	psi
Reservoir thickness	80	ft
Rock compressibility	1×10^{-6}	psi^{-1}
Depth of top layer	1,000	ft
Initial water saturation	40%	–
Water viscosity	0.243	cp
Water density	62.8	lb/ft^3
Reservoir temperature	240	°F
Wellbore radius	0.25	ft
Bottomhole pressure	500	psi
Simulation time	1,000	day

Table 4.7 Parameters for matrix relative permeability curve for Case 7 (Xu et al., 2017).

Parameter	Value
Residual water saturation	20%
Residual oil saturation	10%
Residual gas saturation	10%
Water relative permeability endpoint	0.7
Oil relative permeability endpoint	0.8
Gas relative permeability endpoint	0.8
Water relative permeability exponent	2.0
Oil relative permeability exponent	2.0
Gas relative permeability exponent	2.0

N_2, CH_4, C_2–C_4, C_5–C_7, C_8–C_9, and C_{10+}, and their molar fractions: 0.02%, 0.04%, 25%, 22%, 20%, 13%, and 19.94%, respectively. More details of the fluid properties can be found in Yu et al. (2015). Peng-Robinson equation of state (Peng and Robinson, 1976) is used here to calculate the hydrocarbon phase behavior. The capillary effect is not considered in this study. For matrix gridblocks, a Corey-type relative permeability curve is used for water and gas, and the corresponding parameters are also presented in Table 4.7. Stone's second model is used for three-phase oil relative permeability. For fractures, a straight-line relative permeability model is used for all phases.

Fifteen hydraulic fractures and 200 natural fractures are put into the reservoir (as shown in Fig. 4.32). Each hydraulic fracture is discretized into 11 segments with their width ranging from 0.001 to 0.015 ft. The fracture permeability is constant (10,000 md). The conductivity of natural fractures is 1 md-ft. All natural fractures fully penetrate the reservoir height, with dip angles ranging from 60 to 90 degrees. The hydraulic fractures partially penetrate the reservoir in the vertical direction, with a distance of 16 ft from the top and bottom of the reservoir, respectively.

Fig. 4.33 shows the water, oil, and gas production predicted by the EDFM. Fig. 4.34 shows the profiles of pressure, mole fraction of C8–C9 in the oleic phase, and gas saturation after 1,000 days of production. The pressure profile indicates the pressure interference between hydraulic and natural fractures, which is greatly influenced by fracture geometries. Using the compositional model in RS1 and RS2, the CPU times for Case 7 without fractures are 956 and 549 seconds, respectively. After adding fractures using the EDFM, the CPU times are 3995 seconds (in RS1) and 2921 seconds (in RS2).

(A)

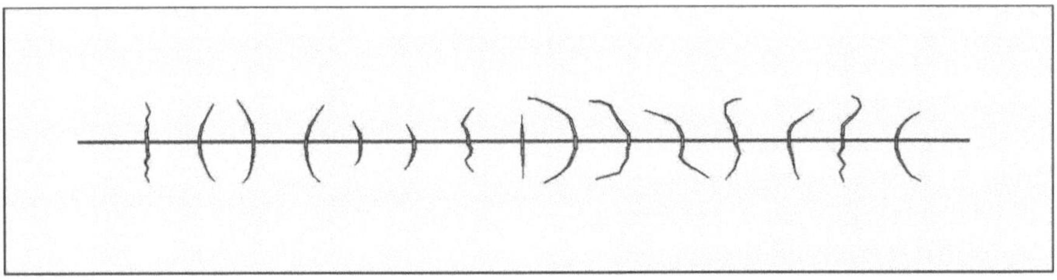

(B)

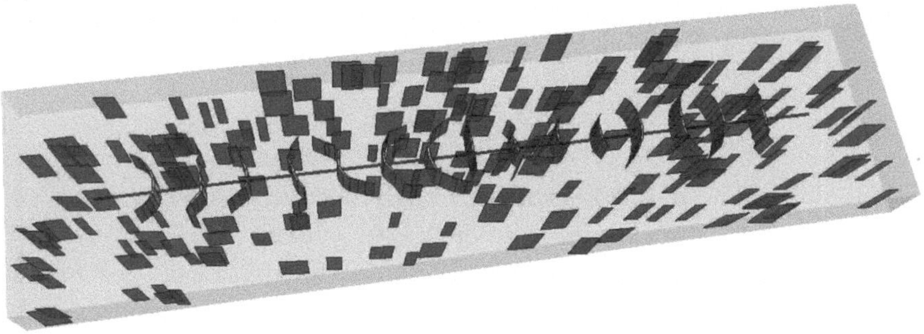

Figure 4.32 ***Reservoir dimensions and fractures in Case 7 (Xu et al., 2017).*** The yellow planes represent hydraulic fractures and the green planes represent natural fractures. (A) Top view of hydraulic fractures. The fracture widths are shown 1,000 times the real size. (B) 3D view of hydraulic fractures and natural fractures.

4.7 Remarks

In this chapter, we presented the EDFM and its application in Cartesian grid. The EDFM was developed to simulate fractures in traditional reservoir simulators without using conforming meshes. As it keeps the structured grids used by conventional simulators, it can be applied to different types of reservoir simulators in a non-intrusive manner.

The accuracy of the EDFM for simulating various types of fracture geometries was confirmed by comparison with the LGR model and a semi-analytical solution. Comparison of computational time showed the advantage of the EDFM over the LGR model. The EDFM is flexible in handling fractures with complex geometries without introducing a large number of extra gridblocks, which guarantees its computational efficiency.

In summary, compared with three traditional methods including the dual-continuum model, the LGR model, and the discrete fracture model

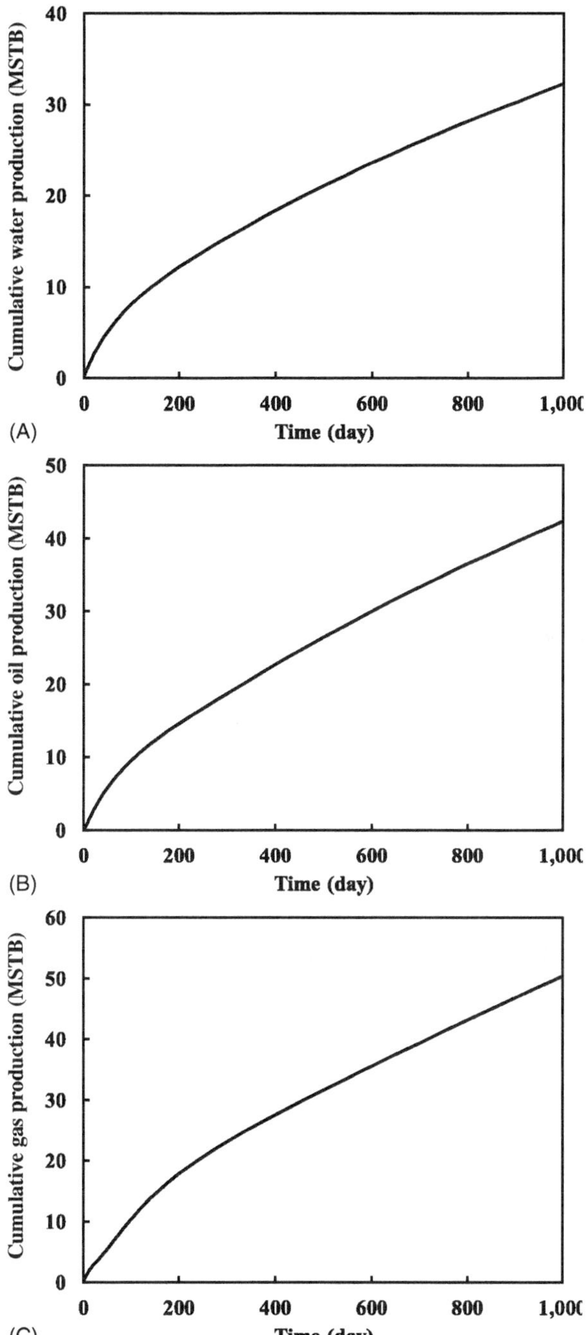

Figure 4.33 Cumulative production predicted by the EDFM for (A) water (B) oil, and (C) gas *(Xu et al., 2017)*.

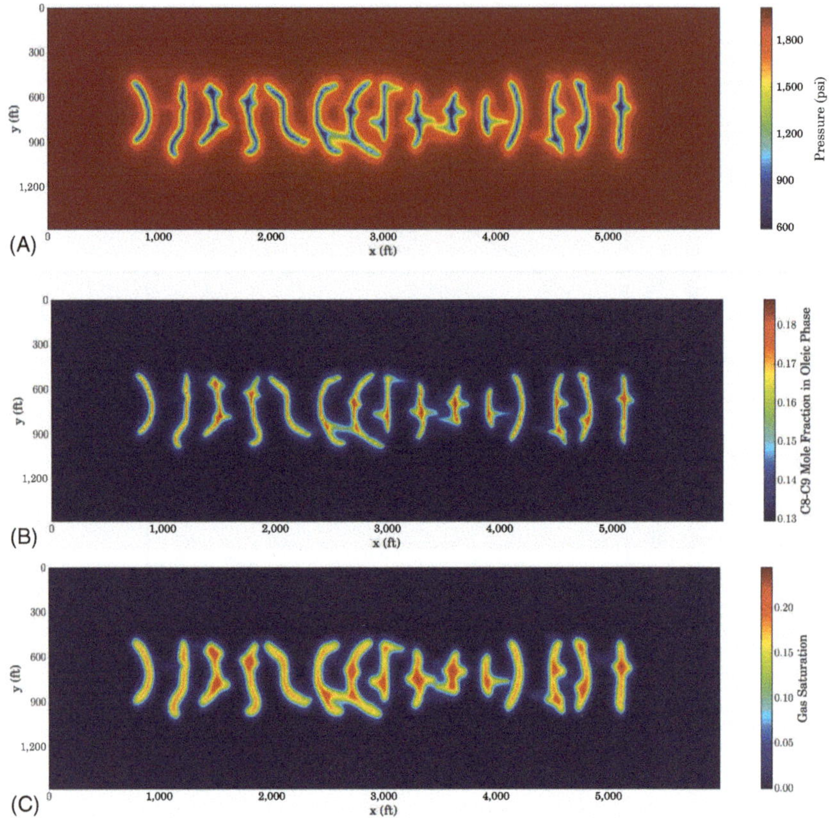

Figure 4.34 Profiles of (A) pressure (B) C8–C9 mole fraction in the oleic phase, and (C) gas saturation after 1,000 days of production predicted by the EDFM (Layer 3) *(Xu et al., 2017)*.

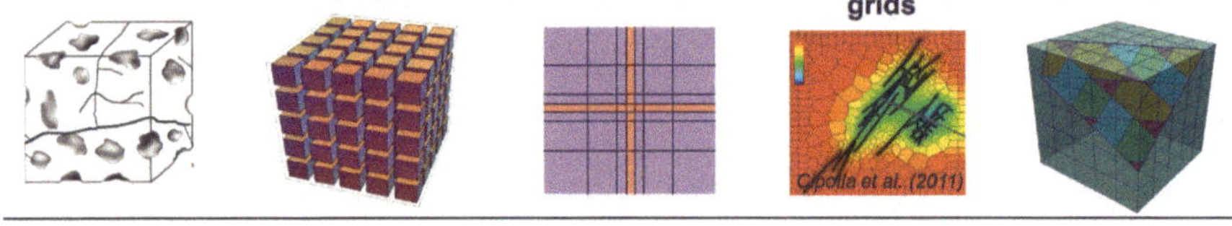

	DPDK	LGR	Unstructured grids	EDFM
Accuracy	Low	High	High	High
Flexibility	Low	Low	High	High
Gridding	Easy	Easy	Difficult	Easy
Computational efficiency	High	Low	Low	High

Figure 4.35 *Comparison of different methods for modeling complex fractures.*

using unstructured grids, the EDFM obtains a balance between accuracy, flexibility, ease of gridding, and computational efficiency for the simulation of complex fractures, as summarized in Fig. 4.35.

Nomenclature

A = area, ft
B = formation volume factor
c = compressibility, psi^{-1}
C_f = fracture conductivity, md-ft
d = average distance, ft
dS = area element, ft^2
dV = volume element, ft^3
H = fracture height, ft
H_s = height of fracture segment, ft
k = reservoir permeability, md
k_f = fracture permeability, md
K = matrix permeability tensor, md
K_α = differential equilibrium portioning coefficient of gas at a constant temperature
L = fracture length, ft
L_{int} = length of fracture intersection line, ft
L_s = length of fracture segment, ft
$\vec{n}$ = normal vector
N = number of nnc
p = pressure, psi
Q = volume flow rate, ft^3/day
re = effective radisu, ft
rw = wellbore radius, ft
R_S = solution gas–oil ratio, scf/STB
T = transmissibility, md-ft or temperature, °F
V = volume, ft^3
V_b = bulk volume, ft^3
V_m = langmuir isotherm constant, scf/ton
w_f = fracture width, ft
$\overline{w_f}$ = average fracture width, ft
WI = well index, md-ft
x = distance, ft
x_f = fracture half-length, ft

Δp = pressure drop, psi
θ = angle between the wellbore axis and the fracture plane
λ = phase mobility, cp^{-1}
μ = viscosity, cp
ρ = density, g/cm^3
ϕ_f = fracture effective porosity

Subscripts and Superscripts
a = adsorbed
b = bulk
c = common face
eff = effective
f = fracture
g = gas
j = phase
L = Langmuir
m = matrix
o = oil
seg = fracture segment
ST = stock tank

References

Abass, H.H., Hedayati, S., Meadows, D.L., 1996. Nonplanar fracture propagation from a horizontal wellbore: experimental study. SPE Prod. Facil. 11 (3), 133–137.

Bahorich, B., Olson, J.E., Holder, J., 2012. Examining the effect of cemented natural fractures on hydraulic fracture propagation in hydrostone block experiments. Paper SPE 160197, SPE Annual Technical Conference and Exhibition. San Antonio, Texas.

Batchelor, G.K., 1970. Slender-body theory for particles of arbitrary cross-section in Stokes flow. J. Fluid Mech. 44 (3), 419–440.

Berkowitz, B., Naumann, C., Smith, L., 1994. Mass transfer at fracture intersections: an evaluation of mixing models. Water Resour. Res. 30 (6), 1765–1773.

Cavalcante Filho, J.S.A., Shakiba, M., Moinfar, A., Sepehrnoori, K., 2015a. Implementation of a preprocessor for embedded discrete fracture modeling in an IMPEC compositional reservoir simulator. Paper SPE 173289, SPE Reservoir Simulation Symposium. Houston, Texas.

Cavalcante Filho, J.S.A., Xu, Y., Sepehrnoori, K., 2015b. Modeling fishbones using the embedded discrete fracture model formulation: sensitivity analysis and history matching. Paper SPE 175124, SPE Annual Technical Conference and Exhibition. Houston, Texas.

Chai, Z., Tang, H., He, Y., Killough, J., Wang, Y., 2018. Uncertainty quantification of the fracture network with a novel fractured reservoir forward model. Paper SPE 191395, SPE Annual Technical Conference and Exhibition. Dallas, Texas.

de Sousa Junior, L.C., dos Santos, L.O.S., de Souza Rios, V., Araújo, C., Celes, W., Sepehrnoori, K., 2016. Methodology for geomechanically controlled transmissibility through active natural fractures in reservoir simulation. J. Pet. Sci. Eng. 147, 7–14.

Ding, D.Y., Farah, N., Bourbiaux, B., Wu, Y.S., Mestiri, I., 2018. Simulation of matrix/fracture interaction in low-permeability fractured unconventional reservoirs. SPE J. 23 (4), 1389–1411.

Ding, D.Y., Wu, Y.S., Jeannin, L., 2014. Efficient simulation of hydraulic fractured wells in unconventional reservoirs. J. Pet. Sci. Eng. 122, 631–642.
Du, S., Liang, B., Yuanbo, L., 2017. Field study: embedded discrete fracture modeling with artificial intelligence in Permian Basin for shale formation. Paper SPE 187202, SPE Annual Technical Conference and Exhibition. San Antonio, Texas.
El Rabaa, W., 1989. Experimental study of hydraulic fracture geometry initiated from horizontal wells. Paper SPE 19720, SPE Annual Technical Conference and Exhibition. San Antonio, Texas.
Flemisch, B., Berre, I., Boon, W., Fumagalli, A., Schwenck, N., Scotti, A., Stefansson, I., Tatomir, A., 2018. Benchmarks for single-phase flow in fractured porous media. Adv. Water Resour. 111, 239–258.
Fumagalli, A., Pasquale, L., Zonca, S., Micheletti, S., 2016. An upscaling procedure for fractured reservoirs with embedded grids. Water Resour. Res. 52 (8), 6506–6525.
Fumagalli, A., Zonca, S., Formaggia, L., 2017. Advances in computation of local problems for a flow-based upscaling in fractured reservoirs. Math. Comput. Simul. 137, 299–324.
Hajibeygi, H., Karvounis, D., Jenny, P., 2011. A hierarchical fracture model for the iterative multiscale finite volume method. J. Comput. Phys. 230 (24), 8729–8743.
Hearn, C.L., Al-Emadi, I.A.A., Worley, P.L.H., Taylor, R.D., 1997. Improved oil recovery in a tight reservoir with conductive faults, ISND Shuaiba, Qatar. Paper SPE 38908, SPE Annual Technical Conference and Exhibition. San Antonio, Texas.
Hui, M.-H., Dufour, G., Vitel, S., Muron, P., Tavakoli, R., Rousset, M., Rey, A., Mallison, B., 2019. A robust embedded discrete fracture modeling workflow for simulating complex processes in field-scale fractured reservoirs. Paper SPE 193827, SPE Reservoir Simulation Conference. Galveston, Texas.
Hustedt, B., Qiu, Y., Zwarts, D., van den Hoek, P.J., 2005. Modeling water-injection induced fractures in reservoir simulation. Paper SPE 95726, SPE Annual Technical Conference and Exhibition. Dallas, Texas.
Jiang, J., Younis, R.M., 2016. Hybrid coupled discrete-fracture/matrix and multicontinuum models for unconventional-reservoir simulation. SPE J. 21 (3), 1009–1027.
Jiang, J., Younis, R.M., 2017. An improved projection-based embedded discrete fracture model (pEDFM) for multiphase flow in fractured reservoirs. Adv. Water Resour. 109, 267–289.
Karimi-Fard, M., Durlofsky, L.J., Aziz, K., 2004. An efficient discrete-fracture model applicable for general-purpose reservoir simulators. SPE J. 9 (2), 227–236.
Langmuir, I., 1918. The adsorption of gases on plane surfaces of glass, mica and platinum. J. Am. Chem. Soc. 40, 1361–1403.
Lee, S.H., Lough, M.F., Jensen, C.L., 2001. Hierarchical modeling of flow in naturally fractured formations with multiple length scales. Water Resour. Res. 37 (3), 443–455.
Li, L., Lee, S.H., 2008. Efficient field-scale simulation of black oil in a naturally fractured reservoir through discrete fracture networks and homogenized media. SPE Reservoir Eval. Eng. 11 (4), 750–758.
Moinfar, A., Sepehrnoori, K., Johns, R.T., Varavei, A., 2013a. Coupled geomechanics and flow simulation for an embedded discrete fracture model. Paper SPE 163666, SPE Reservoir Simulation Symposium. The Woodlands, Texas.
Moinfar, A., Varavei, A., Sepehrnoori, K., John, R.T., 2013b. Development of a coupled dual continuum and discrete fracture model for the simulation of unconventional reservoirs. Paper SPE 163647, SPE Reservoir Simulation Symposium. The Woodlands, Texas.
Moinfar, A., Varavei, A., Sepehrnoori, K., Johns, R.T., 2014. Development of an efficient embedded discrete fracture model for 3D compositional reservoir simulation in fractured reservoirs. SPE J. 19 (2), 289–303.
Olson, J.E., 2008. Multi-fracture propagation modeling: applications to hydraulic fracturing in shales and tight gas sands. Paper ARMA 08-327, 42nd U. S. Rock Mechanics Symposium. San Francisco, California.

Panfili, P., Cominelli, A., 2014. Simulation of miscible gas injection in a fractured carbonate reservoir using an embedded discrete fracture model. Paper SPE 171830, Abu Dhabi International Petroleum Exhibition and Conference. Abu Dhabi.

Panfili, P., Colin, R., Cominelli, A., Giamminonni, D., Guerra, L., 2015. Efficient and effective field scale simulation of hydraulic fractured wells: methodology and application. Paper SPE 175542, SPE Reservoir Characterisation and Simulation Conference and Exhibition. Abu Dhabi, UAE.

Patzek, T.W., Male, F., Marder, M., 2013. Gas production in the Barnett Shale obeys a simple scaling theory. PNAS 110 (49), 19731–19736.

Peaceman, D.W., 1983. Interpretation of well-block pressures in numerical reservoir simulation with nonsquare grid blocks and anisotropic permeability. SPE J. 23 (3), 531–543.

Peng, D.Y., Robinson, D.B., 1976. A new two-constant equation of state. Ind. Eng. Chem. Fundam. 15 (1), 59–64.

Philip, Z.G., Jennings, J.W., Olson, J.E., Laubach, S.E., Holder, J., 2005. Modeling coupled fracture-matrix fluid flow in geomechanically simulated fracture networks. SPE Reservoir Eval. Eng. 8 (4), 300–309.

Ren, G., Jiang, J., Younis, R.M., 2016. A fully coupled XFEM-EDFM model for multiphase flow and geomechanics in fractured tight gas reservoirs. Procedia Comput. Sci. 80, 1404–1415.

Seidle, J.P., Arri, L.E., 1990. Use of conventional reservoir models for coalbed methane simulation. Paper PETSOC 90118, CIM/SPE International Technical Meeting. Calgary, Alberta.

Shakiba, M., 2014. Modeling and simulation of fluid flow in naturally and hydraulically fractured reservoirs using embedded discrete fracture model (EDFM). MS Thesis. The University of Texas at Austin.

Shakiba, M., Sepehrnoori, K., 2015. Using embedded discrete fracture model (EDFM) and microseismic monitoring data to characterize the complex hydraulic fracture networks. Paper SPE 175142, SPE Annual Technical Conference and Exhibition. Houston, Texas.

Tang, H., Chai, Z., Yan, B., Killough, J., 2017. Application of multi-segment well modeling to simulate well interference. Paper URTeC 2668100, SPE/AAPG/SEG Unconventional Resources Technology Conference. Austin, Texas.

Tene, M., Bosma, S.B.M., Al Kobaisi, M.S., Hajibeygi, H., 2017. Projection-based embedded discrete fracture model (pEDFM). Adv. Water Resour. 105, 205–216.

Vitel, S., Souche, L., 2007. Unstructured upgridding and transmissibility upscaling for preferential flow paths in 3D fractured reservoirs. Paper SPE 106483, SPE Reservoir Simulation Symposium. Houston, Texas.

Weijers, L., de Pater, C.J., 1992. Fracture reorientation in model tests. Paper SPE 23790, SPE Formation Damage Control Symposium. Lafayette, Louisiana.

Weng, X., Kresse, O., Cohen, C.E., Wu, R., Gu, H., 2011. Modeling of hydraulic-fracture-network propagation in a naturally fractured formation. SPE Prod. Oper. 26 (4), 368–380.

Wu, K., Olson, J.E., 2015. Simultaneous multifracture treatments: fully coupled fluid flow and fracture mechanics for horizontal wells. SPE J. 20 (2), 337–346.

Xu, Y., 2015. Implementation and application of the embedded discrete fracture model (EDFM) for reservoir simulation in fractured reservoirs. MS Thesis. University of Texas at Austin.

Xu, Y., 2018. Development and application of embedded discrete fracture model for conventional and unconventional reservoir simulation. PhD Dissertation. University of Texas at Austin.

Xu, Y., Cavalcante Filho, J.S.A., Yu, W., Sepehrnoori, K., 2017. Discrete-fracture modeling of complex hydraulic-fracture geometries in reservoir simulators. SPE Reservoir Eval. Eng. 20 (2), 403–422.

Xu, Y., Sepehrnoori, K., 2019. Development of an embedded discrete fracture model for field-scale reservoir simulation with complex corner-point grids. SPE J. 24 (4), 1552–1575.

Yang, D., Xue, X., Chen, J., 2018. High resolution hydraulic fracture network modeling using flexible dual porosity dual permeability framework. Paper SPE 190096, SPE Western Regional Meeting. Garden Grove, California.

Yu, W., 2015. A comprehensive model for simulation of gas transport in shale formation with complex hydraulic fracture geometry. Paper SPE 178747, SPE Annual Technical Conference and Exhibition. Houston, Texas.

Yu, W., Lashgari, H.R., Wu, K., Sepehrnoori, K., 2015. CO2 injection for enhanced oil recovery in Bakken tight oil reservoirs. Fuel 159, 354–363.

Yu, W., Wu, K., Sepehrnoori, K., 2016. A semianalytical model for production simulation from nonplanar hydraulic-fracture geometry in tight oil reservoirs. SPE J. 21 (3), 1028–1040.

CHAPTER FIVE

An extension of the embedded discrete fracture model for modeling dynamic behaviors of complex fractures

5.1 Introduction

Hydraulic fracturing has been a successful technology in the oil and gas industry for several decades. An important aspect of a successful hydraulic fracturing job is the desirable conductivity of hydraulic fractures. In fractured low-permeability reservoirs, fluid will flow dominantly from hydraulic fractures or fracture networks into the wellbore. However, the conductivity of fractures may not remain constant during the whole life of the well. For propped hydraulic fractures, the fracture conductivity primarily depends on proppant pack permeability and proppant pack width (Terracina et al., 2010). Several known mechanisms that may lead to the decline of propped hydraulic fracture conductivity have been studied, including proppant embedment, proppant crushing, fine generation and migration, and proppant flowback (Vincent, 2009; Alramahi and Sundberg, 2012). Extensive field tests and laboratory experiments have been conducted to measure the variation of propped fracture conductivity with stress and time (Alramahi and Sundberg, 2012; Zhang et al., 2014).

For unpropped hydraulic fractures and natural fractures, the flow conductivity is mainly determined by fracture closure under stress (Wu et al., 2017). The conductivity of unpropped fractures is typically several orders of magnitude lower than that of propped hydraulic fractures, and conductivity declines much faster with increasing effective normal stress (Zhang et al., 2014; Wu et al., 2017). Therefore, during long-term reservoir depletion, the geomechanical effects may play a significant role. Several correlations were proposed to relate fracture permeability to stress (Yilmaz et al., 1991; Settari et al., 2005; Bachman et al., 2011). Furthermore, during recovery processes such as water flooding and cold water injection, the stress-regime change in the formation might induce continuously propagating fractures

Embedded Discrete Fracture Modeling and Application in Reservoir Simulation.
http://dx.doi.org/10.1016/B978-0-12-821872-3.00005-8

(Koning and Niko, 1985; Hustedt et al., 2005; Bhardwaj et al., 2016) or shear-induced fractures (Yeo et al., 1998; Zhang et al., 2009; de Sousa Junior et al., 2016). All these geomechanical processes might significantly change the dimensions or conductivity of fractures.

In addition to the long-term stress dependency of fracture permeability, some complex processes in tight reservoirs also involve dynamic fracture behaviors. In recent years, technologies such as refracturing or infill drilling have been gaining more attention from operators as economical methods to further improve recovery in stimulated reservoirs (Vincent, 2011; Morales et al., 2016; Safari et al., 2017; Barree et al., 2017). In these processes, the area and conductivity of fractures can be altered, and new fractures can be created (Vincent, 2011). The orientation of fracture growth and the geometry of new fractures are primarily controlled by the altered stress field after reservoir depletion, which gives rise to complex fracture geometries (Gupta et al., 2012; Safari et al., 2017). In addition, the newly created fractures may, in turn, communicate with existing fractures and impact the conductivity of these fractures, causing interference between fractures and wells (Sardinha et al., 2014; Scott et al., 2015; Morales et al., 2016). In these processes, as fractures provide the main flow path in the near-wellbore region, the dynamic behaviors of complex fractures can dominate the reservoir flow and hydrocarbon recovery.

As the dynamic behaviors of fractures are commonly known in the oil and gas industry, modeling such behaviors has long been an important topic in reservoir simulation. In the simulation of the dynamic processes, the fracture conductivity and/or the fracture geometry need to be changed with time. In addition, processes such as refracturing and infill drilling require new fractures to be added during the simulation. These requirements present challenges for gridding and computational efficiency, and several techniques have been developed in an attempt to solve these challenges (Hustedt et al., 2005; Moinfar et al., 2013; Norbeck et al., 2016). Hustedt et al. (2005) modeled the fracture growth during water injection by dynamically adding special connections between fracture and matrix blocks. Lei et al. (2015) used a dynamic unstructured gridding technique to model the fracture propagation during water flooding, where fractures are explicitly and dynamically gridded during the simulation. Moinfar et al. (2013) developed a coupled fluid flow and geomechanics model based on the embedded discrete fracture model (EDFM) to simulate stress-dependent fracture conductivity; fracture width was correlated with effective stress. Norbeck et al. (2016) developed a framework which coupled fluid flow, geomechanics, and rock failure to study the problem of hydraulic fracturing and shear stimulation. The fluid flow problem was solved with an embedded fracture modeling method in a triangular mesh. de Sousa Junior et al. (2016) implemented a coupled fluid flow and geomechanics model using a simplified version of the EDFM, and activations of natural fractures through shear failure were simulated in their study. Ren et al. (2016) coupled the extended finite element method (XFEM) with the EDFM to simulate two-dimensional multiphase flow in fractured tight reservoirs considering the geomechanical effect. Garipov et al. (2016) coupled a finite-volume method in unstructured triangular mesh with the Galerkin finite-element method to solve fluid flow and poroelasticity equations.

As discussed in Chapter 3, the usage of conforming meshes around fractures might present difficulties in matrix gridding and limit the number of fractures that can be included in the simulation of highly fractured media. More challenges arise for the simulation of complex processes where fractures need to be dynamically added or removed or the fracture geometry changes during the simulation. In such cases, one may need to conduct dynamic re-gridding, which leads to even higher complexity. Several researchers (Moinfar et al., 2013; Ren et al., 2016; de Sousa Junior et al., 2016) used the EDFM to avoid the gridding issues in such cases. However, the usage of in-house simulators or simplified formulations limits the capability of their models to be generalized in conventional simulators.

In this chapter, with modifications of the EDFM formulations, a general method compatible with standard reservoir simulators is introduced to simulate complex dynamic behaviors of hydraulic fractures and natural fractures. This method fully exploits the advantages of the EDFM approach. Different types of dynamic behaviors of complex fractures can be modeled, including pressure-dependent fracture conductivity, growth of existing fractures, creation of new fractures, fracture activations, etc. The method is applied in a commercial reservoir simulator to illustrate its generality. Subsequently, the accuracy of the model is verified against the local grid refinement (LGR) method, and field data are used to perform a history matching study. After that, through four case studies with complex fractures, we demonstrate the applications of the model in both conventional and unconventional reservoirs, including simulations of vertical and horizontal well refracturing, infill drilling, and fracture activation. Key findings from these studies will be discussed.

5.2 Methodology

5.2.1 Half-transmissibility form of the EDFM formulations

The transmissibility formulations of the EDFM are canonically written in a form similar to the Darcy equation (Moinfar et al., 2014; Cavalcante Filho et al., 2015), where the transmissibility factor is a function of the geometry of control volumes (matrix gridblocks and fracture segments), relative position of control volumes, and their permeability as

$$T_{nnc} = \frac{k_{nnc} A_{nnc}}{d_{nnc}} \tag{5.1}$$

In Eq. (5.1), T_{nnc}, k_{nnc}, A_{nnc}, and d_{nnc} are the transmissibility factor, permeability, contact area, and distance related to the non-neighboring connection (NNC). The benefit of writing transmissibility equations in the form of Eq. (5.1) is that it is straightforward and relatively easy to implement in a geometrical calculation code. However, it presents difficulties for the modeling of dynamic fracture behaviors. Because transmissibility factors are calculated before fluid flow simulation, mechanisms such as rock compaction and dilation cannot be simulated with Eq. (5.1) as the parameters in the equation will be altered during the simulation. However, these mechanisms cannot be ignored in particular reservoirs. Some traditional reservoir simulators have the capability of changing gridblock permeability during the simulation to model these mechanisms. In those simulators, as the permeability of a gridblock changes, the corresponding transmissibility factors related to this gridblock are also changed. These dynamic changes reflect the complex processes occurring in the reservoir.

Because each fracture segment is explicitly represented as an active gridblock with modified properties in the EDFM, it is possible to take advantage of the functionalities provided by the above-mentioned simulators. If the NNC transmissibility factors can be linked to gridblock permeability in a loosely coupled manner, the dynamic processes related to fractures can be conveniently modeled in the EDFM by changing fracture gridblock permeabilities. This can be achieved by writing Eq. (5.1) in a half-transmissibility form as

$$\frac{1}{T_{nnc}} = \frac{1}{T_{nnc1}} + \frac{1}{T_{nnc2}}, \tag{5.2}$$

$$T_{nnc1} = \frac{k_1 A_{nnc}}{d_{nnc1}}, T_{nnc2} = \frac{k_2 A_{nnc}}{d_{nnc2}}. \tag{5.3}$$

Therefore,

$$\frac{1}{T_{nnc}} = \frac{d_{nnc1}}{k_1 A_{nnc}} + \frac{d_{nnc2}}{k_2 A_{nnc}}, \tag{5.4}$$

where k_1 and k_2 are the gridblock permeabilities of Control Volumes 1 and 2 (matrix gridblock or fracture segment), respectively, and d_{nnc1} and d_{nnc2} are the equivalent distances from the centroids of control volumes to the common face. Eq. (5.4) has a form similar to the transmissibility equation used for corner-point grid systems (Ponting, 1989). By writing the EDFM formulations in this form, the NNCs in the EDFM can be treated as regular connections in reservoir simulators.

For matrix–fracture intersections (NNC Type I),

$$\begin{aligned}\frac{1}{T_{nnc}} &= \frac{d_{f-m}}{2A_f \vec{n} \cdot (\overline{\overline{K}} \cdot \vec{n})} \approx \frac{d_{f-m}}{(2A_f)\vec{n} \cdot (\overline{\overline{K}} \cdot \vec{n})} + \frac{w_f / 2}{(2A_f)k_f}, \\ &= \frac{d_{f-m} \dfrac{k_m}{\vec{n} \cdot (\overline{\overline{K}} \cdot \vec{n})}}{k_m (2A_f)} + \frac{w_f / 2}{k_f (2A_f)},\end{aligned} \tag{5.5}$$

where A_f is the area of the fracture segment, $\overline{\overline{K}}$ is the permeability tensor of the matrix block, $\vec{n}$ is the normal vector of the fracture plane, k_m is the matrix permeability in a specific direction (the defined direction for the regular connection), k_f is the fracture permeability, w_f is the fracture width, and d_{f-m} is the average normal distance from the matrix to the fracture.

The reason to include the second term on the right-hand side of Eq. (5.5) is that after adding this term, T_{nnc} can be set to zero by simply setting k_f to zero. This provides a convenient method to activate or deactivate fractures in simulators by only manipulating block permeability.

For connections between two fracture segments within an individual fracture (NNC Type II),

$$\frac{1}{T_{nnc}} = \frac{1}{T_1} + \frac{1}{T_2} = \frac{d_{seg1}}{k_f A_c} + \frac{d_{seg2}}{k_f A_c}. \tag{5.6}$$

When the two fracture segments have different permeabilities,

$$\frac{1}{T_{nnc}} = \frac{d_{seg1}}{k_{f1} A_c} + \frac{d_{seg2}}{k_{f2} A_c}, \tag{5.7}$$

where A_c is the area of the common face of these two fracture segments, and d_{seg1} and d_{seg2} are the distances from the centroids of the fracture segments to the common face.

For fracture intersections (NNC Type III),

$$\frac{1}{T_{nnc}} = \frac{d_{f1}}{k_{f1} w_{f1} L_{int}} + \frac{d_{f2}}{k_{f2} w_{f2} L_{int}} = \frac{d_{f1} \frac{\overline{w_f}}{w_{f1}}}{k_{f1}(\overline{w_f} L_{int})} + \frac{d_{f2} \frac{\overline{w_f}}{w_{f2}}}{k_{f2}(\overline{w_f} L_{int})}, \tag{5.8}$$

where L_{int} is the length of the fracture intersection line, d_{f1} and d_{f2} are the average distances from fracture segments 1 and 2 to the intersection line, respectively, and $\overline{w_f}$ is the average fracture width.

The new set of formulations [Eqs. (5.5), (5.7), and (5.8)] takes a form similar to the transmissibility equation in traditional simulators, which indicates that they can be conveniently implemented in these simulators in a non-intrusive manner. Using the new formulations, the fracture segments in the EDFM can be treated as normal control volumes similar to matrix gridblocks, and the EDFM connections can be handled in the same way as regular connections. Table 5.1 summarizes the equivalent parameters required in the modified formulations. Because the NNC transmissibility factors are linked with gridblock permeability, whenever the permeability changes, the transmissibility factors for NNCs will also be modified accordingly. This provides a convenient way to simulate dynamic processes related to fractures.

Table 5.1 Summary of equivalent parameters for the three types of NNCs in the EDFM (Xu et al., 2019).

NNC type	Type of control volumes	Contact area	Equivalent distance to control volume 1	Equivalent distance to control volume 2
Type I	Matrix gridblock and fracture segment	$2A_f$	$d_{f-m} \frac{k_m}{\vec{n} \cdot (\overrightarrow{\overrightarrow{K}} \cdot \vec{n})}$	$\frac{w_f}{2}$
Type II	Fracture segments	A_c	d_{seg1}	d_{seg2}
Type III	Fracture segments	$\overline{w_f} L_{int}$	$d_{f1} \frac{\overline{w_f}}{w_{f1}}$	$d_{f2} \frac{\overline{w_f}}{w_{f2}}$

5.2.2 Modeling matrix- and fracture-permeability change during simulation

A popular approach to simulate geomechanical effects in reservoir simulation is to relate permeability to pore pressure. As a simplified approach, the average effective stress can be correlated with fluid pressure (Settari et al., 2005), which makes it easy and convenient to implement in traditional reservoir simulators. A widely used correlation that relates matrix or fracture permeability with pore pressure is (Yilmaz et al., 1991; Clarkson et al., 2013)

$$k = k_i e^{\gamma(p-p_i)}, \tag{5.9}$$

where k is the absolute permeability at pressure p, k_i is the permeability at a reference pressure p_i, and γ is the permeability modulus for the matrix or fractures. Fractures typically have a higher permeability modulus than the matrix because fracture permeability is more sensitive to stress change than matrix permeability.

Using Eq. (5.9), the change of matrix or fracture permeability with pore pressure can be input into reservoir simulators as a table of permeability multipliers (Settari et al., 2005). The new set of EDFM formulations can be incorporated with Eq. (5.9) to model the variation of fracture permeability and fracture conductivity with pressure.

5.2.3 Modeling of fracture shear failure

The Mohr–Coulomb failure criterion is a convenient method to model fracture shear failure (Zhang et al., 2009; Norbeck et al., 2016; de Sousa Junior et al., 2016). It gives the relationship between the shear strength and the effective normal stress for a fracture plane

$$\tau_{max} = c + \sigma' \tan(\phi), \tag{5.10}$$

where τ_{max} is the shear strength, c is the fracture cohesion, σ' is the effective normal stress, and ϕ is the angle of internal friction for fractures. When the influence of pore pressure is considered in the calculation of effective normal stress,

$$\sigma' = \sigma_n - \beta p, \tag{5.11}$$

where β is the Biot–Willis coefficient, σ_n is the normal stress in the rock, and p is the fluid pore pressure. If the in-situ reservoir stress tensor $\vec{\vec{\sigma}}$ is known, the normal stress σ_n and shear stress τ for a plane with a unit normal vector $\vec{n}$ can be calculated as

$$\sigma_n = (\vec{n} \cdot \vec{\vec{\sigma}}) \cdot \vec{n}, \tag{5.12}$$

$$\tau = \sqrt{(\vec{n} \cdot \vec{\vec{\sigma}})^2 - \sigma_n^2}. \tag{5.13}$$

Therefore, for this fracture plane, if p_c is the critical pore pressure above which shear failure would occur, then

$$\sqrt{(\vec{n} \cdot \vec{\vec{\sigma}})^2 - ((\vec{n} \cdot \vec{\vec{\sigma}}) \cdot \vec{n})^2} = c + ((\vec{n} \cdot \vec{\vec{\sigma}}) \cdot \vec{n} - \beta p_c)\tan(\phi), \tag{5.14}$$

$$p_c = \frac{\left[(\vec{n} \cdot \vec{\vec{\sigma}}) \cdot \vec{n} - \dfrac{\sqrt{(\vec{n} \cdot \vec{\vec{\sigma}})^2 - ((\vec{n} \cdot \vec{\vec{\sigma}}) \cdot \vec{n})^2} - c}{\tan(\phi)} \right]}{\beta}. \tag{5.15}$$

It is important to mention that this study did not use a reservoir geomechanics model to consider rock deformation and effective stress changes during depletion. Instead, we used a simplistic approach to calculate normal and shear stresses on natural fractures based on in-situ reservoir stresses. It means that normal and shear stresses on natural fractures do not change with depletion, and only effective normal stress decreases with pore pressure increases. A more rigorous approach that considers the change of rock stresses in fractured reservoirs is to couple a fluid flow simulator using discrete fracture models with a geomechanics model, which can be more computationally complicated and time-consuming. In this study, because normal and shear stresses are calculated based on in-situ reservoir stresses, the critical pore pressure can be pre-determined. The required inputs for modeling of fracture shear failure include the in-situ stress state of the reservoir such as the vertical stress, the minimum and maximum horizontal stress, the fracture cohesion (c), the friction angle (ϕ), and the Biot–Willis coefficient.

5.3 Model verification and field application

In this section, a case study in a shale-oil reservoir is presented to verify the modified formulations, and a history-matching field case study is presented to demonstrate the field applicability of the approach. As the EDFM method can be applied in general finite-difference or finite-volume simulators in a non-intrusive manner, CMG-IMEX (a black-oil reservoir simulator) and CMG-GEM (a compositional reservoir simulator) are used to perform the case studies. Detailed explanations of the methodology to use EDFM in commercial reservoir simulators can be found in Chapter 4. A bi-wing hydraulic fracture geometry is assumed in the verification case studies to compare the simulation results with the LGR model. In the next section, more complex fractures with dynamic behaviors are presented.

5.3.1 Case 1: Horizontal well with transverse planar hydraulic fractures in the Eagle Ford Shale

The Eagle Ford Shale has been one of the most important shale plays in North America in recent years. This study looks at shale-oil production from the Lower Eagle Ford Shale. The reservoir dimensions are 1,515 × 1,320 × 100 ft, and a uniform 101 × 33 Cartesian grid is used in the horizontal direction, as shown in Fig. 5.1. In the vertical direction, the reservoir is discretized into five layers. The middle three layers have thicknesses of 12 ft, and the thicknesses of the top and bottom layers are both 32 ft. Reservoir parameters obtained from a history-matching study by Simpson et al. (2016) are used, and the key parameters are summarized in Table 5.2. The relative permeability curves used in this study are presented in Fig. 5.2.

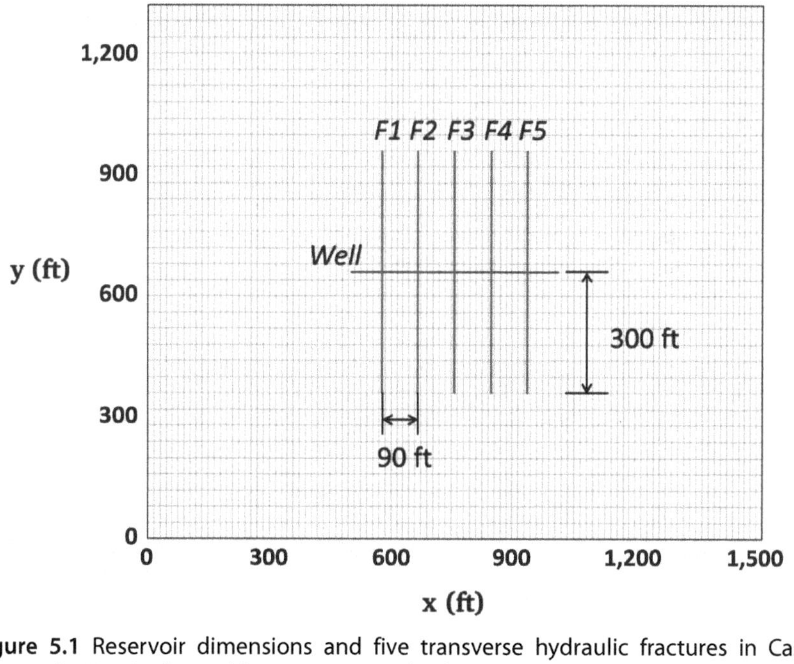

Figure 5.1 Reservoir dimensions and five transverse hydraulic fractures in Case 1 (Xu et al., 2019). The *red lines* represent the fractures. The fracture half-length and spacing are 300 and 90 ft, respectively.

Table 5.2 Simulation parameters for the verification case study Case 1 (Xu et al., 2019).

Parameter	Value	Unit	Parameter	Value	Unit
Initial reservoir pressure	8,125	psi	Initial water saturation within the SRV	0.41	–
Bubble point pressure	3,500	psi	Initial water saturation outside the SRV	0.17	–
Reservoir porosity	0.12	–	Initial oil saturation	0.83	–
Horizontal permeability	0.00047	md	Depth of top layer	10,000	ft
Vertical permeability	0.0000282	md	Bottomhole pressure	3,500	psi
Reservoir temperature	270	°F	Wellbore radius	0.25	ft
Reservoir thickness	100	ft	Simulation time	7,000	day
Rock compressibility	1×10^{-6}	psi^{-1}			

Five planar transverse hydraulic fractures were considered in this study with a spacing of 90 ft. The fractures vertically penetrate the middle three layers of the reservoir, with a distance of 32 ft from both the top and bottom of the reservoir. The fracture width is 0.01 ft, and the initial fracture permeability is 20,000 md. Thus, the initial fracture conductivity is 200 md-ft. Because of this high value of fracture conductivity, non-Darcy flow in fractures is considered.

Three sub-cases were simulated to illustrate different aspects of the methodology. In Case 1a, all five fractures are created at the beginning of the simulation. In Case 1b, only Fractures F1, F3, and F5 are created at $t = 0$. Subsequently, after 1,000 days of depletion, refracturing is performed to reduce the fracture spacing, which creates Fractures F2 and F4. In Case 1c, similar to Case 1a, all fractures are added at $t = 0$, but the geomechanical effects are considered for matrix and fracture permeabilities using Eq. (5.9). A permeability modulus of 1.5×10^{-3} psi^{-1} is chosen for fractures such that large permeability loss is simulated. The permeability modulus for the matrix is 4.0×10^{-4} psi^{-1}. The variation of fracture and matrix permeability with pore pressure is shown in Fig. 5.3. Compaction tables are set up inside the simulator for the matrix and fractures, respectively.

All three sub-cases were simulated using both the EDFM and the LGR model. In the LGR model, each gridblock intersected by fractures is logarithmically refined into seven sub-gridblocks in a direction perpendicular to the fracture plane (as shown in Fig. 5.4). High-permeability gridblocks are used to represent fractures. For Cases 1b and 1c with dynamic fracture behaviors, the permeabilities of these gridblocks are modified with time. Fig. 5.4 illustrates the change of gridblock permeability in the LGR model for Case 1b. In the EDFM, the permeability of fracture gridblocks is modified with time for Cases 1b and 1c.

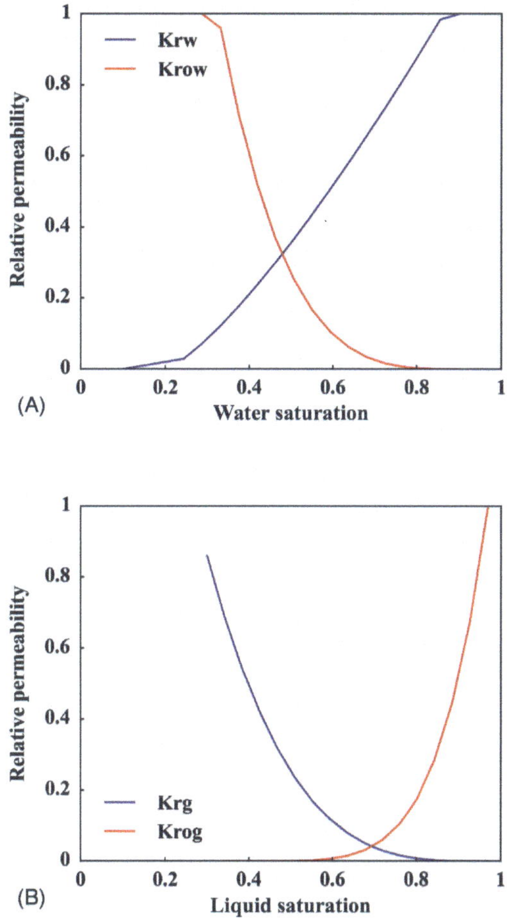

Figure 5.2 (A) Oil–water relative permeability curves and (B) liquid–gas relative permeability curves used in Case 1 (Xu et al., 2019).

Fig. 5.5 presents the field oil production rates predicted by the LGR model and the EDFM for all three sub-cases. Excellent matches between the two methods are obtained for all three cases. For Case 1b, both models capture the sharp oil rate increase after refracturing. For Case 1c, both models predict a faster decline in the oil rate compared with Case 1a. A good match of gas production rates is also obtained for all three cases, which is not shown in Fig. 5.5. These results show the accuracy of the new formulations of the EDFM and the capability of the new methodology to model dynamic fracture behaviors. It is also important to mention that the gridblock size around fractures in this case study is adequate for the simulation of long-term flow behaviors. If an accurate simulation of transient flow behaviors around fractures within a

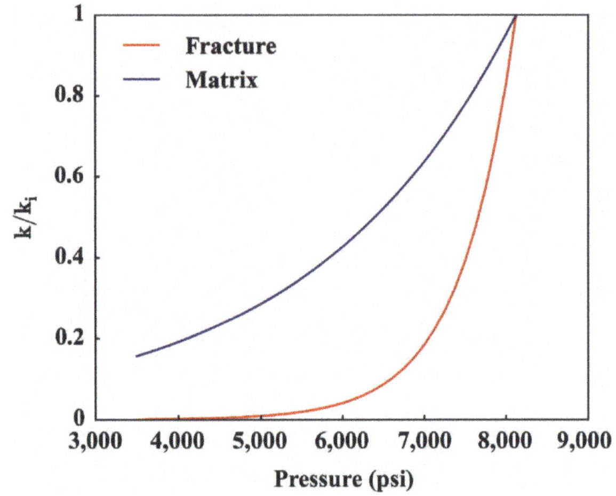

Figure 5.3 Variation of the matrix and fracture permeability with pore pressure in Case 1c (Xu et al., 2019).

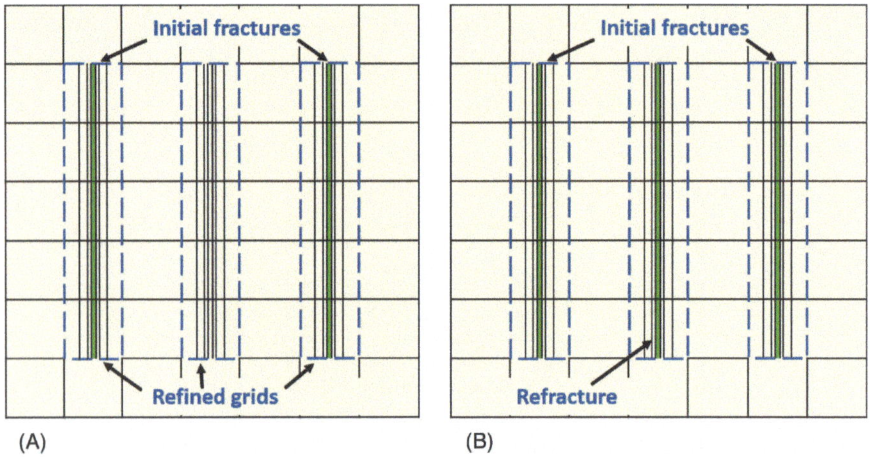

Figure 5.4 Illustration of the LGR model for Case 1b (Xu et al., 2019). All matrix gridblocks containing fractures are refined into seven sub-gridblocks at the beginning of the simulation. Fractures are represented as high-permeability sub-gridblocks, as shown in green. The size of "green" sub-gridblocks is enlarged in the direction perpendicular to the fracture plane for illustration purposes. The actual size used in the simulation is the same as the fracture width (0.01 ft). (A) Gridblock permeability before refracturing; (B) gridblock permeability after refracturing.

short period of production is desired, a smaller gridblock size around fractures (as shown in the LGR model) might be required (Panfili et al., 2015).

5.3.2 Case 2: History matching of an Eagle Ford shale-oil well

An actual Eagle Ford shale-oil well with 583 days of oil, gas, and water production data was selected to perform history matching using the dynamic

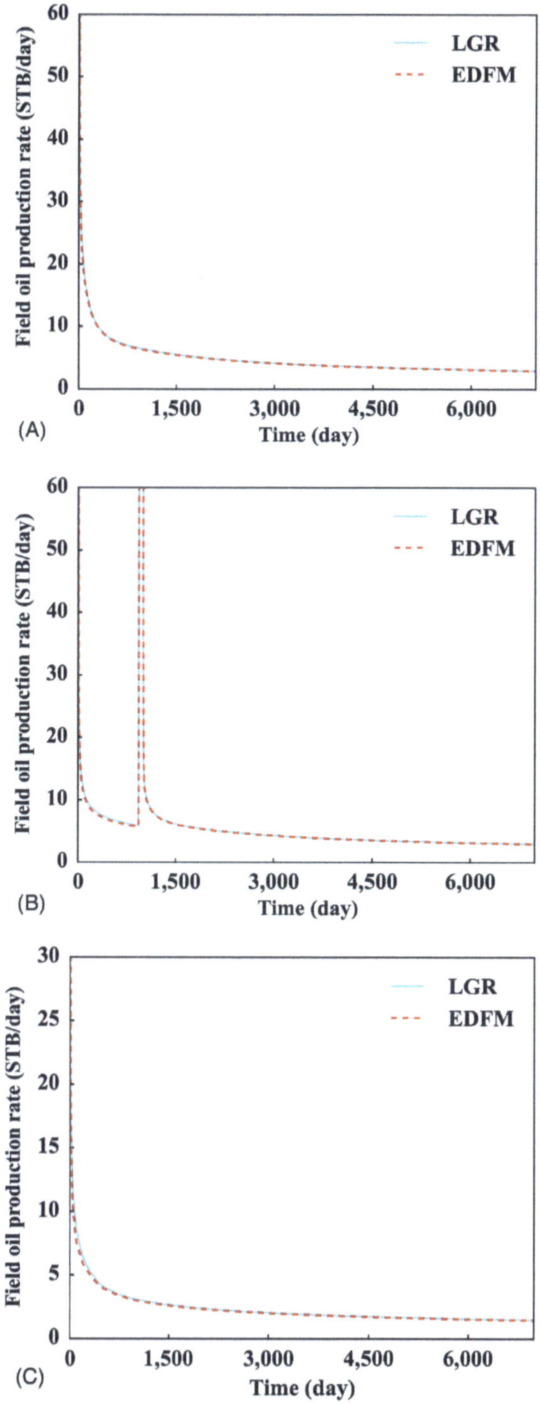

Figure 5.5 Field oil production rate for Cases (A) 1a, (B) 1b, and (C) 1c predicted by the LGR model and the EDFM (Xu et al., 2019).

EDFM method. A 3D reservoir model including 76 planar hydraulic fractures with dimensions of 7,780 × 1,320 × 100 ft is built, as shown in Fig. 5.6. A uniform 778 × 65 Cartesian grid is used in the horizontal direction, and three layers are considered in the vertical direction. The middle layer has a thickness of 26 ft, and the top and bottom layers have a thickness of 37 ft. A compositional model with four pseudo-components is used. Tables 5.3 and 5.4 summarize the key properties and binary interaction coefficients of these four pseudo-components. Under the reservoir temperature of 270 °F, the oil gravity is 40° API, the GOR is 2,400 scf/STB, the oil formation volume factor is 1.65 RB/STB, and the bubble point pressure is 3,446 psi. The relative permeability curves presented in Fig. 5.2 are used. In addition, the pressure-dependent matrix and fracture permeabilities are considered with a permeability modulus of 1.0×10^{-4} for the matrix and 2.0×10^{-4} for fractures.

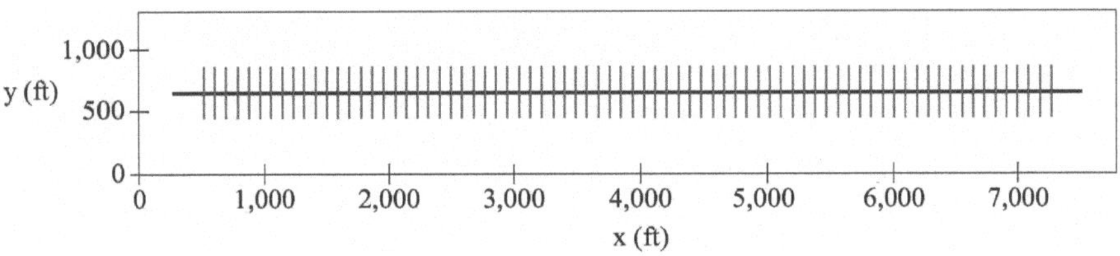

Figure 5.6 Reservoir dimensions and 76 transverse hydraulic fractures in Case 2 (Xu et al., 2019). The *red lines* represent fractures. The fracture spacing is 90 ft.

Table 5.3 Key properties of the four pseudo-components used for history matching (Xu et al., 2019).

Component	Molar fraction	Critical pressure (atm)	Critical temperature (K)	Critical volume (L/mol)	Molar weight (g/gmol)	Acentric factor	Parachor coefficient
C_1	0.12724	45.40	190.60	0.0990	16.04	0.0080	77.00
C_2–C_5	0.26599	33.05	275.47	0.2255	51.08	0.1691	168.37
C_6–C_{10}	0.38089	22.09	430.83	0.3941	102.94	0.2838	297.22
C_{11+}	0.22588	16.05	701.56	0.8870	267.15	0.6716	661.44

Table 5.4 Binary interaction coefficients between the four pseudo-components (Xu et al., 2019).

Component	C_1	C_2–C_5	C_6–C_{10}	C_{11+}
C_1	0	0.0168	0.0462	0.1110
C_2–C_5	0.0168	0	0.0077	0.0454
C_6–C_{10}	0.0462	0.0077	0	0.0163
C_{11+}	0.1110	0.0454	0.0163	0

Table 5.5 Key parameters used for obtaining a good history match in Case 2 (Xu et al., 2019).

Parameter	Value	Unit	Parameter	Value	Unit
Initial reservoir pressure	8,125	psi	Initial water saturation within SRV	0.33	–
Reservoir temperature	270	°F	Initial water saturation outside SRV	0.17	–
Reservoir porosity	0.12	–	Fracture height	26	ft
Horizontal permeability	0.0009	md	Fracture half-length	210	ft
Vertical permeability	0.000054	md	Fracture spacing	90	ft
Rock compressibility	1×10^{-6}	psi^{-1}	Initial fracture conductivity	40	md-ft
Depth of top layer	10,000	ft	Wellbore radius	0.25	ft
Initial oil saturation	0.83	–	Simulation time	583	day

The oil flow rate is used as a history matching constraint. The produced gas is simply solution gas because reservoir pressure is above the bubble point pressure during production. Fracture height, fracture half-length, fracture conductivity, matrix permeability, and initial water saturation within the stimulated reservoir volume (SRV) are the main parameters to be tuned to match the water flow rate and flowing bottomhole pressure (BHP). In addition, the dimensions of the SRV are defined as the distance between the first and the last fracture × fracture length × fracture height. A larger water saturation within the SRV is assumed to represent some of the injected fracturing fluid. Table 5.5 summarizes the key parameters used for achieving a good history match. A comparison of field production data and simulation results is shown in Fig. 5.7. It can be observed that good agreements were obtained. As the reliability of the numerical model forms the basis for the history matching study, the good agreements again verify the model and demonstrate its applicability. The results also show that the model can be applied in compositional simulation considering pressure-dependent matrix and fracture permeabilities.

5.4 Case studies with complex fractures

In this section, several case studies with complex fractures are presented. These studies consider different types of problems involving dynamic fracture behaviors in both conventional and unconventional reservoirs. The simulator used in Case 1 is also used for these simulations. All these studies present challenges when using traditional simulators without using the methodology presented in this chapter.

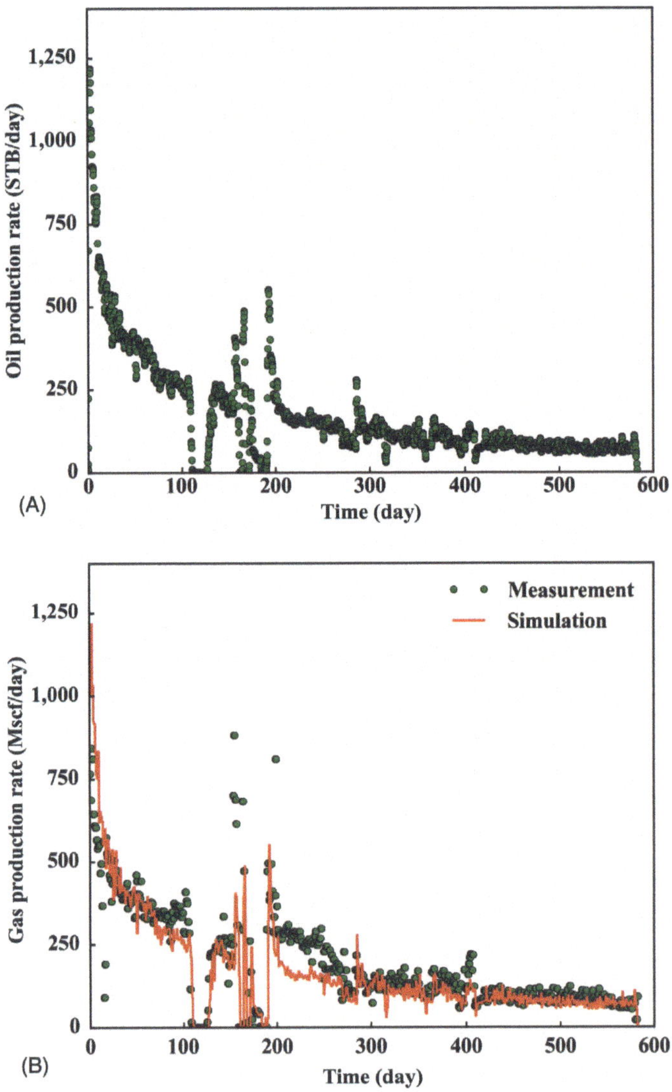

Figure 5.7 History matching results using the EDFM (Xu et al., 2019). (A) Oil production rate used as a history matching constraint. (B) Measured and simulated gas production rate. (C) Measured and simulated water production rate. (D) Measured and simulated well BHP.

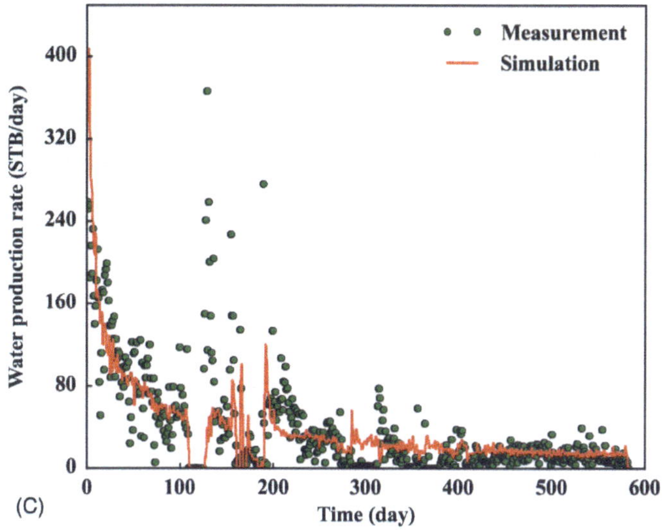

(C)

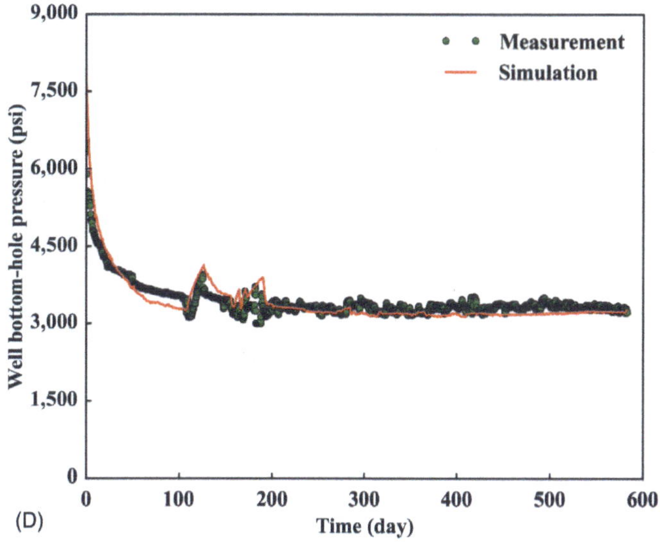

(D)

Figure 5.7 **(*Cont.*)**

5.4.1 Case 3: Vertical well refracturing in tight gas reservoir

Refracturing vertical wells in tight reservoirs has been studied for a long time in the oil and gas industry. By 1970, 35% of the fracturing jobs had been restimulated (Coulter and Menzie, 1973). The created fractures further enhance the connectivity between the well and the reservoir, and they improve the drainage efficiency of understimulated zones. To demonstrate the capability of the EDFM to simulate realistic refracturing cases, a modified case with fracture geometry measured by surface tiltmeter mapping (Wolhart et al., 2007) in the Codell Formation in the Wattenberg Field is simulated in this study. The orientations of initial fractures and refractures are available from Wolhart et al. (2007).

Some of the reservoir properties for the Codell Formation are available in Wolhart et al. (2007) and Roussel and Sharma (2012), and they are summarized in Table 5.6. Fig. 5.8 shows the orientations of fractures measured

Table 5.6 Simulation parameters for Case 3 (Xu et al., 2019).

Parameter	Value	Unit	Parameter	Value	Unit
Reservoir temperature	200	°F	Rock compressibility	3×10^{-6}	psi^{-1}
Reservoir porosity	0.14	–	Depth of top layer	7,500	ft
Reservoir permeability	0.05	md	Bottomhole pressure	1,000	psi
Initial reservoir pressure	4000	psi	Wellbore radius	0.25	ft
Reservoir thickness	20	ft	Simulation time	7,000	day

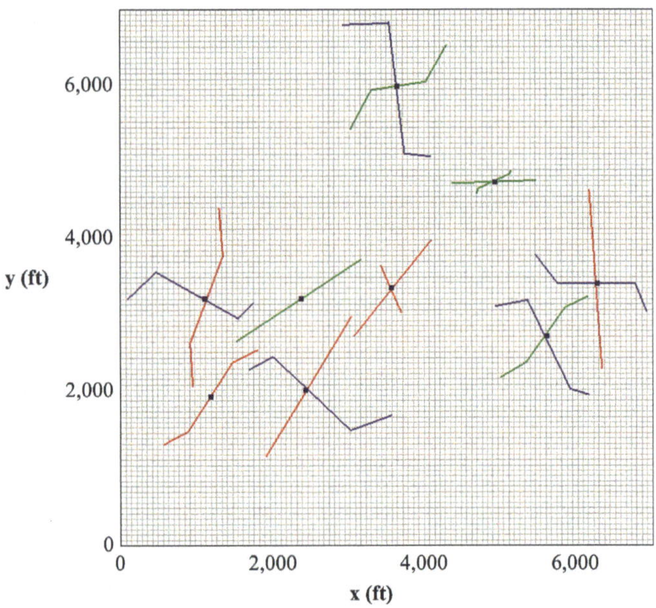

Figure 5.8 Reservoir simulation grid, well locations, and fracture geometries in Case 3 (Xu et al., 2019). All wells start producing from $t = 0$. Fractures created at different times are shown in different colors. The *green lines* represent initial fractures. The *red lines* represent refractures created after 1,000 days. The *blue lines* represent refractures created after 2,000 days. The total lengths of "red" fractures and "blue" fractures are the same.

by surface tiltmeter mapping (Wolhart et al., 2007), and the initial fractures and refractures are shown in green and red, respectively. It can be observed that the refractures have different orientations versus the initial fractures as a result of the changed pore pressure and altered stress field after depletion. In Wolhart et al. (2007), only one refracturing treatment was performed for several wells, which turned out to be a success, and the authors mentioned the potential of a "third-time" fracturing. Therefore, in this study, based on the measured fracture orientations, we assume one more refracturing treatment is performed for several wells. Considering the local redistribution of pore pressure (Siebrits et al., 2000), the orientations of new fractures are assumed to be nearly perpendicular to the existing fractures, as represented by *blue lines* in Fig. 5.8. Nonplanar geometries of these fractures are also assumed by taking into account the extent of stress reversal (Siebrits et al., 2000; Roussel and Sharma, 2012). All fractures have the same width of 0.0328 ft and permeability of 10,000 md.

A 100 × 100 × 1 matrix grid is used for the simulation with gridblock sizes of 70 × 70 × 20 ft. Only gas flow is simulated because the detailed reservoir-fluid-property data are not available. Pressure/volume/temperature (PVT) data of methane are used for the input of the gas PVT table. All wells start producing from the beginning of the simulation. The first refracturing treatment is performed at 1,000 days, and the second refracturing treatment is performed at 2,000 days. For comparison, the total length of fractures created during the second refracturing treatment ("second-time" refractures) is set to be the same as the total length of the refractures created during the first treatment ("first-time" refractures).

To evaluate the incremental recovery brought by each refracturing treatment, three sub-cases (3a, 3b, and 3c) were simulated. In Case 3a, only initial fractures were simulated; in Case 3b, the initial fractures and "first-time" refractures were considered; and in Case 3c, all fractures were simulated.

Fig. 5.9 presents the gas production rate and cumulative gas production for all three sub-cases. After each treatment, a large increase in the gas production rate can be observed. It can also be found that although the two refracturing treatments create the same length of fractures, the "first-time" refractures bring a much higher increase in gas production than the "second-time" refractures. Regarding long-term recovery (for 7,000 days), the cumulative gas production in Case 3b is 3,608 MMscf higher than that in Case 3a, while the difference between Case 3c and Case 3b is 1,074 MMscf. The pressure profiles in Fig. 5.10 shows the drainage area of each fracture, and they partly explain the reason. It can be observed that the "first-time"

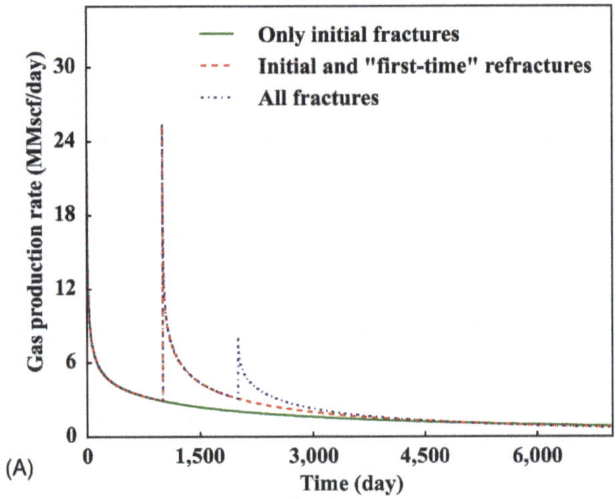

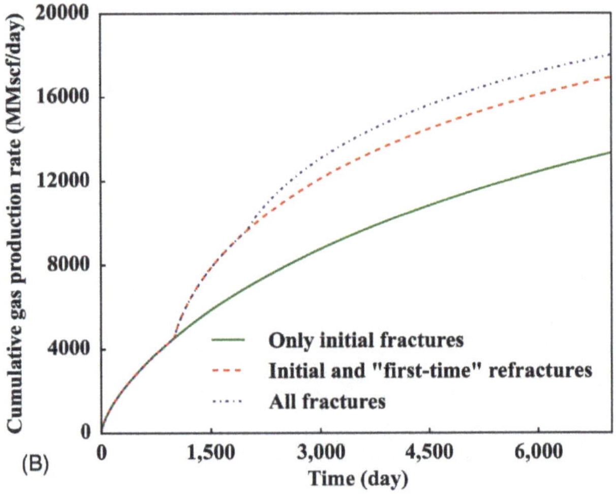

Figure 5.9 Comparison of (A) gas production rate and (B) cumulative gas production for three sub-cases in Case 3 (Xu et al., 2019).

refractures significantly expand the depleted reservoir area, while the "second-time" refractures mainly grow in zones that are already depleted. These results illustrate that the key to a successful refracturing work is to create fractures that connect the wellbore to areas not well-drained by existing fractures.

Wolhart et al. (2007) have stated that stress reorientations might be a beneficial factor for the success of restimulation treatment. Geomechanical modeling is typically required to determine the geometry of refractures.

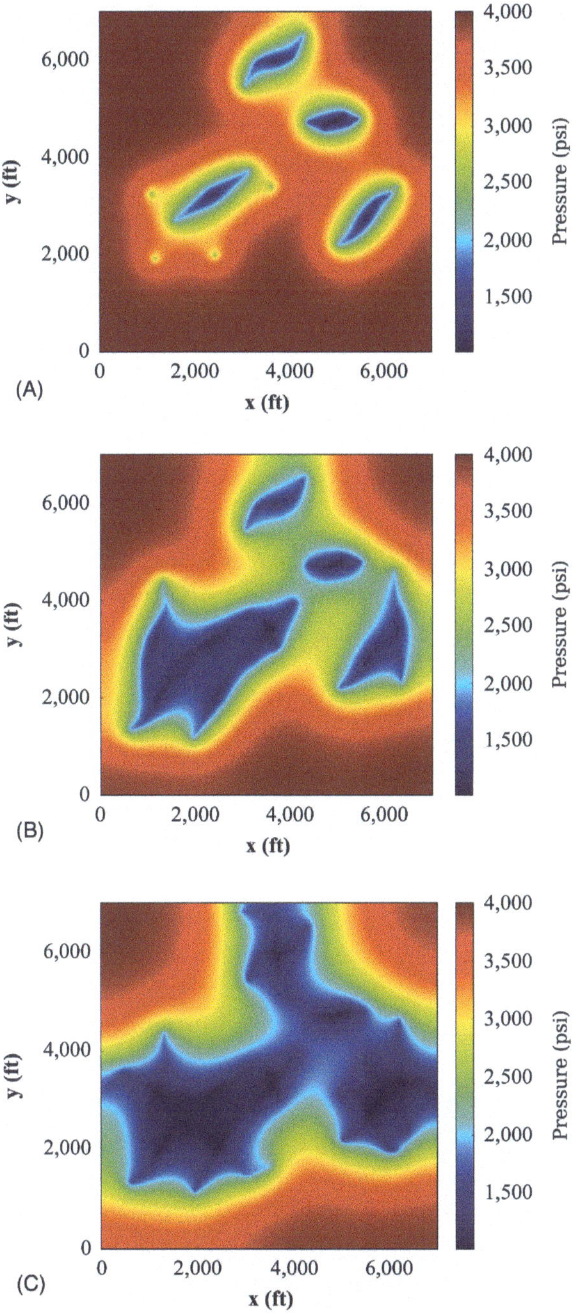

Figure 5.10 Matrix pressure profiles after (A) 1,000 days, (B) 2,000 days, and (C) 3,000 days for Case 3c predicted by the EDFM (Xu et al., 2019).

However, a prerequisite for accurate modeling of fracture reorientation is an accurate simulation of pore pressure change, as fracture orientations can be significantly influenced by reservoir depletion. In addition, accurate and efficient reservoir simulations are valuable tools for finding undrained zones, which provide guidance for the refracturing work, and performing "what-if" studies to evaluate the economics of a refracturing job. The refracturing processes can be affected by many factors (Roussel and Sharma, 2012), and therefore tend to introduce complexities in both the orientation and the geometry of newly created fractures, which poses challenges for reservoir simulation. The EDFM is an efficient and convenient method to simulate complex fractures, which makes the method introduced in this chapter suitable for modeling such processes.

5.4.2 Case 4: Horizontal well refracturing with nonplanar fractures

Refracturing horizontal wells in shale plays has attracted much attention from shale operators in recent years because of its relatively low cost compared with drilling a new well (Vincent, 2011; Gupta et al., 2012). Vincent (2011) summarized the mechanisms contributing to the success of a refracturing treatment, including the creation of new fractures, restoration of fracture conductivity, enlarged fracture geometry, increased fracture height. Case 1 investigated a scenario in which new fractures are created between the original fractures; therefore, this case study will focus on the effects of other mechanisms.

The reservoir dimensions in the simulation model are 4,000 × 2,980 × 100 ft, and a 400 × 149 × 7 matrix grid is used. In the vertical direction, the reservoir is discretized into seven layers, with thicknesses of 23.54, 8.46, 12, 12, 12, 8.46, and 23.54 ft from the top to the bottom, respectively. Most reservoir and simulation parameters are the same as those used in Case 1. Ten nonplanar fractures are considered, as shown in Fig. 5.11. They all penetrate the middle three layers of the reservoir. Because the influence of varying fracture conductivity is not a focus in this study, a constant fracture width of 0.01 ft and fracture permeability of 20,000 md are used for all fractures.

A refracturing treatment is performed after 1,500 days of simulation. On the basis of the simulation model, three scenarios were considered to investigate different mechanisms of refracturing:

1. **Case 4a**: The variation of fracture permeability with pore pressure is simulated using Eq. (5.9) with a permeability modulus of 1.5×10^{-3} psi^{-1}. After refracturing, the permeability of all fractures is restored to 20,000 md.

2. Case 4b: The change of fracture permeability with pressure is not considered. After refracturing, all fractures grow in the horizontal direction, as shown in Fig. 5.11A and B.

3. Case 4c: The change of fracture permeability with pressure is not considered. After refracturing, all fractures grow in the vertical direction, as shown in Fig. 5.11C.

The objective of simulating Case 4a is to study the effect of enhanced fracture conductivity after refracturing, while Cases 4b and 4c are simulated to investigate the impact of fracture growth. For comparison, the total area of fracture growth is the same in Cases 4b and 4c.

Fig. 5.12 presents the cumulative oil production at 7,000 days for Cases 4a, 4b, and 4c. As references, the production curves for corresponding cases without refracturing are also plotted. For all three cases, the refracturing treatment significantly improves recovery. By comparing the results of Cases 4b and 4c, it can be observed that with the same increase in fracture area, fracture growth in the horizontal direction can be more beneficial than growth in the vertical direction. The pressure profiles in Fig. 5.13 show the depleted zone in two layers of the reservoir for Cases 4b and 4c at 7,000 days. For Case 4b, although the fractures do not grow into Layer 2, the pressure profile in Fig. 5.13A shows good drainage in this layer, indicating good vertical connectivity in the formation. In contrast, the drainage

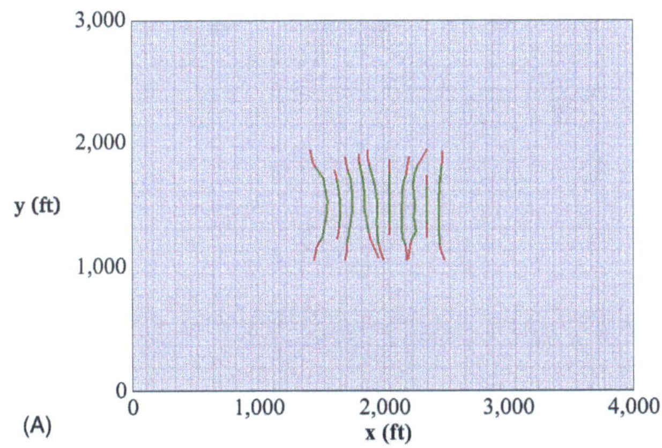

Figure 5.11 Reservoir dimensions and fracture geometries in Case 4 (Xu et al., 2019). (A) The top view; (B, C) closer 3D views of fractures, in which the lengths in the *x*- and *z*-directions are shown twice and 10 times the real size, respectively. The initial hydraulic fractures are shown in green, and the newly created fracture areas during the refracturing treatments in Cases 4b and 4c are shown in pink and blue, respectively. The total fracture areas shown in "pink" and "blue" are the same.

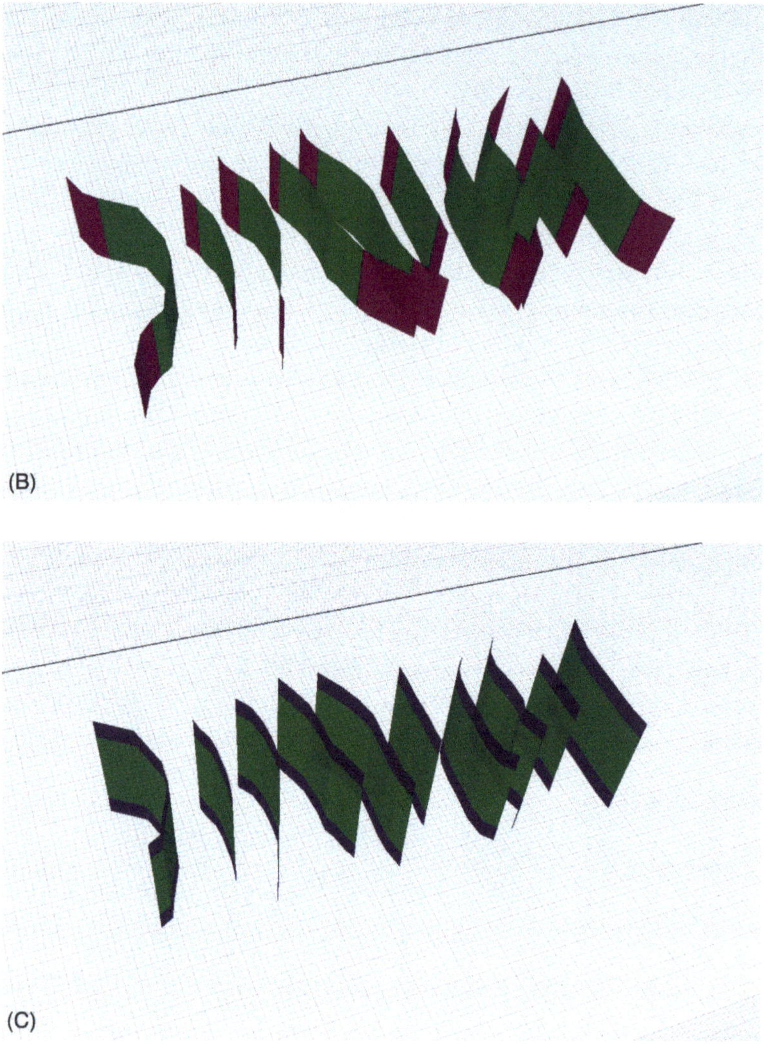

Figure 5.11 ***(Cont.)***

area in the horizontal direction in Case 4c is obviously smaller than that in Case 4b, as shown in Fig. 5.13. Therefore, the fracture growth in the horizontal direction is more desirable than growth in the vertical direction because of good interlayer connectivity.

To confirm the importance of vertical reservoir connectivity, we simulated another two cases, where the ratio of vertical/horizontal matrix permeability (k_v / k_h) for Cases 4b and 4c is changed from 0.06 to 0.005. For

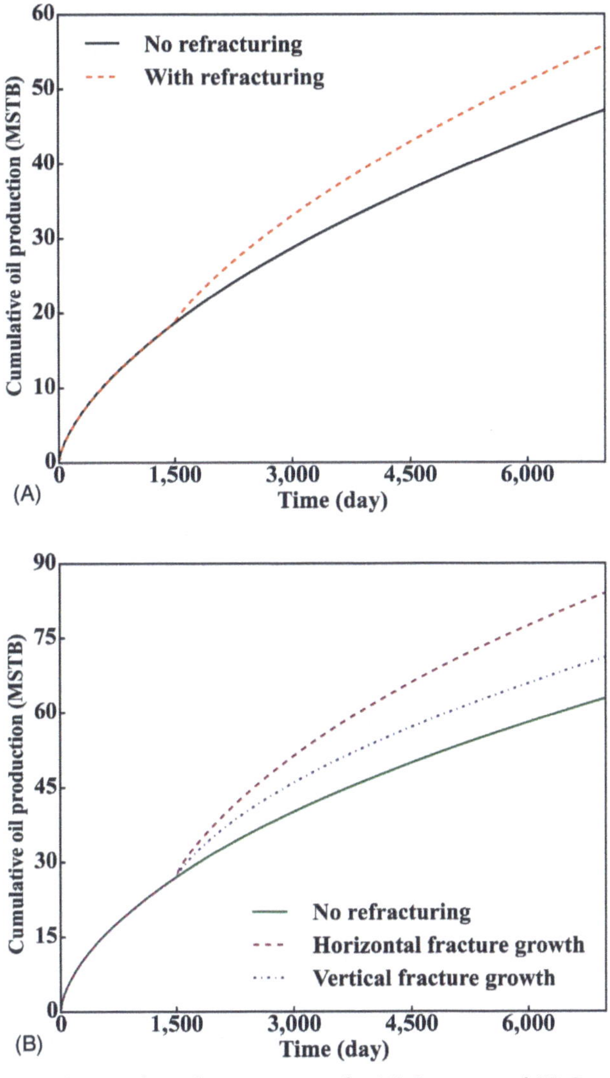

Figure 5.12 Cumulative oil production curves for (A) Case 4a and (B) Case 4b and Case 4c (Xu et al., 2019). In each plot, the cumulative oil production for the case without refracturing is included as a reference.

this vertical permeability value, the two cases reach a similar recovery, as shown in Fig. 5.14. These results show that the relative importance of the two mechanisms discussed in Cases 4b and 4c depends on the contrast between matrix permeability in corresponding directions.

The pressure profiles in Fig. 5.13 also show that the zones between fractures are well-drained. In this case, if new clusters are created, these clusters will mainly grow in the well-stimulated zones, making the treatment relatively less attractive in terms of economics. An accurate reservoir simulation can provide guidance for such decision-making processes. This case study shows that the enhancement of fracture conductivity and the growth of existing fractures can both play a critical role in a successful refracturing treatment. When quantifying the influence of these mechanisms, a dynamic simulation technique for fractures needs to be used. It should also be noted that although the three different mechanisms were considered independently in the three sub-cases, in real life, the mechanisms discussed in Cases 4a, 4b, and 4c might need to be considered simultaneously. Owing to the complexity of proppant transport mechanisms, the transportation of proppants into initially unpropped zones will unavoidably result in changes within the formerly propped zones. The coupling effects of these mechanisms can also be quantitatively studied using the dynamic EDFM approach in this chapter.

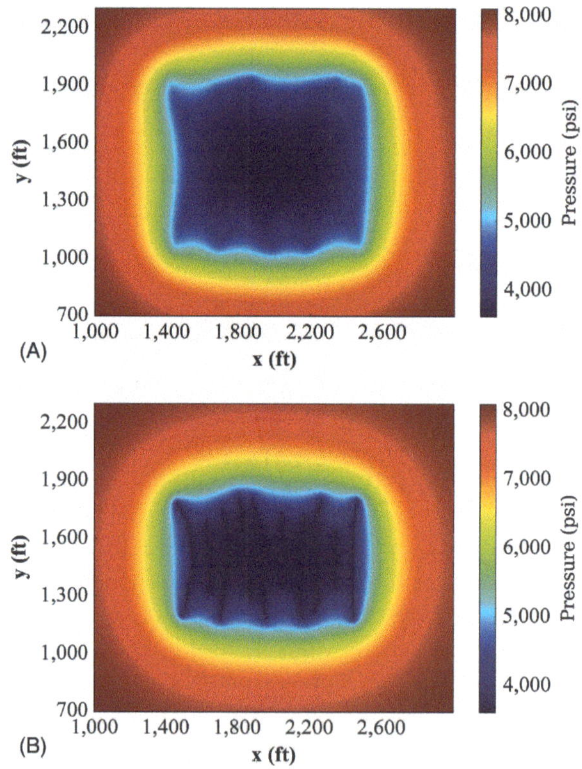

Figure 5.13 Pressure profiles at 7,000 days for (A) Case 4b, Layer 2; (B) Case 4c, Layer 2; (C) Case 4b, Layer 4; and (D) Case 4c, Layer 4 (Xu et al., 2019). Layer 2 is the layer immediately above the fractured layers; Layer 4 is the middle layer of the reservoir.

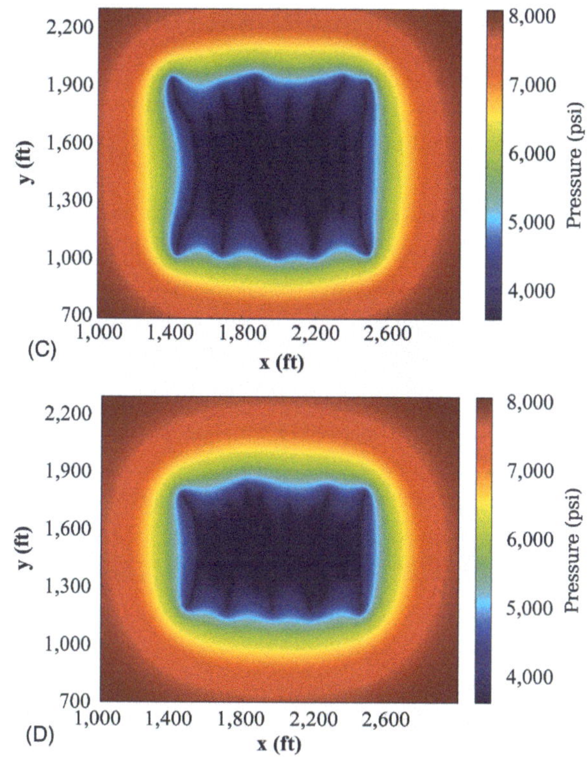

Figure 5.13 (*Cont.*)

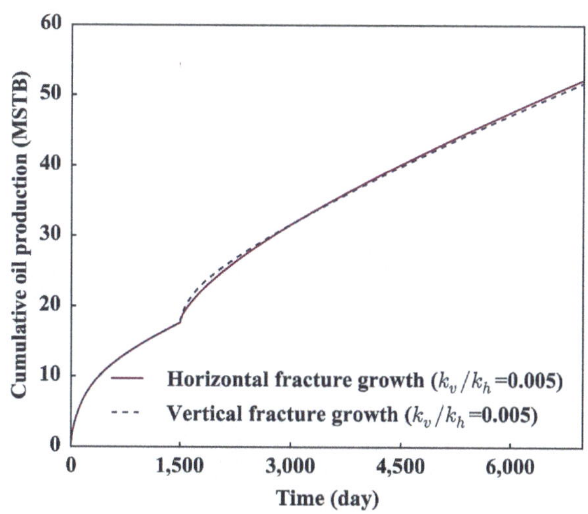

Figure 5.14 Cumulative oil production curves for Cases 4b and 4c with a lower vertical permeability (Xu et al., 2019).

5.4.3 Case 5: Multiple-well interference through fracture hits

In shale reservoirs, the drilling and fracturing of new wells may impact the production of old, adjacent wells, which is referred to as well interference. A common reason for well interference is "fracture hits," in which the hydraulic fractures of different wells are connected to each other. Fracture hits occur very frequently in cases with small well spacing or infill drilling. It can be beneficial, but most of the time, it should be avoided because it may cause loss of original fracture conductivity, production of fracturing fluid from parent wells, or damage to tubing and casing (Awada et al., 2016; Bommer et al., 2017). Here, a synthetic case study in a shale-oil reservoir with infill drilling is presented to investigate the influence of fracture hits.

Most of the simulation parameters are the same, as listed in Table 5.2. The reservoir dimensions are 9,000 × 6,000 × 100 ft, and the gridblock sizes in the x- and y-directions are 60 and 20 ft, respectively. In the vertical direction, the reservoir is discretized into seven layers with thicknesses of 20, 12, 12, 12, 12, 12, and 20 ft from the top to the bottom. Six horizontal wells in the y-direction are drilled and hydraulically fractured at different times, as shown in Fig. 5.15. Well-2 and Well-5 are infill wells. All four parent

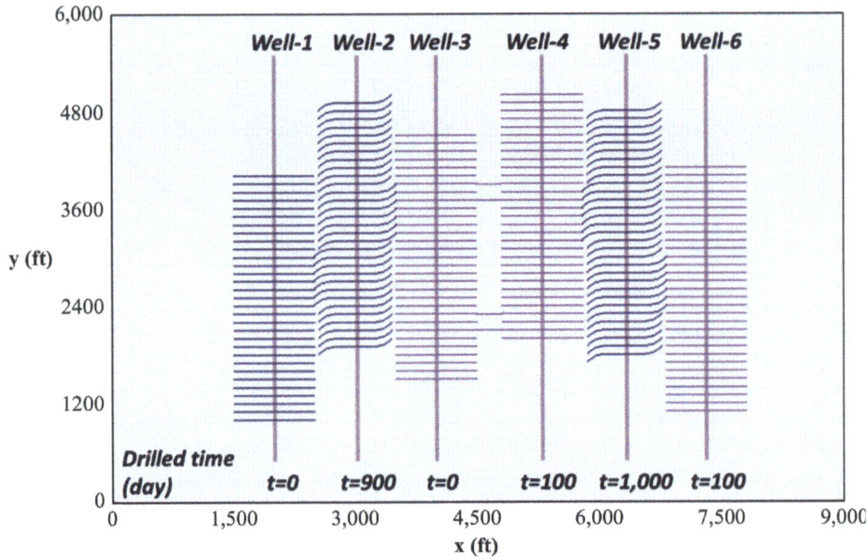

Figure 5.15 Top view of reservoir dimensions, well locations, and fracture geometries in Case 5 (Xu et al., 2019). Fractures in different colors have different depths. The drilling schedule is also shown. The horizontal distances between Well-1 and Well-2, Well-2 and Well-3, Well-3 and Well-4, Well-4 and Well-5, and Well-5 and Well-6 are 1,030; 970; 1,300; 1,030; and 970 ft, respectively.

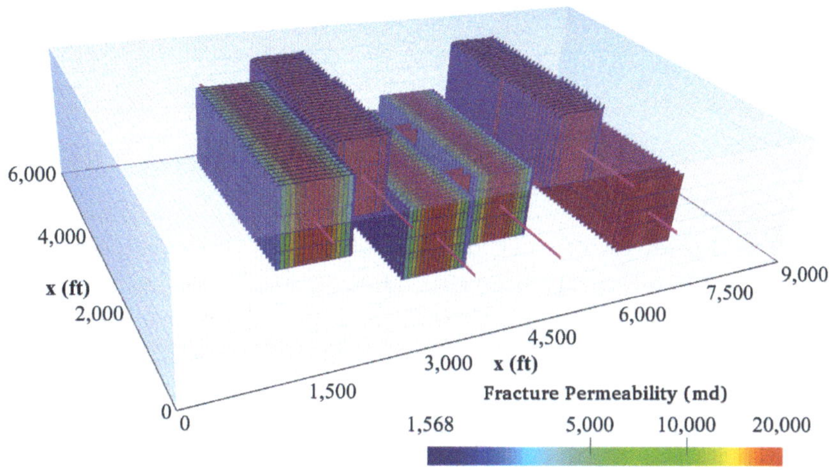

Figure 5.16 Three-dimensional (3D) view of reservoir dimensions, fracture depth, and fracture permeability in Case 5 (Xu et al., 2019). The length in the *z*-direction is shown 20 times the real size. All fractures have the same height (56 ft), but they penetrate different layers of the reservoir. Well-1, Well-2, and Well-5 have the same depth (10,028 ft) and Well-3, Well-4, and Well-6 have the same depth (10,072 ft). Fractures of Well-1, Well-2, and Well-5 fully penetrate Layers 1, 2, 3, and 4, and fractures of Well-3, Well-4, and Well-6 penetrate Layers 4, 5, 6, and 7. The fractures of Well-1, Well-3, and Well-4 have varying permeability along the length as represented by different colors (shown on the logarithmic scale).

wells (Well-1, Well-3, Well-4, and Well-6) have the same producing time of 900 days before infill drilling. Each well has 31 hydraulic fractures along the *x*-direction, with a spacing of 100 ft. The fracture heights for all fractures are the same, but fractures have different depths, as shown in Figs. 5.15 and 5.16. The fracture shapes of Well-1, Well-3, Well-4, and Well-6 are assumed to be rectangular, with a fracture half-length of 500 ft. Regarding the fracture geometry of infill wells (Well-2 and Well-5), studies have shown that fracture curving might occur because of the altered stress field after depletion (Gupta et al., 2012; Safari et al., 2017). Determination of the geometry of infill well fractures is not an easy task because it is affected by various factors such as local stress, pore pressure, inter-well spacing, and pre-existing natural fractures (Marongiu-Porcu et al., 2016; Safari et al., 2017). For simplicity, all fractures of Well-2 and Well-5 are assumed to have similar shapes (shown in Fig. 5.15) as modeled by Safari et al. (2017), and the lengths of these fractures range from 965 to 1034 ft. The width of all fractures is 0.01 ft. For Well-2, Well-5, and Well-6, a constant permeability of 20,000 md is used for all hydraulic fractures. To investigate the influence of varying fracture permeability on the analysis of fracture hits, a more realistic treatment of fracture permeability is used for the fractures of Well-1, Well-3, and Well-4. For these fractures, the permeability changes along the length with a correlation

$$k_f(x) = k_{\max}\left[1 - \left(\frac{x}{x_f}\right)^2\right], \tag{5.16}$$

where $k_{\max}$ is the fracture permeability near the wellbore, x is the distance from the wellbore along the fracture length, and x_f is the fracture half-length. In this study, $k_{\max}$ is 20,000 md and x_f is 500 ft.

As can be seen in Fig. 5.15, in this synthetic case, fracture hits occur between each pair of adjacent wells. Well-1 and Well-2, Well-2 and Well-3, Well-4 and Well-5, and Well-5 and Well-6 are connected by seven fractures, and Well-3 and Well-4 are connected by four fractures. Therefore, each time a new well is fractured, the production of its adjacent wells will be affected through connecting fractures. In addition, as has been discussed in the literature (Morales et al., 2016; Bommer et al., 2017), fracture hits can damage parent-well fractures because of the migration of solids into parent wells or removal of proppants from the near-wellbore region of the parent wells. Therefore, it is assumed that after Well-5 is fractured, the permeabilities of Well-4 fractures that are connected to Well-5 suddenly drop to 20 md.

In a typical procedure to identify fracture hits between multiple wells, all wells are first shut in for a period to establish a pressure buildup, and then wells are brought back at different times to observe if the opening of a well can influence the buildup trend of other wells (Awada et al., 2016). Many studies have been conducted on monitoring or simulating the pressure response (Sardinha et al., 2014; Scott et al., 2015; Yu et al., 2017). Here, instead of simulating the variation of well BHP for shut-in wells, this study focuses on the impact of fracture hits on producing wells. All wells are operated under a constant BHP of 3,500 psi throughout the whole simulation.

The predicted pressure profiles at different layers after 3,000 days of production are shown in Fig. 5.17, and the simulated oil rates for all wells are presented in Fig. 5.18. Each time the hydraulic fractures of a new well hit the fractures of other wells, a quick response in the oil flow rate can be observed in corresponding wells. When fracture-conductivity damage in parent wells is not considered (Well-1, Well-3, and Well-6), the oil rates of these wells experience a sudden increase (when fracture hits occur) and a decrease (when the child well starts producing at a low BHP). When severe conductivity damage is taken into account (Well-4), the fracture hit can cause a sudden drop in oil rate and permanent damage to parent-well productivity. Furthermore, several factors might affect the severity of interwell interference. Well-3 is connected to both Well-2 and Well-4 at different times, but a much stronger response in its oil rate can be observed at 900 days (when Well-2 is fractured) compared with that at 100 days (when Well-4 is fractured), and the main difference is the number of connecting fractures. Also, by comparing the oil rate responses of Well-1 and Well-3 at 900 days, it can be found that when fractures are at different depths (Well-2

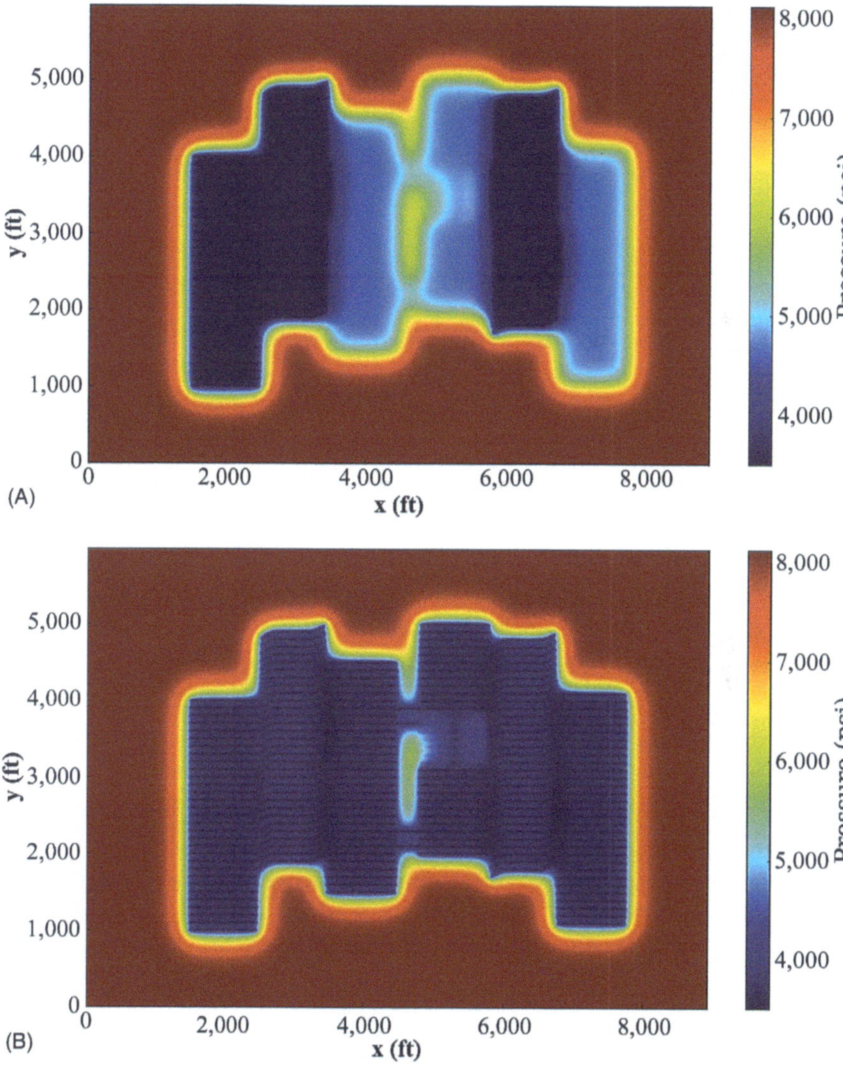

Figure 5.17 Pressure profiles in (A) Layer 2, (B) Layer 4, and (C) Layer 6 at 3,000 days predicted by the EDFM simulation (Xu et al., 2019).

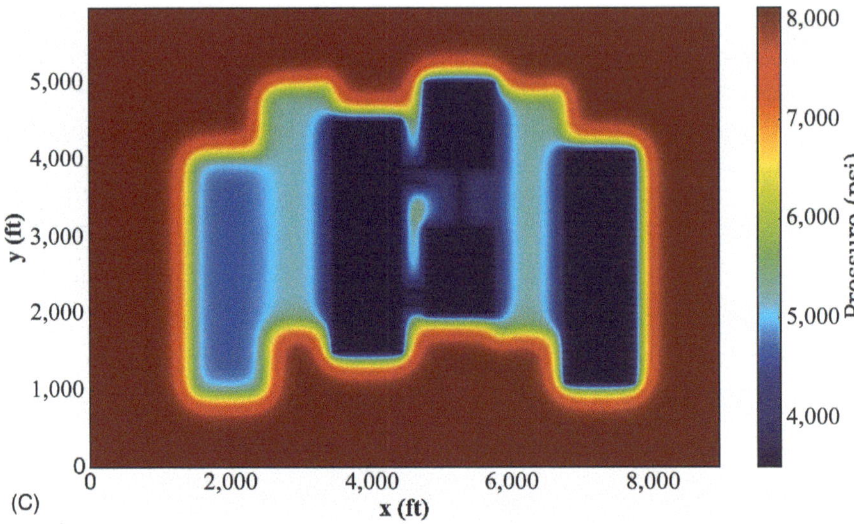

(C)

Figure 5.17 ***(Cont.)***

and Well-3), the inter-well interference can be relatively mitigated. The influence of varying fracture conductivity can also be seen when the response of Well-3 at 900 days is compared with that of Well-6 at 1,000 days. With all other parameters being the same, a much stronger response is predicted with the constant-conductivity fracture model (Well-6) than the varying-conductivity model (Well-3).

The results of this study show the influence of various factors that affect well interference, including the number of connecting fractures, fracture depth, damage to parent-well fracture conductivity, and variation of fracture conductivity along the length. Because the first three factors have also been discussed in previous studies (Awada et al., 2016; Yu et al., 2017), we would like to highlight the impact of fracture-conductivity variation along the length. In this study, because of the high value of initial fracture conductivity, the cumulative oil production of Well-1, Well-3, Well-4, and Well-6 before infill drilling has a negligible difference. In this stage, the varying conductivity does not have a big influence. However, when fracture hits occur, the varying-conductivity model predicts a much weaker response in the oil rate than does the constant-conductivity fracture model. In typical studies of primary production from hydraulically fractured wells, the description of fracture conductivity along with the length is sometimes simplified, and constant conductivity in the fracture plane is assumed because the fluid flow is mainly controlled by near-wellbore fracture conductivity.

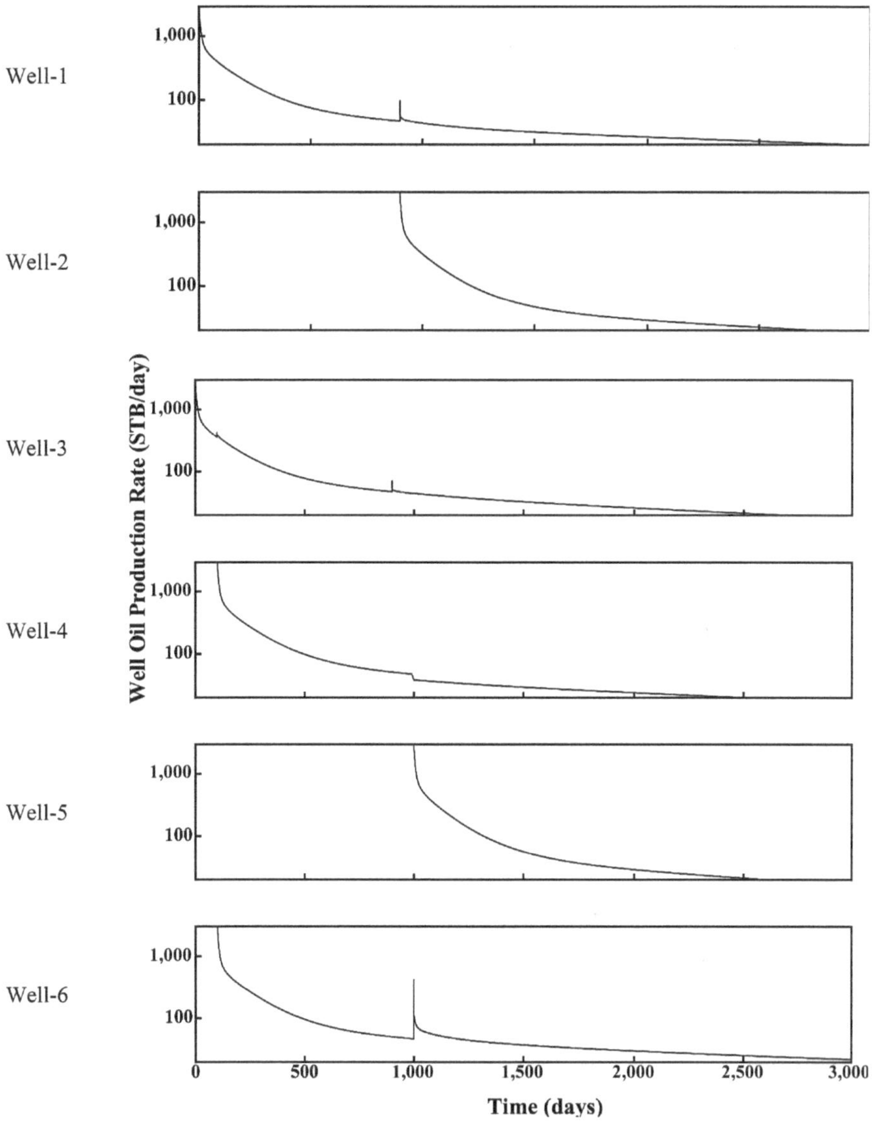

Figure 5.18 Oil production rate curves for all six wells (Xu et al., 2019). A logarithmic scale is used for the *y*-axis.

However, in the analysis of well interference, because the communicating fractures are connected near the tip, an accurate representation of fracture conductivity along the length becomes necessary. It should also be noted that the decline of fracture conductivity over time is not considered in this study. Because the fracture conductivity near the tip is typically low, if the stress-dependent effect is considered, the connectivity at fracture hits will decrease very rapidly, and the inter-well interference might experience a large downgrade over time. These dynamic effects should also be considered in well-interference studies.

5.4.4 Case 6: Natural fracture activation during water injection

In naturally fractured reservoirs, because the fluid flow is significantly affected by the existence of conductive fractures, an accurate characterization of natural fractures is of great importance. However, oftentimes even if the fracture distribution can be characterized from outcrops or field observation, whether these fractures are mechanically open cannot be easily determined. Moreover, during water injection, the increase in pore pressure might even induce shear slippage in the fracture plane (Hossain et al., 2002; Zhang et al., 2009), which changes the status of a fracture from closed to open (known as fracture activation). In this study, we simulate such a process.

The reservoir has dimensions of 3,010 × 3,010 × 50 ft, and it is uniformly discretized using a 301 × 301 × 1 matrix grid. The key simulation parameters are summarized in Table 5.7. A Corey-type relative permeability model is used, and the parameters for water/oil relative permeability curves are also provided in Table 5.7. The initial water saturation is the same as the residual water saturation. The capillary pressure is neglected in this study. An inverted five-spot pattern is simulated where four oil producers (Prod-1, Prod-2, Prod-3, and Prod-4) are in the corners, and a water injector (Inj-1) is in the center of the reservoir, as shown in Fig. 5.19. The in-situ stress state of the reservoir is also given in Table 5.7. Here, x is the direction of minimum horizontal stress, and y is the direction of maximum horizontal stress.

Three sets of randomly generated natural fractures (Sets 1, 2, and 3) with different orientations are placed inside the reservoir. The length of natural fractures ranges from 50 to 150 ft. Sets 1 and 2 each include 600 vertical fractures, randomly distributed in the whole reservoir. Set 3 has 1,200 fractures with a dip angle of 60 degrees, and the fractures are placed in only part of the reservoir, as shown in Fig. 5.19. The angles between the three sets of fractures and the x-axis are 75, 30, and 60 degrees, respectively. The conductivity of all fractures is 20,000 md-ft when they become activated.

Eq. (5.15) is used to relate fracture activation with pore pressure. A fracture cohesion (c) of zero and a friction angle (ϕ) of 30 degrees are used in Eq. (5.15) because they are classical values for these parameters (Hossain et al., 2002; de Sousa Junior et al., 2016). The pore pressure is assumed to be fully effective, and hence the Biot–Willis coefficient (β) is

Table 5.7 Simulation parameters for Case 6 (Xu et al., 2019).

Parameter	Value	Unit	Parameter	Value	Unit
Reservoir temperature	200	°F	Residual water saturation	0.2	–
Reservoir porosity	0.18	–	Residual oil saturation	0.2	–
Horizontal permeability	200	md	Water rel. perm. endpoint	0.21	–
Vertical permeability	20	md	Oil rel. perm. endpoint	0.70	–
Initial reservoir pressure	6,000	psi	Water rel. perm. exponent	1.5	–
Reservoir thickness	50	ft	Oil rel. perm. exponent	2.5	–
Rock compressibility	3×10^{-6}	psi^{-1}	Vertical stress	7,600	psi
Initial water saturation	0.2	–	Minimum horizontal stress	7,100	psi
Initial oil saturation	0.8	–	Maximum horizontal stress	7,300	psi
Water viscosity	0.79	cp	Injection rate of water	2,000	STB/day
Oil viscosity	6.3	cp	Producer bottomhole pressure	5,500	psi
Water compressibility	1×10^{-6}	psi^{-1}	Wellbore radius	0.48	ft
Oil compressibility	1×10^{-6}	psi^{-1}	Simulation time	3,000	day
Depth of top layer	8,000	ft			

unity. All fractures are initially non-active. The normal and shear stresses on all fractures are evaluated using Eqs. (5.12) and (5.13) depending on the orientation of fractures and the in-situ reservoir stress state. The critical pore pressure is calculated using Eq. (5.15). When the shear stress on a fracture plane exceeds the shear strength (i.e., the pore pressure exceeds the critical pore pressure), the fracture is considered to be activated and conductive. As a simplified approach, the entire fracture will be activated once the critical pore pressure is reached within the fracture. Through the setting of several compaction tables for different sets of fractures, this scenario can be simulated without any restarting. It should also be noted that shear failure is irreversible. After a fracture becomes activated, it will remain conductive, even if the fluid pressure in the fracture drops at a later time.

Two additional scenarios in which the dynamic fracture activations are not taken into account were also simulated to investigate the influence of

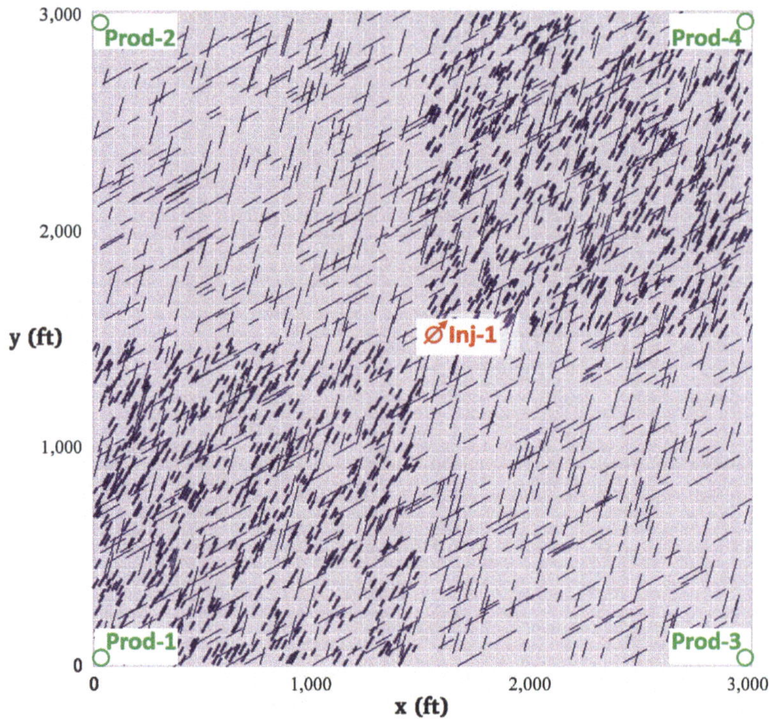

Figure 5.19 Top view of matrix grid, locations of wells and distribution of fractures in Case 6 (Xu et al., 2019). There exist three sets of natural fractures in different orientations in the reservoir, and the total number of fractures is 2,400.

dynamic fracture behaviors on production forecasting. In the first scenario, all fractures are considered non-conductive (i.e., they have no influence on the production); in the second scenario, all fractures are conductive from the beginning of the simulation.

Fig. 5.20 presents the simulated water–oil ratio for all four producers. The water-breakthrough time for each producer can be found in the plot. For the case where all fractures are non-conductive, the water breakthrough time for all four producers is the same; when all fractures are conductive from the beginning, the water phase has an earlier breakthrough at Prod-1 and Prod-4 than at Prod-2 and Prod-3 because of the high fracture density in corresponding regions; when fracture activations are considered, the predicted water-breakthrough time for all producers is between the preceding two extremes. It should also be mentioned that if capillary pressure is considered in this study, the predicted water breakthrough time will be delayed (for water-wet rock) because of the imbibition process, as studied previously

(Karimi-Fard et al., 2004; Moinfar et al., 2014). Fig. 5.21 shows the water saturation profiles for the two cases with conductive fractures, where the impact of fractures can be clearly observed.

The results show that when fracture activations are considered (Fig. 5.21A), the water phase moves much slower compared with the case

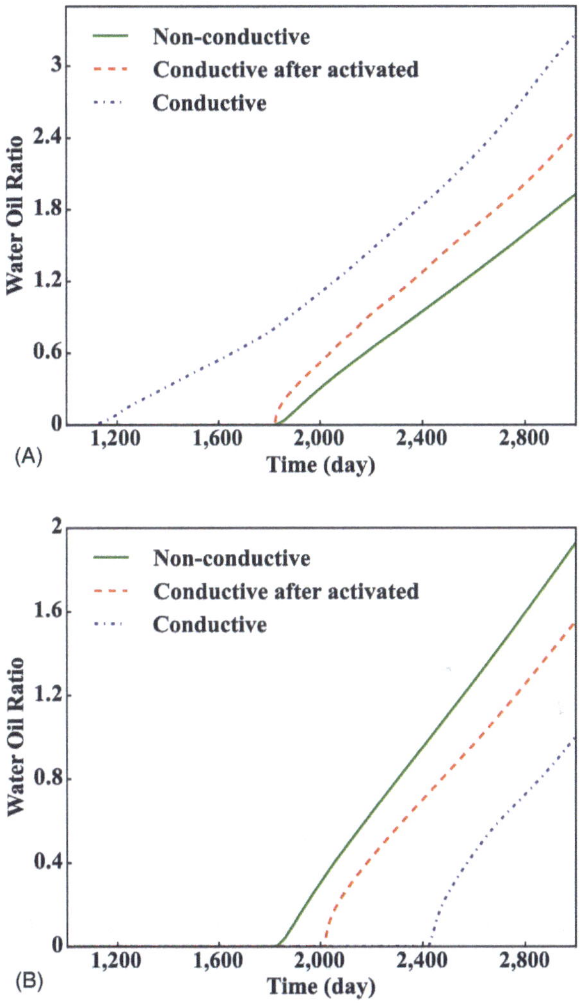

Figure 5.20 Water–oil ratio of producers (A) Prod-1, (B) Prod-2, (C) Prod-3, and (D) Prod-4 for three scenarios (Xu et al., 2019). Scenario 1 is "non-conductive", in which all fractures are non-conductive; Scenario 2 is "Conductive after activated", in which fractures are initially non-conductive, but they become conductive after shear failure; Scenario 3 is "Conductive", in which all fractures are conductive from the beginning.

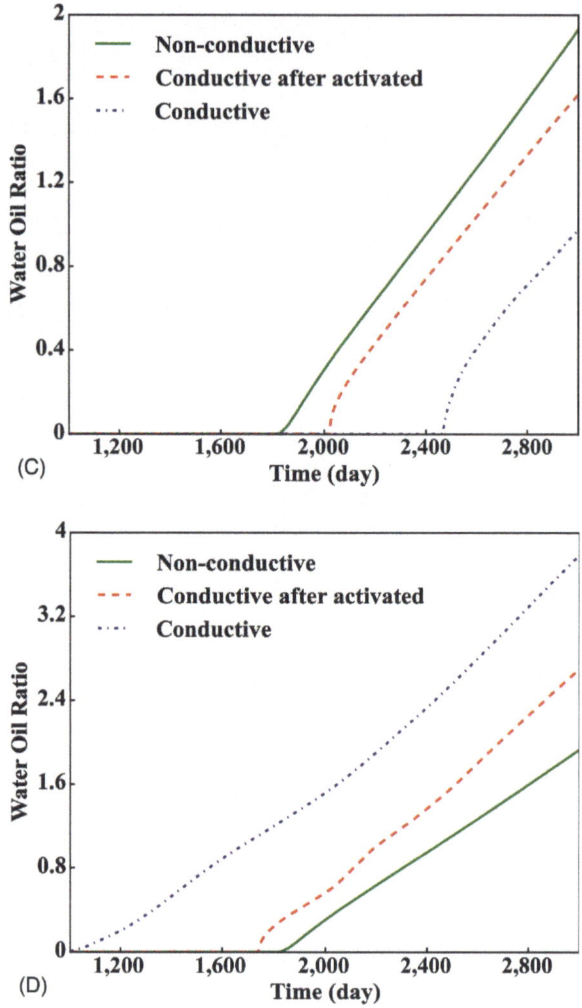

Figure 5.20 (*Cont.*)

in which all fractures are conductive (Fig. 5.21B). A primary reason is that the fractures are not effectively connected to form a network. In Fig. 5.22, the activated fractures at four different times are presented. Although most fractures in Set 3 are activated, because of the relatively high critical pore pressure for fractures in Sets 1 and 2, the activated fractures are not well connected, and the water phase cannot move through an effective fracture network from the injector to producers. Therefore, the activated fractures mainly enhance the equivalent matrix permeability in corresponding

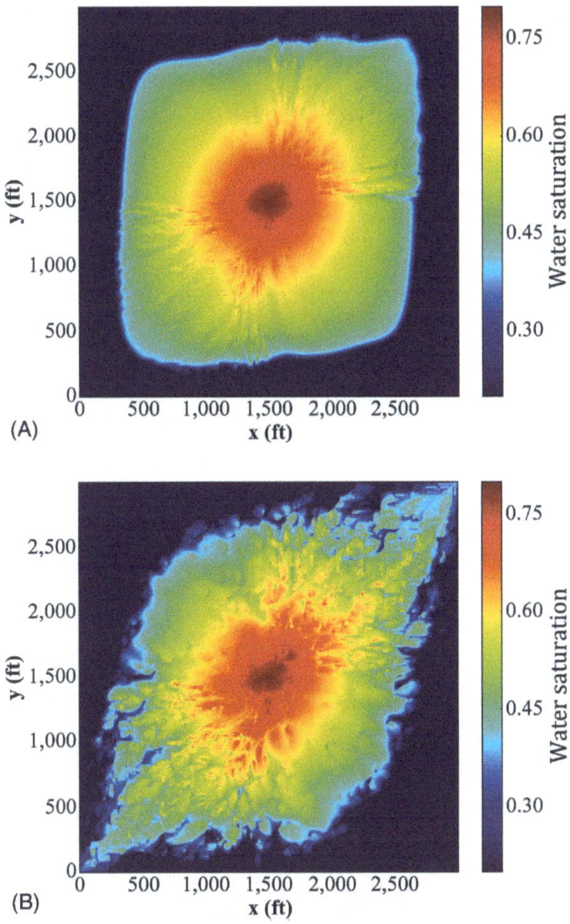

Figure 5.21 Matrix water saturation profiles at 1,500 days (Xu et al., 2019). (A) Fractures are considered conductive after shear failure. (B) All fractures are conductive.

regions, and their impact is less significant compared with a well-connected fracture network.

The results of this study show the importance of considering dynamic fracture behaviors in naturally fractured reservoirs. One might either overestimate or underestimate the influence of fractures when fracture activations are not considered. In the above simulations, it means that the predicted water-breakthrough time will be either too early or too late. This example also demonstrates the capability of the EDFM to simulate complex systems with a large number of fractures. Because of the separate gridding of the matrix and fractures, the EDFM approach avoids the difficulty of

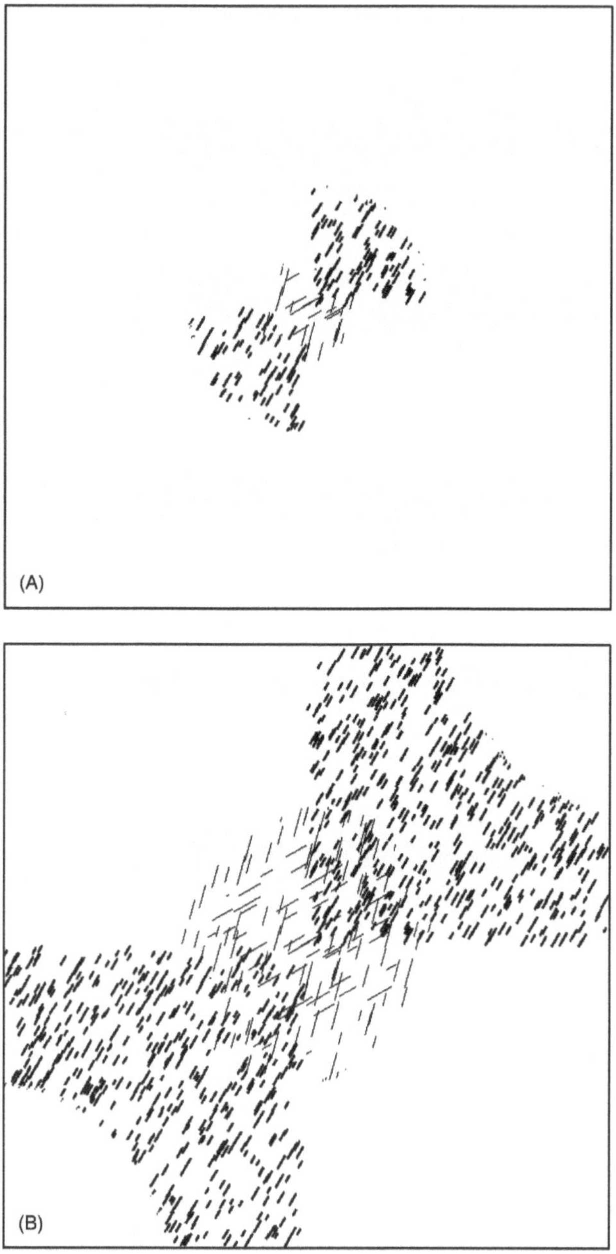

Figure 5.22 Activated fractures at (A) 50 days, (B) 200 days, (C) 2,000 days, and (D) 3,000 days (Xu et al., 2019).

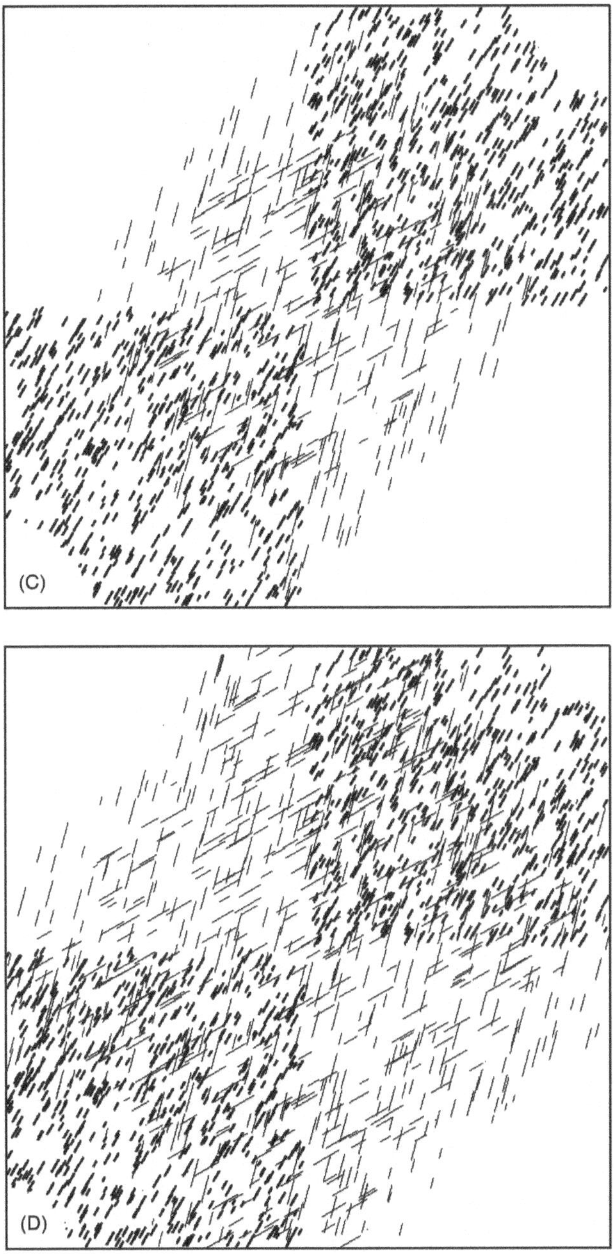

Figure 5.22 (*Cont.*)

mesh generation around complex fractures, making it a convenient method for large-scale cases.

Nomenclature

A area, ft^2
c fracture cohesion, psi
d distance, ft
k reservoir permeability, md
$\overrightarrow{\overrightarrow{K}}$ matrix permeability tensor, md
k_f fracture permeability, md
k_h horizontal reservoir permeability, md
k_m matrix permeability in a specific direction, md
k_v vertical reservoir permeability, md
L_{int} length of fracture intersection line, ft
$\vec{n}$ unit normal vector
p pore pressure, psi
p_c critical pore pressure for fracture activation, psi
T transmissibility factor, md-ft
w_f fracture width, ft
$\overline{w_f}$ average fracture width, ft
x distance, ft
x_f fracture half-length, ft
β Biot–Willis coefficient
γ permeability modulus, psi^{-1}
σ stress tensor, psi
σ' effective normal stress, psi
σ_n normal stress, psi
τ shear stress, psi
τ_{max} shear strength, psi
ϕ angle of internal friction, degree

References

Alramahi, B., Sundberg, M.I. 2012. Proppant embedment and conductivity of hydraulic fractures in shales. Paper ARMA 2012-291, 46th U.S. Rock Mechanics/Geomechanics Symposium, Chicago, Illinois.
Awada, A., Santo, M., Lougheed, D., Xu, D., Virues, C., 2016. Is that interference? A work flow for identifying and analyzing communication through hydraulic fractures in a multiwell pad. SPE J. 21 (5), 1554–1566.

Bachman, R.C., Sen, V., Khalmanova, D., Mangha, V.O., Settari, A. 2011. Examining the effects of stress dependent reservoir permeability on stimulated horizontal Montney gas wells. Paper SPE 149331, Canadian Unconventional Resources Conference, Calgary, Alberta, Canada.
Barree, R.D., Miskimins, J.L., Svatek, K.J. 2017. Reservoir and completion considerations for the refracturing of horizontal wells. Paper SPE 184837, SPE Hydraulic Fracturing Technology Conference and Exhibition, The Woodlands, Texas.
Bhardwaj, P., Hwang, J., Manchanda, R., Sharma, M.M. 2016. Injection induced fracture propagation and stress reorientation in waterflooded reservoirs. Paper SPE 181883, SPE Annual Technical Conference and Exhibition, Dubai, UAE.
Bommer, P., Bayne, M., Mayerhofer, M., Machovoe, M., Staron, M. 2017. Re-designing from scratch and defending offset wells: case study of a six-well Bakken zipper project, McKenzie County, ND. Paper SPE 184851, SPE Hydraulic Fracturing Technology Conference and Exhibition, The Woodlands, Texas.
Cavalcante Filho, J.S.A., Shakiba, M., Moinfar, A., Sepehrnoori, K. 2015. Implementation of a preprocessor for embedded discrete fracture modeling in an IMPEC compositional reservoir simulator. Paper SPE 173289, SPE Reservoir Simulation Symposium, Houston, Texas.
Clarkson, C.R., Qanbari, F., Nobakht, M., Heffner, L., 2013. Incorporating geomechanical and dynamic hydraulic-fracture-property changes into rate-transient analysis: example from the Haynesville Shale. SPE Reservoir Eval. Eng. 16 (3), 303–316.
Coulter, G.R., Menzie, D.E., 1973. The design of refrac treatments for restimulation of subsurface formations. Paper SPE 4400, SPE Rocky Mountain Regional Meeting, Casper, Wyoming.
de Sousa Junior, L.C., dos Santos, L.O.S., de Souza Rios, V., Araújo, C., Celes, W., Sepehrnoori, K., 2016. Methodology for geomechanically controlled transmissibility through active natural fractures in reservoir simulation. J. Pet. Sci. Eng. 147, 7–14.
Garipov, T.T., Karimi-Fard, M., Tchelpi, H.A., 2016. Discrete fracture model for coupled flow and geomechanics. Comput. Geosci. 20 (1), 149–160.
Gupta, J., Zielonka, M., Albert, R.A., El-Rabaa, A.M., Burnham, H.A., Choi, N.H. 2012. Integrated methodology for optimizing development of unconventional gas resources. Paper SPE 152224, SPE Hydraulic Fracturing Technology Conference, The Woodlands, Texas.
Hossain, M.M., Rahman, M.K., Rahman, S.S., 2002. A shear dilation stimulation model for production enhancement from naturally fractured reservoirs. SPE J. 7 (2), 183–195.
Hustedt, B., Qiu, Y., Zwarts, D., van den Hoek, P.J. 2005. Modeling water-injection induced fractures in reservoir simulation. Paper SPE 95726, SPE Annual Technical Conference and Exhibition, Dallas, Texas.
Karimi-Fard, M., Durlofsky, L.J., Aziz, K., 2004. An efficient discrete-fracture model applicable for general-purpose reservoir simulators. SPE J. 9 (2), 227–236.
Koning, E.J.L., Niko, H. 1985. Fractured water-injection wells: a pressure falloff test for determining fracture dimensions. Paper SPE 14458, SPE Annual Technical Conference and Exhibition, Las Vegas, Nevada.
Lei, Z., Gong, B., Wang, F., Wang, T., Li, Q. 2015. A dynamic discrete fracture model for fluid flow in fractured low-permeability reservoirs. Paper SPE 175554, SPE Reservoir Characterisation and Simulation Conference and Exhibition, Abu Dhabi, UAE.
Marongiu-Porcu, M., Lee, D., Shan, D., Morales, A., 2016. Advanced modeling of interwell-fracturing interference: an Eagle Ford shale-oil study. SPE J. 21 (5), 1567–1582.
Moinfar, A., Sepehrnoori, K., Johns, R.T., Varavei, A. 2013. Coupled geomechanics and flow simulation for an embedded discrete fracture model. Paper SPE 163666, SPE Reservoir Simulation Symposium, The Woodlands, Texas.
Moinfar, A., Varavei, A., Sepehrnoori, K., Johns, R.T., 2014. Development of an efficient embedded discrete fracture model for 3D compositional reservoir simulation in fractured reservoirs. SPE J. 19 (2), 289–303.

Morales, A., Zhang, K., Gakhar, K., Porcu, M.M., Lee, D., Shan, D., Malpani, R., Pope, T., Sobernheim, D., Acock, A., 2016. Advanced modeling of interwell fracturing interference: an Eagle Ford Shale oil study - refracturing. Paper SPE 179177, SPE Hydraulic Fracturing Technology Conference, The Woodlands, Texas.
Norbeck, J.H., McClure, M.W., Lo, J.W., Horne, R.N., 2016. An embedded fracture modeling framework for simulation of hydraulic fracturing and shear stimulation. Comput. Geosci. 20 (1), 1–18.
Panfili, P., Colin, R., Cominelli, A., Giamminonni, D., Guerra, L. 2015. Efficient and effective field scale simulation of hydraulic fractured wells: methodology and application. Paper SPE 175542, SPE Reservoir Characterisation and Simulation Conference and Exhibition, Abu Dhabi, UAE.
Ponting, D.K. 1989. Corner point geometry in reservoir simulation. 1st European Conference on the Mathematics of Oil Recovery, Cambridge, UK.
Ren, G., Jiang, J., Younis, R.M., 2016. A fully coupled XFEM-EDFM model for multiphase flow and geomechanics in fractured tight gas reservoirs. Procedia Comput. Sci. 80, 1404–1415.
Roussel, N.P., Sharma, M.M., 2012. Role of stress reorientation in the success of refracture treatments in tight gas sands. SPE Prod. Oper. 27 (4), 346–355.
Safari, R., Lewis, R., Ma, X., Mutlu, U., Ghassemi, A., 2017. Infill-well fracturing optimization in tightly spaced horizontal wells. SPE J. 22 (2), 582- 295.
Sardinha, C.M., Petr, C., Lehmann, J., Pyecroft, J.F., Merkle, S. 2014. Determining interwell connectivity and reservoir complexity through frac pressure hits and production interference analysis. Paper SPE 171628, SPE/CSUR Unconventional Resources Conference - Canada, Calgary, Alberta, Canada.
Scott, K.D., Chu, W.C., Flumerfelt, R.W. 2015. Application of real-time bottom-hole pressure to improve field development strategies in the Midland Basin Wolfcamp Shale. Paper URTeC 2154675, Unconventional Resources Technology Conference, San Antonio, Texas. https://doi.org/10.15530/URTEC-2015-2154675.
Settari, A.T., Bachman, R.C., Walters, D.A. 2005. How to approximate effects of geomechanics in conventional reservoir simulation. Paper SPE 97155, SPE Annual Technical Conference and Exhibition, Dallas, Texas.
Siebrits, E., Elbel, J.L., Hoover, R.S., Diyashev, I.R., Griffin, L.G., Demetrius, S.L., Wright, C.A., Davidson, B.M., Steinsberger, N.P., Hill, D.G. 2000. Refracture reorientation enhances gas production in Barnett Shale tight gas wells. Paper SPE 63030, SPE Annual Technical Conference and Exhibition, Dallas, Texas.
Simpson, M.D., Patterson, R., Wu, K. 2016. Study of stress shadow effects in Eagle Ford Shale: insight from field data analysis. Paper ARMA 2016-190, 50th US Rock Mechanics/Geomechanics Symposium, Houston, Texas.
Terracina, J.M., Turner, J.M., Collins, D.H., Spillars, S. 2010. Proppant selection and its effect on the results of fracturing treatments performed in shale formations. Paper SPE 135502, SPE Annual Technical Conference and Exhibition, Florence, Italy.
Vincent, M.C. 2009. Examining our assumptions-have oversimplifications jeopardized our ability to design optimal fracture treatments? Paper SPE 119143, SPE Hydraulic Fracturing Technology Conference, The Woodlands, Texas.
Vincent, M.C., 2011. Restimulation of unconventional reservoirs: when are refracs beneficial? J. Can. Pet. Technol. 50 (5), 36–52.
Wolhart, S.L., McIntosh, G.E., Zoll, M.B., Weijers, L. 2007. Surface tiltmeter mapping shows hydraulic fracture reorientation in the Codell Formation, Wattenberg Field, Colorado. Paper SPE 110034, SPE Annual Technical Conference and Exhibition, Anaheim, California.
Wu, W., Kakkar, P., Zhou, J., Russell, R., Sharma, M.M. 2017. An experimental investigation of the conductivity of unpropped fractures in shales. Paper SPE 184858, SPE Hydraulic Fracturing Technology Conference and Exhibition, The Woodlands, Texas.

Xu, Y., Yu, W., Sepehrnoori, K., 2019. Modeling dynamic behaviors of complex fractures in conventional reservoir simulators. SPE Reservoir Eval. Eng. 22 (3), 1110–1130.

Yeo, I.W., de Freitas, M.H., Zimmerman, R.W., 1998. Effect of shear displacement on the aperture and permeability of a rock fracture. Int. J. Rock Mech. Min. Sci. 35 (8), 1051–1070.

Yilmaz, O., Nur, A., Nolen-Hoeksema, R., 1991. Pore pressure profiles in fractured and compliant rooks. SPE 22232 unsolicited.

Yu, W., Xu, Y., Weijermars, R., Wu, K., Sepehrnoori, K. 2017. Impact of well interference on shale oil production performance: a numerical model for analyzing pressure response of fracture hits with complex geometries. Paper SPE 184825, SPE Hydraulic Fracturing Technology Conference and Exhibition, The Woodlands, Texas. https://doi.org/10.2118/184825-MS.

Zhang, J., Kamenov, A., Hill, A.D., Zhu, D., 2014. Laboratory measurement of hydraulic-fracture conductivities in the Barnett Shale. SPE Prod. Oper. 29 (3), 216–227.

Zhang, X., Jeffrey, R.G., Thiercelin, M., 2009. Mechanics of fluid-driven fracture growth in naturally fractured reservoirs with simple network geometries. J. Geophys. Res. Solid Earth 114 (B12), B12406.

CHAPTER SIX

Field-scale applications of the embedded discrete fracture model

6.1 Field-scale case study in the Piceance Basin Niobrara Formation

With the rapid development of unconventional oil and gas reservoirs, a large number of horizontal wells have been drilled and fractured. Simultaneously, the issue of well interference has become more critical (Lawal et al., 2013; King and Valencia, 2016). In field operations, the shut-in and build-up pressure test is often utilized to identify the degree of inter-well communication vertically and areally (Portis et al., 2013; Sardinha et al., 2014; Sani et al., 2015; Scott et al., 2015; Liang et al., 2017). Well interference might occur through the matrix, connecting hydraulic fractures, natural fractures, or a combination of these mechanisms (Yu et al., 2018c). The occurrence of well interference may adversely affect well productivity in multi-well pads (Yaich et al., 2014; Malpani et al., 2015; Liang et al., 2017; Tang et al., 2017). Well interference also plays a vital role in the optimization of well spacing.

Generally, the geometries and locations of fractures in multi-well pads are highly uncertain and complex. Given the impacts of complex fracture systems, this presents a great challenge for efficient history matching with actual production data and pressure responses of multiple wells from different pads. Several analytical models have been developed to model the pressure response or determine the degree of well interference (Awada et al., 2016; Yu et al., 2016b; Molina and Zeidouni, 2017). However, there are difficulties for analytical and semi-analytical solutions in capturing the effects of heterogeneous reservoir properties, multiple layers, multiphase flow, complex fracture networks, and natural fractures. Thus, advanced numerical models are needed to address these challenges.

Li et al. (2017) built a three-dimensional, two-phase numerical reservoir model with six horizontal wells from two well pads with planar fracture hits. The traditional local grid refinement (LGR) approach was applied to handle hydraulic fractures. A good match between the simulation result and

Embedded Discrete Fracture Modeling and Application in Reservoir Simulation.
http://dx.doi.org/10.1016/B978-0-12-821872-3.00006-X

the actual data was obtained using 11 hydraulic fracture hits across the laterals with high conductivity. However, the LGR approach with structured gridding has difficulty capturing complex hydraulic and natural fracture geometries. Marongiu-Porcu et al. (2016) used a reservoir simulator with unstructured-gridding capability to simulate a synthetic well-interference case with complex fracture geometries predicted by the fracture modeling. However, the usage of unstructured grids may have complicated gridding issues around the fractures and generate a high computational cost, especially for cases with a large number of hydraulic and natural fractures, as mentioned in Chapter 3. Furthermore, it is challenging to use this approach to perform efficient history matching for multi-well pads with fracture hits, because the fracture geometry needs to be frequently altered and simulations need to be repeatedly run to obtain a good history match. These issues can be addressed by using the embedded discrete fracture model (EDFM).

This study will demonstrate the application of the EDFM in a field-scale study with multiple wells in a specific shale reservoir. In this study, the EDFM is used to model a gas reservoir in the Mancos Shale of the Piceance Basin, United States. First, a history matching study is conducted to match the calculated well bottomhole pressure (BHP) of six horizontal wells by varying fracture properties while honoring other diagnostic test results. The interference between wells, through connecting hydraulic fractures, is considered to obtain a good match. The high computational efficiency of the EDFM is illustrated through its comparison with the LGR approach. Subsequently, the advantages of the EDFM are fully exploited through sensitivity studies. We investigate the impacts of hydraulic fracture half-length, natural fracture density, and natural fracture conductivity on long-term gas production. The influence of natural fractures on well interference is also studied.

6.1.1 Completions of six subject wells

Six horizontal wells are located in the Rangely Bench of the Niobrara Formation in western Colorado, which is in the lower portion of the Mancos Shale in the Piceance Basin (Li et al., 2016). The thickness of the Niobrara Formation is ~1,200 ft with interbedded calcareous shale and shaly limestone facies. The operator applied a mechanical earth model (MEM) using all available data and estimated that the preferred fracture orientation is in the NE direction and that there is no barrier to fracture growth in the vertical direction. Borehole images showed that two sets of natural fractures are present in the formation. One set is oriented

~15 degrees to the maximum-horizontal-stress direction. The difference between the maximum and minimum horizontal stresses was evaluated to be around 350 psi. More details of the geological description of the formation can be found in Li et al. (2016, 2017).

Six horizontal wells were drilled and completed from two pads in the SE direction. The 3D view of the six wells is displayed in Fig. 6.1. First, Well 1 was completed with 2,370,661 lbs of 100 mesh, 7,045,191 lbs of 40/70 white sand, and 529,561 bbls of slickwater. Subsequently, Well 3 was completed with 2,302,729 lbs of 100 mesh, 8,127,509 lbs of 40/70 white sand, and 545,687 bbls of slickwater. Well 2 was lastly completed with 2,754,880 lbs of 100 mesh, 13,027,220 lbs of 40/70 white sand, and 545,687 bbls of slickwater. After 6 months, Wells 4, 5, and 6 were completed. Well 4 was completed with 2,894,269 lbs of 100 mesh, 10,422,224 lbs of 40/70 white sand, and 660,475 bbls of slickwater. Well 6 was completed with 1,695,380 lbs of 100 mesh, 9,376,189 lbs of 40/70 white sand, and 527,992 bbls of slickwater. Well 5 was lastly completed with 2,368,080 lbs of 100 mesh, 12,066,504 lbs of 40/70 white sand, and 634,932 bbls of slickwater. The detailed completion information is summarized in Table 6.1. Microseismic fracture mapping was performed for all six wells. However,

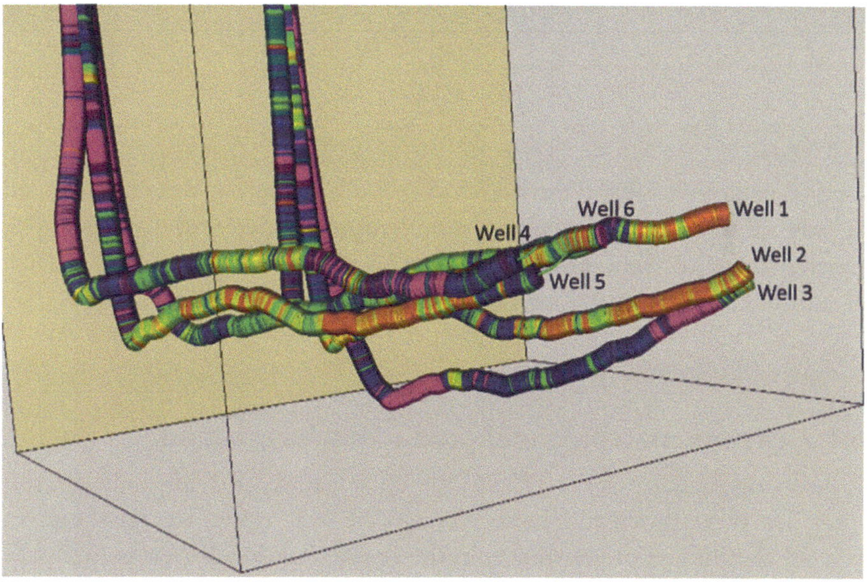

Figure 6.1 ***Three-dimensional view of six horizontal wells in the target area.*** *(Modified from Li et al., 2016, reprinted by permission of the AAPG whose permission is required for further use).*

Table 6.1 Detailed completion information for the six horizontal wells (Xu et al., 2018).

Well	Stage spacing (ft)	Number of stages	Proppant per ft (lbs/ft)	Max proppant concentration (ppa)	Volume per ft (bbls/ft)
Well 1	200	42	973	1.50	55
Well 2	200	50	1,571	1.75	72
Well 3	200	46	1,150	1.50	60
Well 4	200	49	1,367	2.00	68
Well 5	200	46	1,621	2.00	71
Well 6	200	47	1,332	2.00	64

Table 6.2 Detailed comparison of CPU time between the EDFM and the LGR model for the history matching and production forecasting stages. In Columns 3 and 5, "history" refers to the history matching stage of 413 days, and "forecasting" refers to the production forecasting stage of 40 years (Xu et al., 2018).

Simulation model	Number of gridblocks	Number of time steps	CPU time (seconds)	The ratio of CPU time (LGR/EDFM)
EDFM	291,900	733 (history)	1,589	13.7 (history)
		670 (forecasting)	1,763	19.1 (forecasting)
LGR	855,750	1,420 (history)	21,720	
		1,378 (forecasting)	33,619	

the toe-half of Well 2 and the heel-half of Wells 4, 5, and 6 were not mapped because of the observation distance limitation of microseismic events. The whole laterals of Wells 1 and 3 were mapped. Fig. 6.2 shows a map view of all microseismic events, which clearly illustrates the significant overlap between wells. Hence, there is a high chance that well interference due to complex fracture hits occurred.

6.1.2 History matching and production forecasting

A reservoir model using a 125 × 300 × 7 matrix grid was built for this study. The dimensions of the reservoir and the well trajectories are shown in Fig. 6.3. A uniform discretization in the horizontal direction is used. In the vertical direction, each layer of the matrix grid represents a geological unit in the formation. For hydraulic fractures, the fracture spacing was 40–50 ft during completion. However, fracture modeling results and production

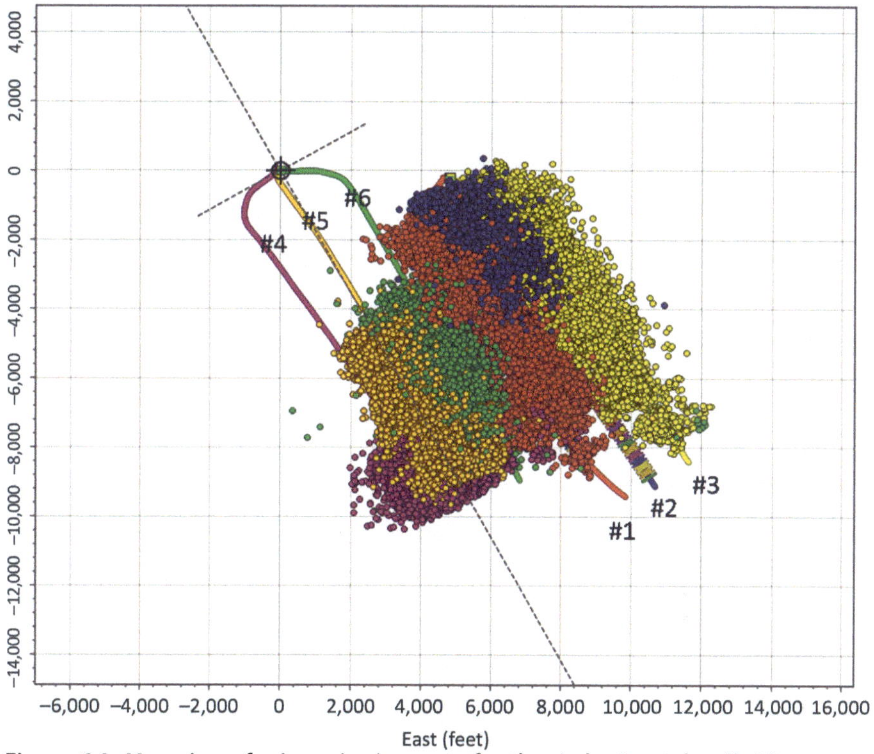

Figure 6.2 ***Map view of microseismic events for the six horizontal wells.*** The events for different wells are shown in different colors. "#1" represents "Well 1", and similarly for "#2", "#3", "#4", "#5", and "#6". *(Li et al., 2016, reprinted by permission of the AAPG whose permission is required for further use).*

logging tool (PLT) data showed that the cluster efficiency was around 30%. Therefore, the effective fracture spacing was greater than 100 ft, and this value was used in the model. The fractures were simulated using the EDFM.

A history matching study was performed for the six wells. Two-phase (gas and water) flow was simulated using a black-oil commercial simulator (CMG, 2017b). The gas flow rate was used as a constraint for the history matching, as shown in Fig. 6.4. Simulations were performed to match the BHP, which was calculated from surface pressure, and it matched well with the measured data from PLT (Li et al., 2017). The BHP of all six wells was matched simultaneously.

On the basis of the availability of measurement data, the hydraulic fracture geometry and fracture orientations have less uncertainty. The fracture number and fracture length used in this study were similar to those in Li et al. (2017). The microseismic data indicate that the fractures are mainly planar, and the micro-deformation data show that the "deformed" fracture height is between 750 and 950 ft (Li et al., 2016). The fracture width used in the model is 0.01 ft. For history matching, the most important unknowns are fracture conductivities. In Li et al. (2017), different fracture permeabilities in different directions were used. In this study, we assumed that the fracture permeabilities in all directions were the same. In addition, updated formation porosity and permeability values were used. To limit the number of unknowns, the conductivities of fractures for each well were assumed to be similar. It is also worth mentioning that the fracture conductivity changed after several well operations. These dynamic fracture conductivity behaviors need to be considered to obtain a good history matching. Another important factor to consider is the interference between wells. It had been observed, in production data, that communication between wells exists (Li et al., 2017). As an example, at 328 days, Well 6 was shut-in for ~60 days. A continuous increase in gas rate and BHP was then observed in Well 5 (Fig. 6.4E). This indicates strong communication between Wells 5 and 6 through fracture hits. Therefore, several long fractures were added to the model to represent well communication, as shown in Fig. 6.3.

The results of the history matching are presented in Fig. 6.5. A good match was obtained for all wells. The pressure profiles in the matrix and fractures at

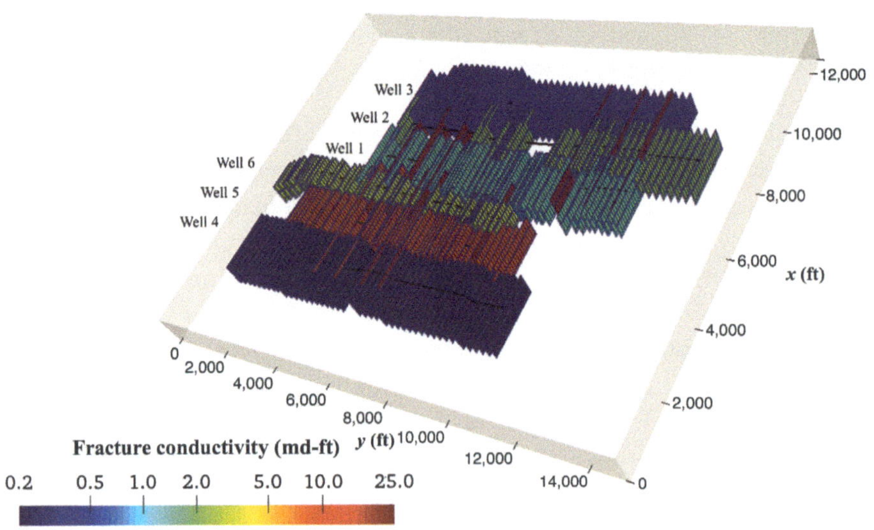

Figure 6.3 ***Reservoir dimensions, well trajectories, fracture locations, and fracture conductivity at the end of the history matching stage (Xu et al., 2018).*** The fracture conductivity is shown on the logarithmic scale.

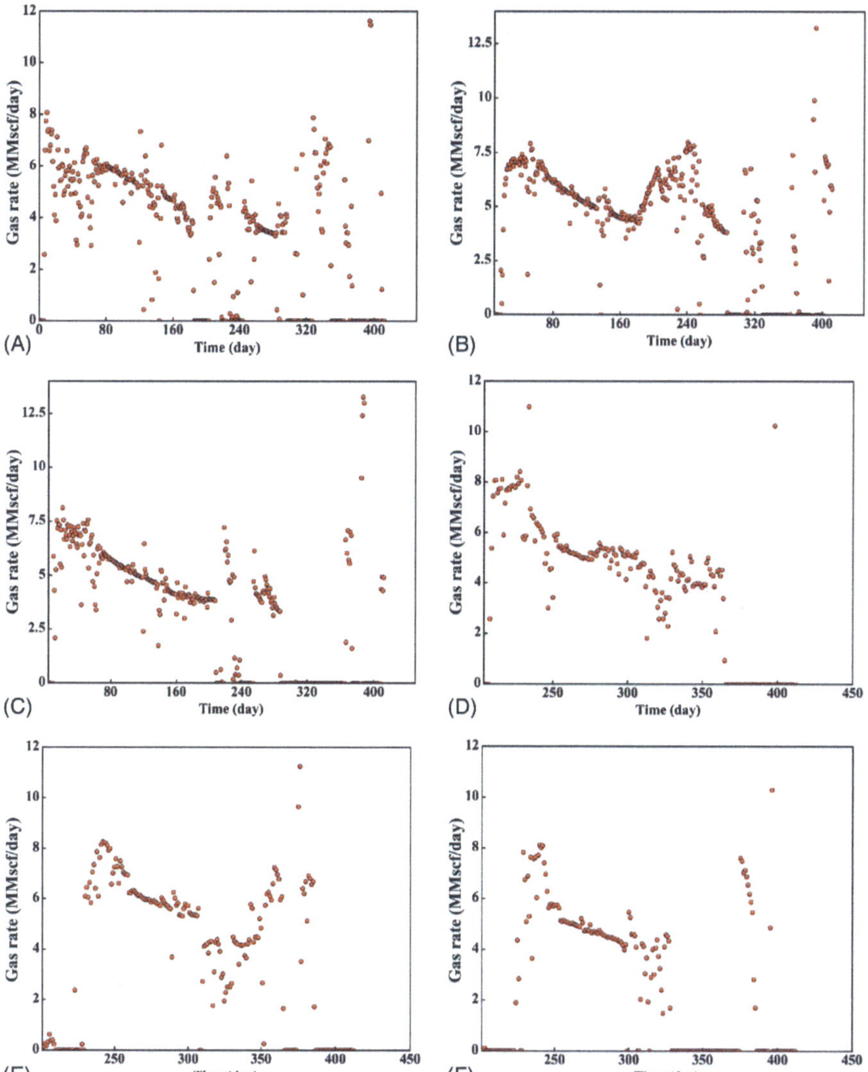

Figure 6.4 ***Gas-flow-rate constraint used for history matching (Xu et al., 2018).*** (A) through (F) show the gas flow rate of Wells 1 through 6, respectively.

the end of the history matching stage are shown in Fig. 6.6. The fracture conductivity obtained from history matching is shown in Fig. 6.3, and the difference among wells can be clearly observed. It should be noted that like all other history matching studies, the fracture parameters obtained from this study can be non-unique. For example, it might be possible to get alternative matches with higher fracture conductivities and fewer transverse fractures.

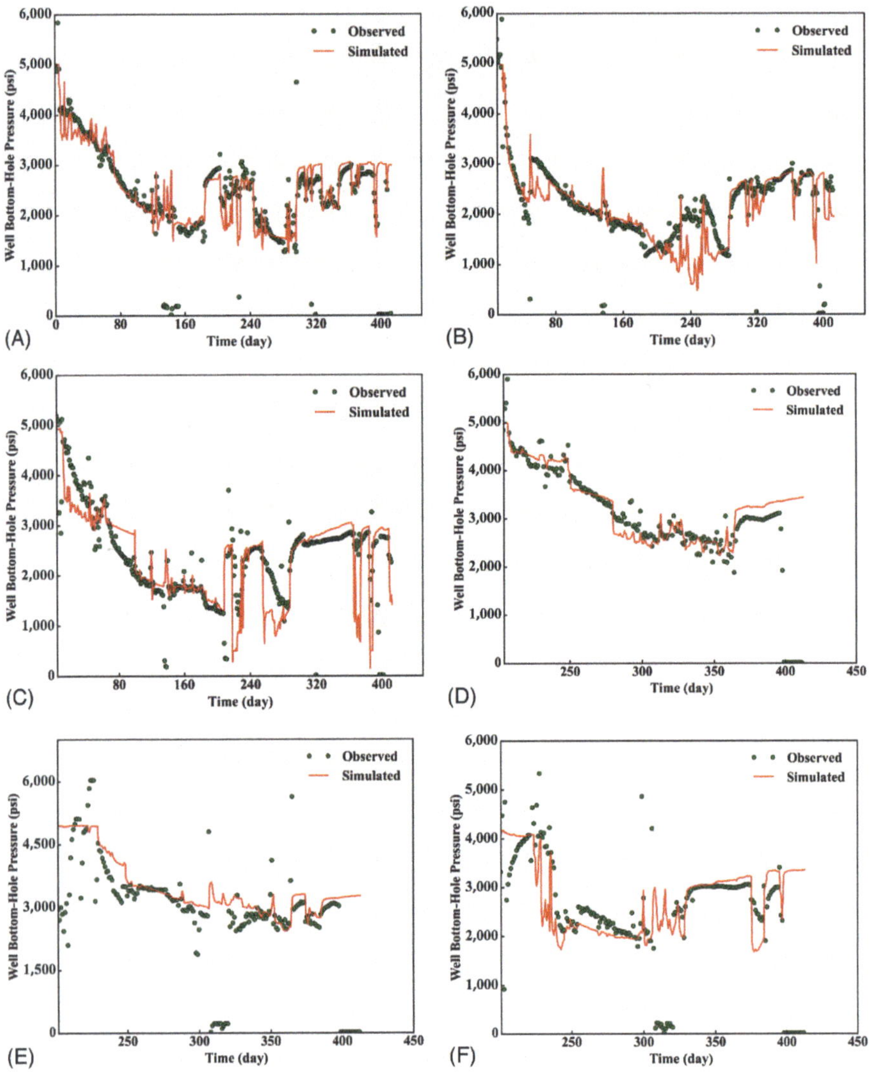

Figure 6.5 ***History matching results of the BHP using the EDFM (Xu et al., 2018).*** (A) through (F) show the history matching results for Wells 1 through 6, respectively.

With the fracture properties obtained from history matching, production forecasting was also conducted to evaluate the well performance for 40 years. A minimum BHP of 500 psi was used for all wells in the forecasting stage. Fig. 6.7 presents the predicted pressure profile after 40 years of production. The stimulated reservoir volume (SRV) and the pressure interference between wells can be clearly seen from the pressure profile.

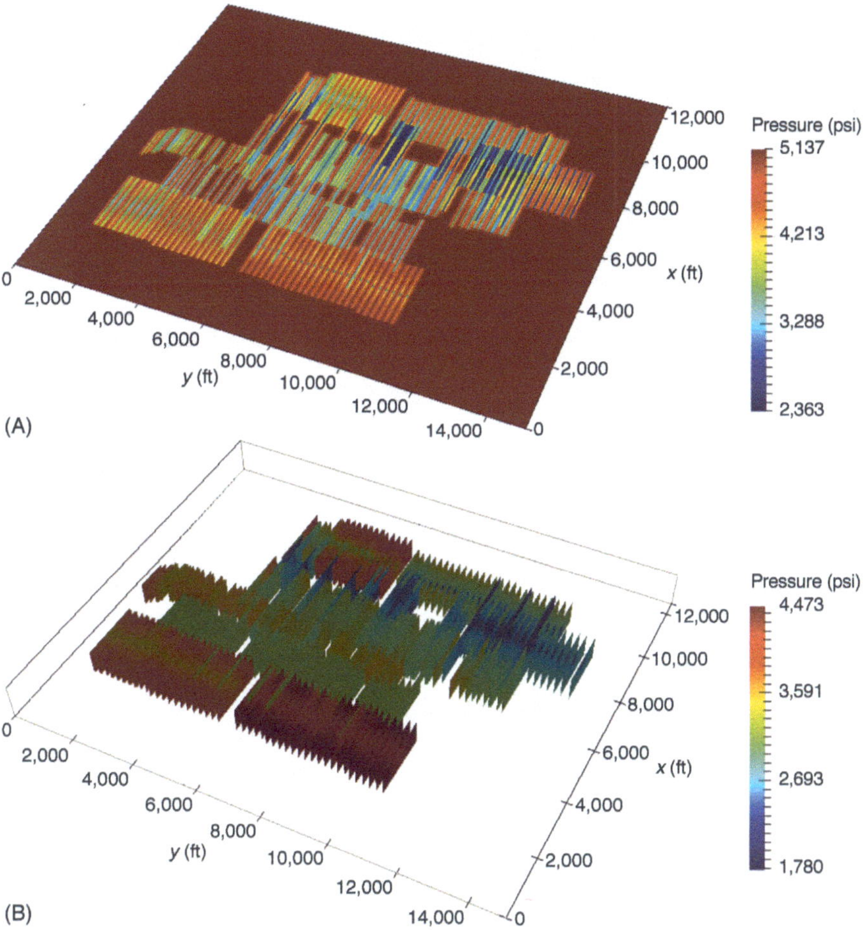

Figure 6.6 ***Predicted pressure profiles in the matrix and fractures after the history matching stage (413 days) (Xu et al., 2018).*** (A) The pressure in the bottom layer of the matrix. (B) The pressure in the fractures.

Computational efficiency is one of the most important aspects to consider when evaluating a numerical model. Compared with the LGR model with 855,750 gridblocks by Li et al. (2017), the EDFM model built in this work has 291,900 gridblocks. Correspondingly, using a single processor, for the history matching stage with a maximum time step of 1.0 day, the central processing unit (CPU) time of the EDFM simulation is 26 minutes, and the CPU time using the LGR model is 362 minutes. For the production forecasting stage with a maximum time step of 30 days, the CPU time using the EDFM is 29 minutes, while the CPU time using the LGR model

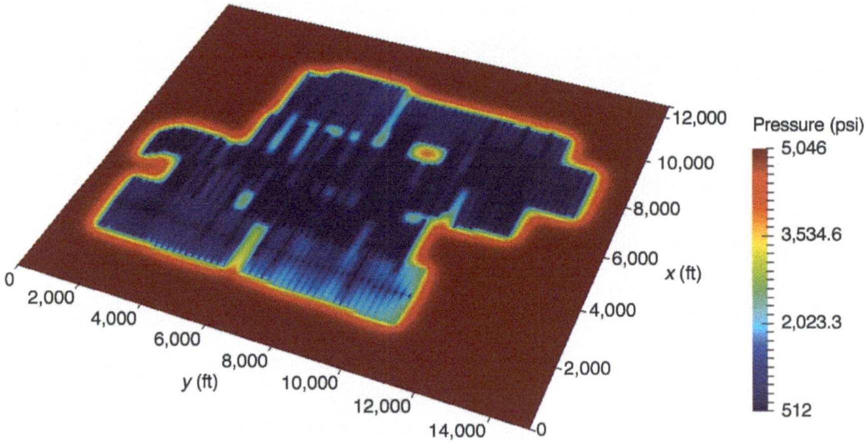

Figure 6.7 ***Predicted pressure profile at the bottom layer of the matrix after 40 years of production (Xu et al., 2018).***

is 560 minutes. A detailed comparison is shown in Table 6.2. Although the two cases have differences in fracture properties, this comparison illustrates the advantages of the EDFM regarding memory usage and computational efficiency. Furthermore, to the best of our knowledge, this is the first time that the efficiency of the EDFM and the LGR is compared using a complex, real-life reservoir model with multiple wells.

6.1.3 Impact of propped fracture length

The effective length of fractures is a key factor impacting the pressure interference between wells. The history matching results show that the fracture half-length is ~1,000 ft, and well communication can be observed due to overlapping fractures. However, fracture modeling results show that the propped fracture length is shorter than the total fracture length (Li et al., 2016). If the unpropped portion of the hydraulic fractures is assumed to have a negligible effect on long-term production, the actual fracture length used for production forecasting should be lower than that obtained from history matching. Therefore, the half-length of fractures was varied to investigate the influence of fracture length on long-term production, and the results are shown in Fig. 6.8. In Fig. 6.8, "base model" refers to the model obtained from history matching. The *y*-axis represents the total cumulative gas production of six wells. It can be seen that fracture half-length has a significant influence on cumulative production. If the fracture half-length is only a quarter of that in the base model, the cumulative gas production after 40 years is only 47.0%

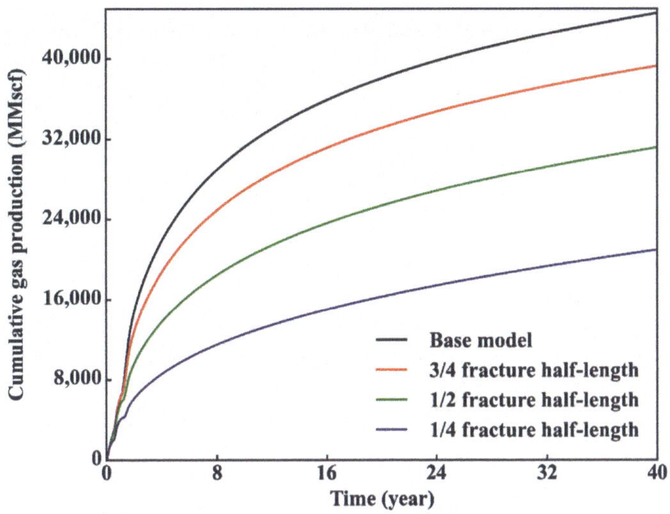

Figure 6.8 ***Impact of fracture half-length on cumulative gas production (Xu et al., 2018).*** "3/4 fracture half-length" means that the half-length of each fracture is set to be 3/4 of that obtained from history matching. Similar definitions apply to "1/2 fracture half-length" and "1/4 fracture half-length."

of that predicted by the base model. Therefore, a propped fracture length close to the well spacing is vital for high well performance. In addition, although the closure of propped hydraulic fractures under stress is not considered in this study, it is important to maintain the fracture length and fracture conductivity over the long term to maximize gas production.

6.1.4 Impact of natural fractures

The geological analysis results indicated the existence of two sets of natural fractures with different orientations in the formation. The orientation of one set of open natural fractures is close to the direction of the maximum horizontal stress, and the open natural fractures are expected to increase the hydraulic fracture complexity. Another set of natural fractures are mineralized, and their orientation is almost perpendicular to the open natural fractures. In the previous simulations, the influence of the open natural fractures was not considered because their orientation was similar to hydraulic fractures. In addition, the mineralized set of fractures was assumed to remain closed and have no influence on well performance. Here, using the EDFM, a sensitivity study was conducted to investigate the influence of natural fractures. A height similar to that of hydraulic fractures is assumed for natural fractures. Scenarios with different fracture density and fracture conductivity were simulated.

We first investigated the influence of the mineralized set of natural fractures. The orientation of this set of fractures is nearly perpendicular to the orientation of hydraulic fractures, as indicated by the geological data. Therefore, these natural fractures could help to create a complex fracture network. Three simulations with 500, 2,000, and 5,000 randomly generated natural fractures, as shown in Fig. 6.9, were conducted to study the impact

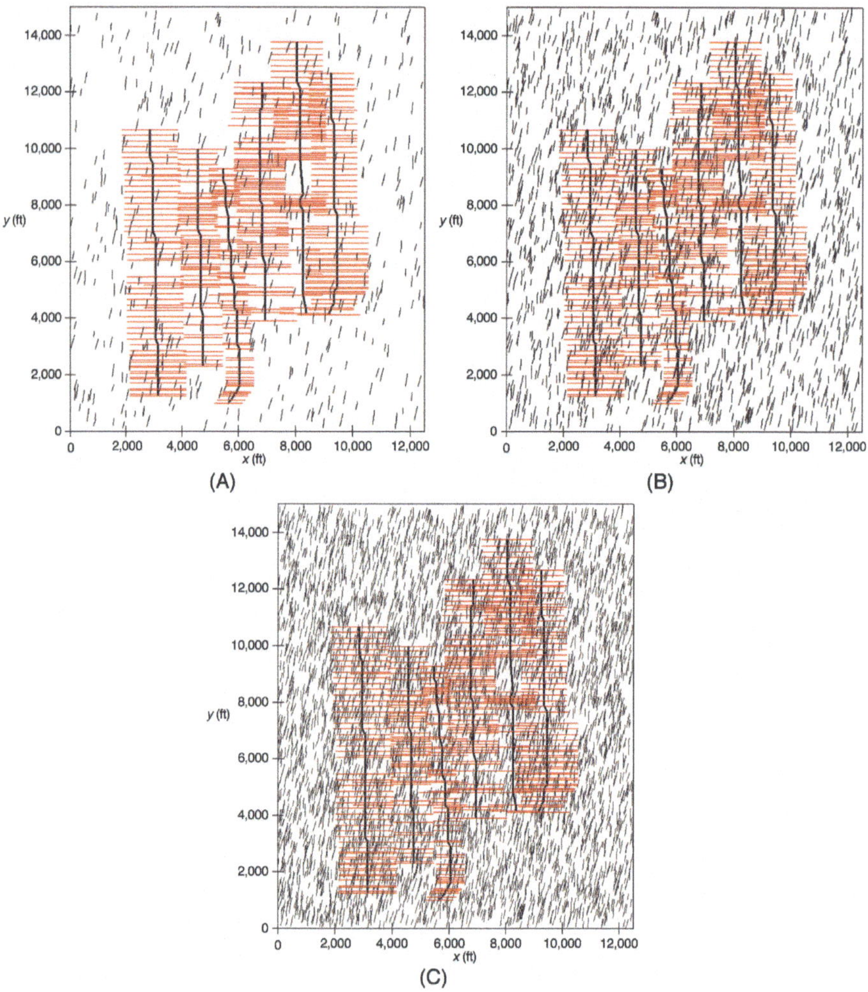

Figure 6.9 Map view of three case studies with (A) 500, (B) 2,000, and (C) 5,000 natural fractures (Xu et al., 2018). The hydraulic fractures are shown in red, and the natural fractures are shown in gray.

of fracture density. All fractures are vertical, and the natural fracture conductivity is 0.002 md-ft. The length of the natural fractures ranges from 200 to 500 ft. Fig. 6.10 presents the cumulative gas production for 40 years with different numbers of natural fractures. It can be observed that the existence of natural fractures has a high impact on long-term production. With 5,000 natural fractures, the cumulative gas production increases by 8.6% compared with the case without natural fractures. With 500 and 2,000 natural fractures, the contribution of natural fractures is 1.1% and 4.6%, respectively. The pressure profiles after 40 years of production with 500 and 5,000 natural fractures are shown in Fig. 6.11, where the difference in the SRV can be clearly observed. The case with 5,000 natural fractures has a larger SRV. Also, it should also be noted that because of the tight spacing between wells and the pressure interference between hydraulic fractures, the contribution of natural fractures is diminished because the increase in SRV by the natural fractures is less significant compared with cases with a large well spacing.

The influence of natural fracture conductivity was also investigated. For the case with 2,000 natural fractures, the conductivity of natural fractures was varied from 0.0002 to 0.02 md-ft. Fig. 6.12 shows the predicted cumulative gas production with different natural fracture conductivities. It can be observed that gas production is very sensitive to natural fracture conductivity. When the fracture conductivity increases, the relative contribution from

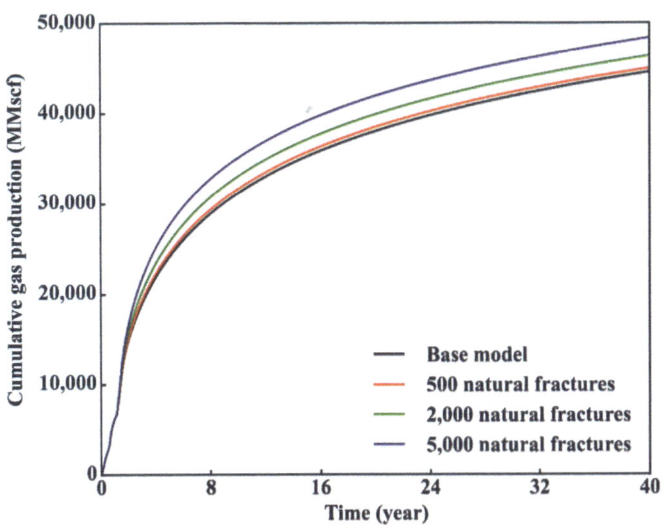

Figure 6.10 ***Impact of natural fracture density on cumulative production (Xu et al., 2018).*** The "base model" refers to the model obtained from history matching.

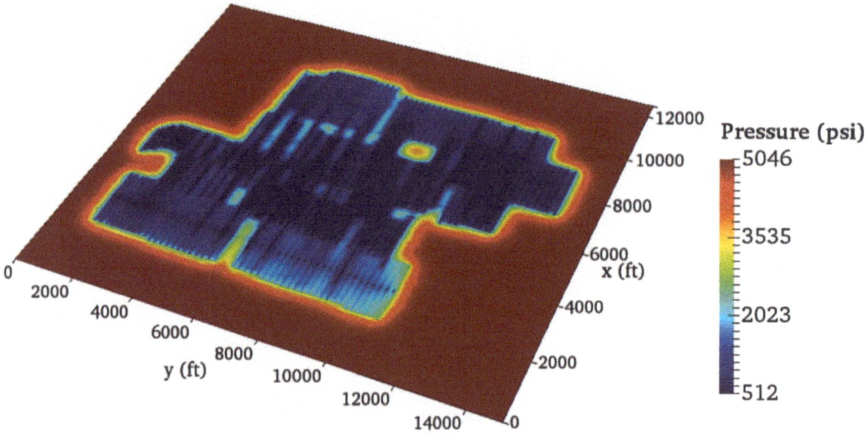

(A)

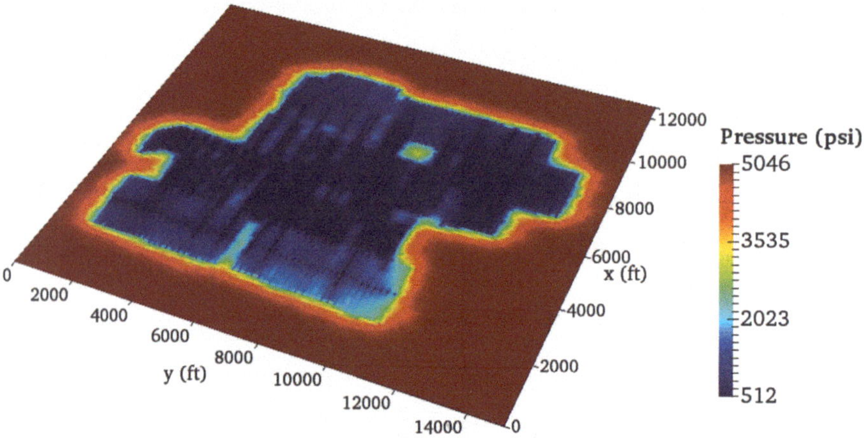

(B)

Figure 6.11 ***Predicted pressure profiles at the bottom layer of the matrix after 40 years of production (Xu et al., 2018).*** (A) With 500 natural fractures. (B) With 5,000 natural fractures.

natural fractures also increases. Compared with the case without natural fractures, the case with natural fracture conductivity of 0.02 md-ft results in 8.7% more cumulative gas production after 40 years. When the natural fracture conductivity is as low as 0.0002 md-ft, the contribution of natural fractures is 1.3%, which is almost negligible.

The impact of the other set of natural fractures was also investigated through simulations. As mentioned before, these natural fractures are oriented 15 degrees to the hydraulic fractures. Therefore, besides the 2,000 fractures in the mineralized set, another 2,000 natural fractures were added into the simulation, as shown in Fig. 6.13. In Fig. 6.13, the surface areas of the

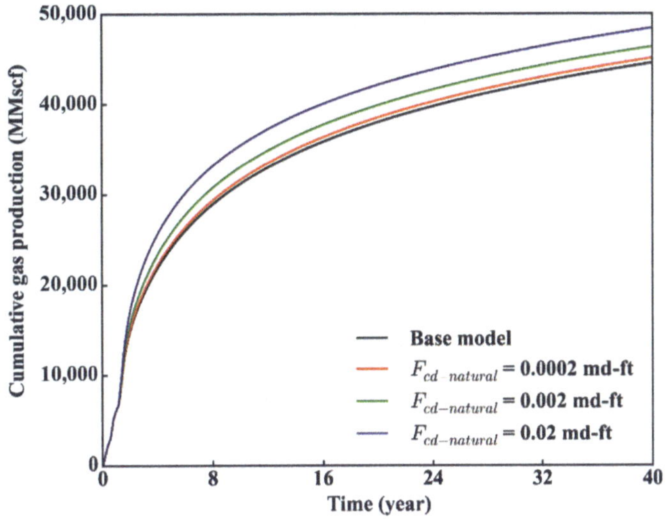

Figure 6.12 ***Impact of natural fracture conductivity on cumulative production (Xu et al., 2018).*** The "base model" refers to the model obtained from history matching. $F_{cd\text{-}natural}$ is the natural fracture conductivity. The number of natural fractures is 2,000.

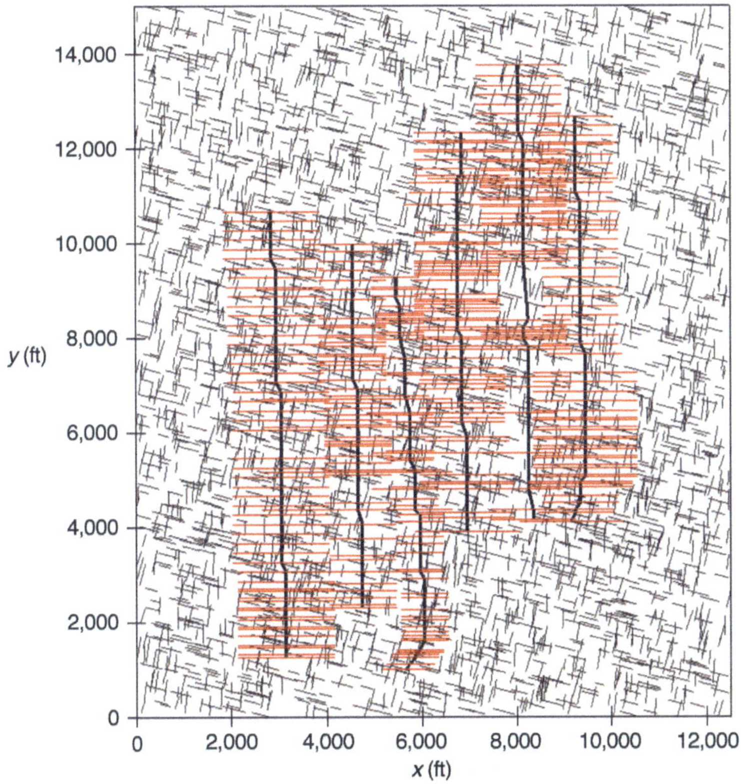

Figure 6.13 ***Map view of the two sets of natural fractures with different orientations (Xu et al., 2018).*** Each fracture set contains 2,000 natural fractures.

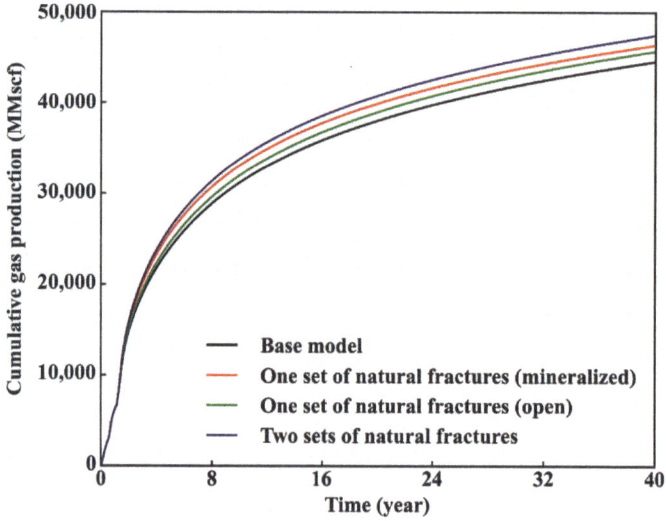

Figure 6.14 ***Impact of different sets of natural fractures (Xu et al., 2018).*** The "base model" refers to the model obtained from history matching.

two sets of natural fractures are close. The conductivity of all natural fractures is 0.002 md-ft. Fig. 6.14 compares the contribution of each set of natural fractures in cumulative gas production. The results show that fracture orientation has a significant influence on production. For the same fracture surface area, the natural fractures whose orientations are close to the *y*-direction have more contribution than the fractures with orientations close to the *x*-direction. The reason is that more natural fractures can be connected to the hydraulic fractures to form complex networks when their orientation is very different from that of the hydraulic fractures. When both sets of fractures are considered, a more complex network could be created, which results in a larger SRV and better well performance. Fig. 6.15 presents the predicted pressure profile after 40 years of production for this scenario. A 6.4% increase in cumulative production is observed with the two sets of natural fractures.

6.1.5 Impact of well spacing with and without natural fractures

Finding optimal well spacing is a key aspect of unconventional reservoir development. Larger well spacing typically results in less interference between hydraulic fractures, which improves single-well performance. However, there may be a tradeoff between the performance of a standalone well versus all wells in full-field development. This makes it necessary to

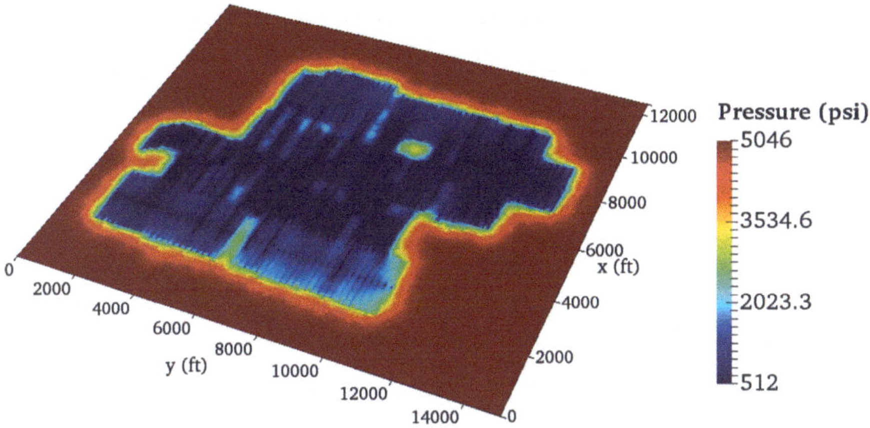

Figure 6.15 ***Predicted pressure profile at the bottom layer of the matrix after 40 years of production for the case with two sets of natural fractures (Xu et al., 2018).***

find an optimal well spacing. In this section, the impact of well spacing in addition to natural fractures is investigated.

A case with double well spacing compared with the base model was simulated. This was achieved by removing three wells (Wells 2, 4, and 6) from the model. In an effort to study the influence of natural fractures, cases with the two sets of natural fractures, as shown in Fig. 6.13, are also simulated. The fracture conductivity was varied from 0.002 to 0.02 md-ft. Table 6.3 summarizes the cumulative gas production after 40 years. A much better single-well performance was obtained with the new well spacing, as the three wells produce more than 64% of the gas produced from six wells with the tighter spacing. This indicates that the current well spacing may be too close. Furthermore, the natural fractures with 0.002 md-ft

Table 6.3 Influence of well spacing on cumulative gas production for cases with and without natural fractures (Xu et al., 2018). In the first column, "HF" refers to hydraulic fractures, and "NF" refers to natural fractures.

Fractures in the simulation	Number of wells	Cumulative gas production (MMscf)	The ratio of gas production with three wells to that with six wells)
Only HF	3	28,552	64.1%
	6	44,541	
HF + 4,000 NF	3	31,166	65.7%
(0.002 md-ft)	6	47,412	
HF + 4,000 NF	3	34,667	68.1%
(0.02 md-ft)	6	50,931	

conductivity improve the gas production by 9.2%, compared with the 6.4% increase when six wells are considered. With a larger well spacing, the relative importance of natural fractures on long-term production increases.

Another observation from Table 6.3 is that when natural fractures are considered, the benefit of having a larger well spacing is more significant, as shown in the last column of Table 6.3. The reason is that the existence of natural fractures intensifies the pressure interference between wells. Fig. 6.16 shows the pressure profiles after 40 years for the cases with and without natural fractures. With natural fractures, the interference between

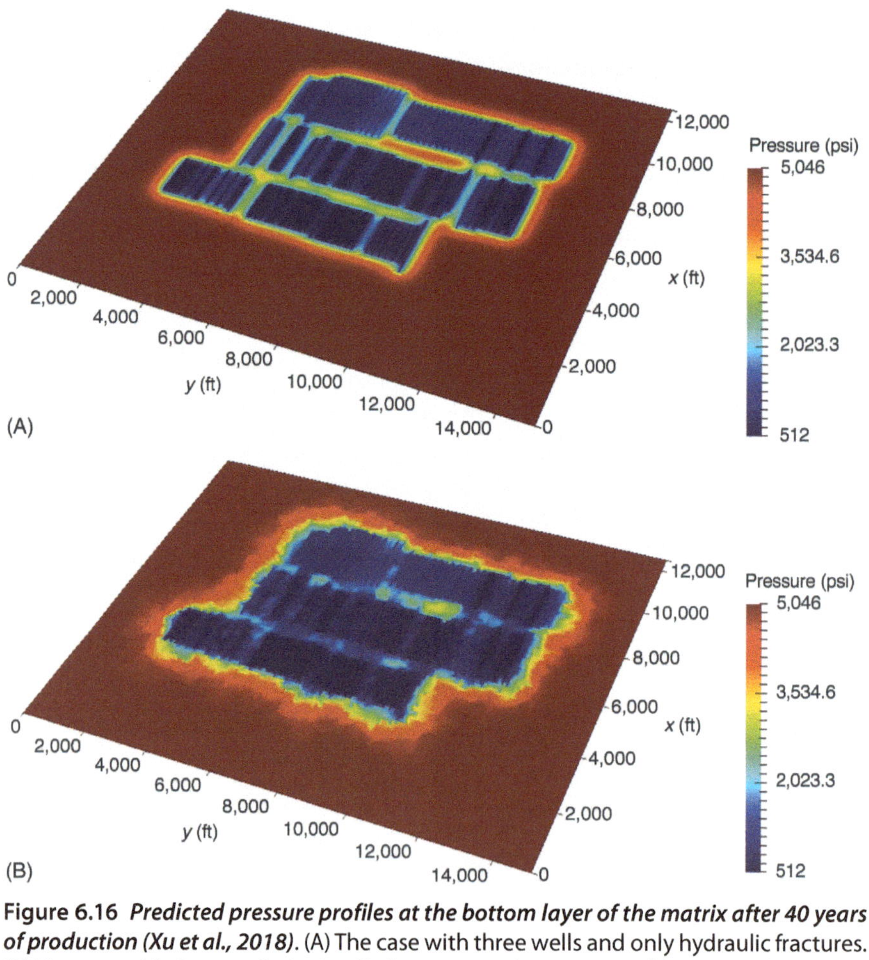

Figure 6.16 ***Predicted pressure profiles at the bottom layer of the matrix after 40 years of production (Xu et al., 2018).*** (A) The case with three wells and only hydraulic fractures. (B) The case with three wells, hydraulic fractures, and 4,000 natural fractures. The natural fracture conductivity is 0.02 md-ft.

wells is much stronger. Therefore, when the existence of natural fractures is taken into account, the optimal well spacing should be larger compared with the cases without natural fractures. A more detailed characterization of natural fractures will help to find the optimal well spacing.

6.1.6 CPU time comparison

The history matching and sensitivity studies have shown the capability and flexibility of the EDFM for field-scale simulations. The comparison of CPU time with the LGR model also demonstrates the computational efficiency of the EDFM. However, when the total number of fractures increases, the number of fracture segments in the EDFM also increases, which affects CPU time. The influence of the number of fracture segments on EDFM computational performance has seldom been studied. Therefore, here, we provide the CPU time for several of the above runs with different numbers of fractures, as shown in Table 6.4. The CPU time in the table is the summation of CPU time for the history matching stage (413 days, with a maximum time step of 1.0 day) and production forecasting stage (40 years, with a maximum time step of 30 days). The adaptive-implicit scheme was used for all the simulations when recording the CPU time.

Table 6.4 Comparison of CPU time for cases with different numbers of hydraulic fractures and natural fractures (Xu et al., 2018). In the first column, "HF" refers to hydraulic fractures, "NF (Set 1)" refers to the mineralized natural fractures, and "NF (Set 2)" refers to the open natural fractures. For all the simulations listed here, the conductivity of natural fractures is 0.002 md-ft. A single processor was used for the simulations when recording the CPU time.

Simulation model	Number of fracture segments	Number of time steps	CPU Time (second)
Base model	28,323	1,403	3,352
HF + 500 NF (Set 1)	49,683	1,398	4,054
HF + 2,000 NF (Set 1)	113,378	1,365	5,577
HF + 5,000 NF (Set 1)	240,273	1,342	9,399
HF + 2,000 NF (Set 2)	89,693	1,409	5,002
HF + 2,000 NF (Set 1) + 2,000 NF (Set 2)	174,748	1,360	7,315
HF (only three wells)	15,034	1,366	2,523
HF (only three wells) + 2,000 NF (Set 1) + 2,000 NF (Set 2)	161,459	1,331	6,452

It can be observed that all simulations summarized in Table 6.4 take a similar number of time steps, and the main difference is the CPU time per time step. Although the number of fracture gridblocks is typically less than the number of matrix gridblocks, the CPU time of an EDFM simulation is mainly determined by fracture gridblocks because of their small pore volumes. Fig. 6.17 plots the CPU time versus the number of fracture segments. The simulations with three wells and six wells are shown in the figure with different colors. An important finding is that for all the simulations with six wells, the CPU time roughly follows a linear relationship with the number of fracture segments. The two simulations with three wells follow a similar trend. However, both simulations lie below the fitting line. The reason is that these two cases have fewer high-conductivity hydraulic fractures, and the relatively high conductivity of hydraulic fractures makes them more important than natural fractures when it comes to computational efficiency. As a rule of thumb, when the matrix grid and matrix properties are the same, the CPU time of an EDFM simulation T_{sim} is mainly affected by the total number of fracture segments N_s. In this study, a linear relationship between T_{sim} and N_s was observed. The fracture conductivity also has a noticeable impact on CPU time.

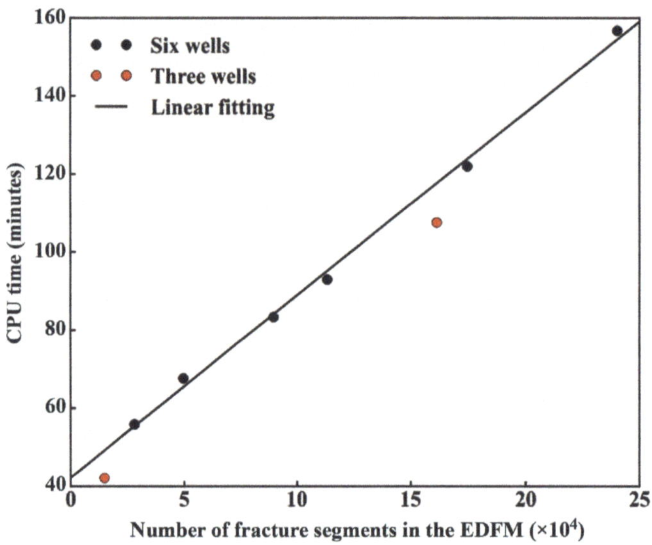

Figure 6.17 ***Relationship between simulation CPU time and the number of fracture segments in the EDFM (Xu et al., 2018).***

6.1.7 Discussions

The results of this case study show that the EDFM is a practical approach to be used with history matching and sensitivity studies for complex fracture systems. First, engineers can easily alter the number, location, geometry, and properties of the fractures; these values need to be frequently modified in history matching and sensitivity studies. Second, when modeling complex fractures, the EDFM has a relatively high computational efficiency compared with some traditional methods such as LGR. Third, the EDFM maintains a high computational performance when the number of fractures increases. In this study, the CPU time required to perform an EDFM simulation was found to increase linearly with the number of fracture segments. The properties of the fractures also had a noticeable influence on the CPU time.

This study also shows that it is extremely helpful to combine engineering measurements such as MEM, microseismic, micro-deformation, and PLT with discrete fracture models to reduce the uncertainty in the number, location, and geometry of fractures. This will greatly speed up the reservoir model building and the history matching processes.

6.2 Field-scale case study in the Marcellus Shale

Several critical processes for gas transport from shale formation with complex fractures to wellbore have been documented by Vanorsdale and Boring (1987): first, free gas is flushed from the fractures, then adsorbed gas streams off the fracture surfaces, and finally absorbed gas is released slowly from the shale matrix. In addition, the presence of a large number of nanopores in shale rock results in important gas transport mechanisms, such as gas slippage and Knudsen diffusion, and it increases overall ultimate gas recovery, especially at lower reservoir pressure during long-term depletion (Javadpour et al., 2007; Wang and Reed, 2009; Civan et al., 2011; Sakhaee-Pour and Bryant, 2012; Akkutlu and Fathi, 2012; Yu et al., 2016a). Understanding gas flow in complex fractures, gas desorption, and gas slippage in nanopores plays an important role in adequately evaluating stimulation effectiveness and accurately predicting long-term shale gas production.

Detailed numerical reservoir simulation is the rigorous approach to properly model shale gas production through fully capturing the complexity of fracture systems and complex gas transport physics in nanopores. Although there are many existing reservoir simulation studies for shale gas reservoirs (Cipolla et al., 2010; Wu et al., 2014; Sun et al., 2015; Li et al., 2015; Fung and Du, 2016), the impacts of complex hydraulic and natural fractures and complex gas transport mechanisms on well productivity have not been well captured. Yu et al. (2018b) developed a numerical model using the EDFM to investigate the effects of such complexities on shale-gas well performance based only on synthetic case studies. However, they did not couple the fracture-propagation model with the EDFM to examine the influences of complex fracture geometries on well performance based on field shale-gas production data. In particular, the comparison of history matching and production forecasting for simple bi-wing fractures and complex fractures was not considered. In addition, the accuracy of modeling gas slippage effects using the empirical correlation should be improved.

In this study, we apply the EDFM in conjunction with a commercial compositional reservoir simulator (CMG, 2017a) to efficiently simulate shale gas production when considering complex fracture geometries. The main reason to choose a compositional reservoir simulator here is that it has the capability to accurately model the gas desorption effect. An overall workflow diagram for this study is shown in Fig. 6.18. We compare the simulated production profiles and computational efficiency of the EDFM with small and large block sizes against the LGR method for simple bi-wing hydraulic

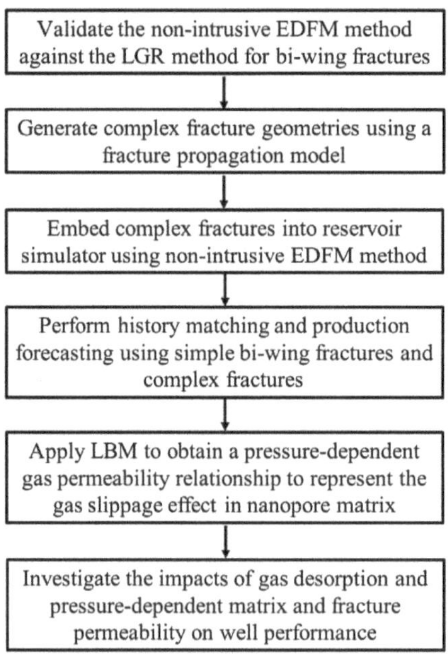

Figure 6.18 ***An overall workflow diagram for shale gas case study (Yu et al., 2018a).***

fractures. Subsequently, an actual shale-gas well from Marcellus is used to perform history matching and production forecasting with simple and complex fractures. A complex fracture-propagation model is used to predict the fracture geometry. The lattice Boltzmann method (LBM) is used to model the gas slippage phenomenon. In addition, the impacts of fracture geometry, gas desorption, gas slippage, pressure-dependent matrix permeability, and fracture conductivity on gas recovery are systematically investigated.

6.2.1 Modeling gas slippage in nanopores using LBM

The lab-generated Klinkenberg gas permeability relationship is not available. Hence, the LBM is applied to simulate the gas slippage phenomenon in nanopores in this study, which can improve the accuracy of the empirical method. The LBM has a solid foundation for modeling gas transport in nanopores in shale with a high Knudsen number (K_n) because of the strict derivation from the Boltzmann equation (Wang et al., 2018). Hence, it can predict the gas slippage effect more accurately than many empirical models such as the one proposed by Florence et al. (2007) (Chen et al., 2015; Li and Sultan, 2017; Wang et al., 2018). The basic fundamentals of LBM are briefly discussed here. More details about the LBM derivations can be found in the work by Liu et al. (2010). The entire LBM simulation process consists of two important parts: the collision process and the streaming process (Liu et al., 2010). The general LBM equation can be given by

$$f_i(\mathbf{x}+\mathbf{e}_i\Delta t, t+\Delta t) - f_i(\mathbf{x},t) = -\frac{1}{\tau}\left[f_i(\mathbf{x},t) - f_i^{eq}(\mathbf{x},t)\right], \tag{6.1}$$

where f_i is the distribution function, which is a function of space, velocity and time. If the velocity field is discretized at different directions, $\mathbf{e}_i$ is used to represent the discretization velocities. The distribution function can be simplified to $f_i(\mathbf{x},t)$. In general, the D2Q9 model was used to model gas slippage, as shown in Fig. 6.19.

The equilibrium function is

$$f_i^{eq} = w_i\rho\left[1 + 3(\mathbf{e}_i\cdot\mathbf{u}) + \frac{9(\mathbf{e}_i\cdot\mathbf{u})^2}{2} - \frac{3\mathbf{u}^2}{2}\right], \tag{6.2}$$

where $w_0 = 4/9$, $w_{1,2,3,4} = 1/9$, and $w_{5,6,7,8} = 1/36$. The fluid density and velocity can be obtained below:

$$\sum_i f_i^{eq} = \rho, \tag{6.3}$$

$$\sum_i \mathbf{e}_i f_i^{eq} = \rho \mathbf{u}. \tag{6.4}$$

To accurately handle gas slippage in nanopores, attention should be paid to two key points: one is the change of boundary conditions, and the other is the modification of the relaxation time.

(1) Boundary conditions:

$$f_i = (1-r) f_b + r f_s. \tag{6.5}$$

The mixing bounce-back and specular reflection boundary conditions are used here, which provides a free adjustable parameter to model slip flow phenomena (Zhang et al., 2005).

(2) Relaxation factor:

The relaxation factor τ should be corrected based on the Knudsen number for fluid flow in nanopores. The relaxation factor is related to the bulk Knudsen number and the effective mean free path as follows (Tang et al., 2008):

$$\tau^* = \tau \frac{2}{\pi} \arctan\left(\sqrt{2} K_n^{-3/4}\right), \tag{6.6}$$

where the Knudsen number (K_n) can be obtained using the following expression:

$$K_n = \frac{\lambda}{H} = \frac{k_B T}{\sqrt{2}\pi P d^2} \frac{1}{H}, \tag{6.7}$$

where λ is the molecular mean free path, k_B is the Boltzmann constant, d is the molecular diameter, P is the fluid pressure, T is the ambient temperature, and H is the characteristic length of flow. Then the dynamic viscosity can be obtained by

$$\nu = \frac{1}{3} \frac{\Delta x^2}{\Delta t} (\tau - 0.5), \tag{6.8}$$

where τ is the relaxation factor, $\tau = 1/\Omega$, and Ω is the relaxation time. Using the above LBM method, the fluid flow in nanopores can be simulated. When the flow in the nanopore is steady, the flow flux Q and pressure drop ΔP can be obtained; then the apparent permeability k_a can be calculated by

$$k_a = \frac{Q}{\Delta P} \frac{\mu L}{A}, \tag{6.9}$$

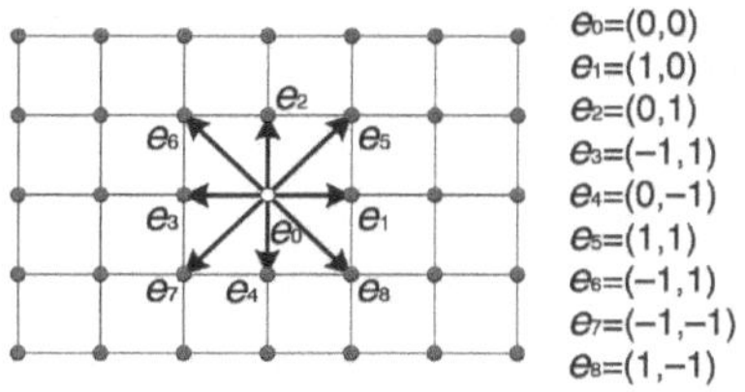

Figure 6.19 ***Schematic of the classic D2Q9 model used in LBM simulation (Yu et al., 2018a).***

where L is flow length, A is the cross-section area, and μ is the kinematic viscosity. The intrinsic permeability k_i can be determined by

$$k_i = \frac{R^2}{8} \tag{6.10}$$

where R is the radius of the nanopore. Finally, the ratio γ of apparent permeability over intrinsic permeability can be obtained by

$$\gamma = k_a / k_i. \tag{6.11}$$

When examining the effect of gas slippage on shale-gas well performance, the LBM equations are initialized and solved. Inputs of nanopore size, reservoir temperature, and pressure are needed. A nanopore size of 10 nm is considered in this study. The LBM-based simulation then generates a pressure-dependent gas permeability relationship, which can effectively represent the gas slippage effect in the nanopore matrix. Correspondingly, the pressure-dependent gas permeability relationship is transferred to a pressure-dependent matrix permeability table and assigned to the compositional reservoir simulator as an input to capture the gas slippage effect.

6.2.2 EDFM verification with LGR for simple bi-wing fractures

First, to verify the EDFM method against the LGR method, we built a basic shale-gas reservoir model with dimensions of 2,168 × 1,000 × 100 ft. A compositional numerical reservoir simulator is used. The reservoir model includes a single horizontal well with a lateral length of 1,968 ft and 21 simple bi-wing hydraulic fractures.

Each fracture is assumed to fully penetrate the reservoir thickness. A block size of 16.9375 × 20 × 100 ft is used in the LGR model. Each matrix block containing a hydraulic fracture is logarithmically refined into 5 × 1 × 1 cells to appropriately simulate long-term transient gas flow behavior from the tight matrix to hydraulic fractures. The five locally refined cell sizes are 8.2938, 0.2012, 0.01, 0.2012, and 8.2938 ft, respectively. The middle cell represents a hydraulic fracture with a width of 0.01 ft. The same block size (referred to as small block size) is also used in the EDFM simulation, as shown in Fig. 6.20A and B. In addition, another EDFM simulation with a larger block size (54.2 × 50 × 100 ft, referred to as large block size) is conducted for comparison purposes, as displayed in Fig. 6.20C and D.

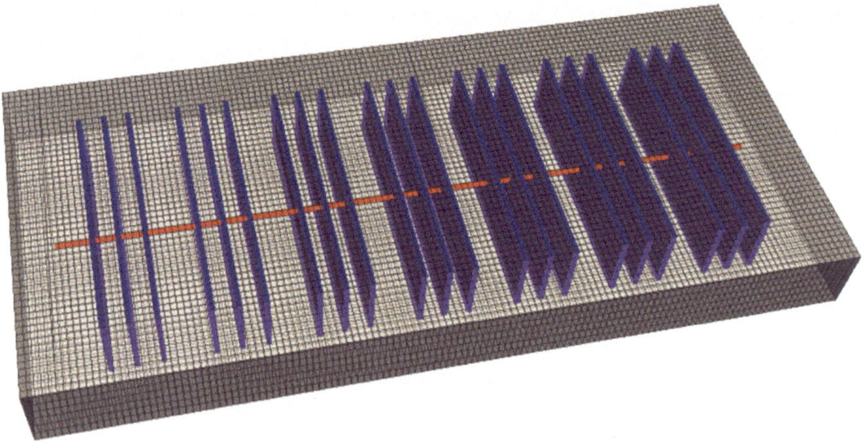

(A)

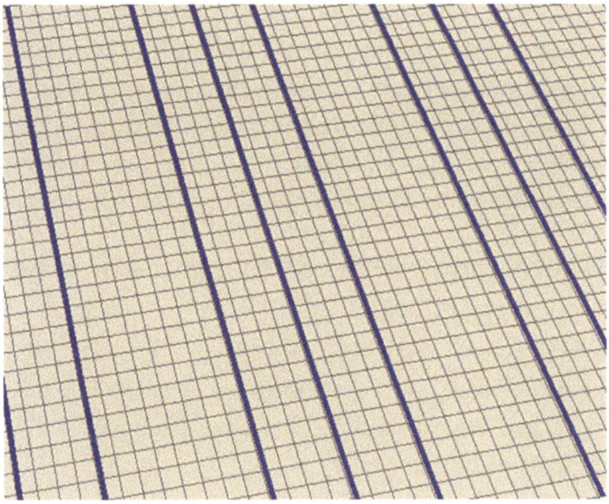

(B)

Figure 6.20 ***Two reservoir models with different block sizes (Yu et al., 2018a). Both models contain a horizontal well and 21 bi-wing hydraulic fractures.*** (A) A basic shale-gas reservoir model with small block size. (B) A small portion of the reservoir model with small block size. (C) A basic shale-gas reservoir model with large block size. (D) A small portion of the reservoir model with large block size.

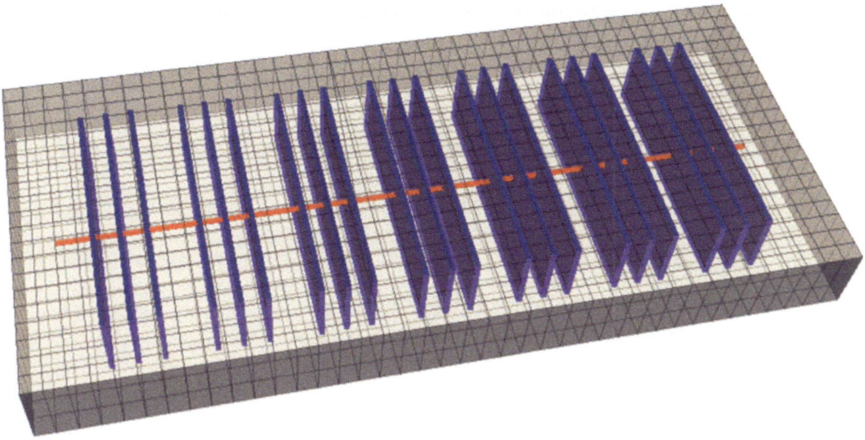

(C)

(D)

Figure 6.20 (*Cont.*)

Because only a single layer is considered in the model, the fractures are 2.5D. Seven stages and three clusters per stage are considered. The cluster spacing is 68 ft, and the separation distance between two adjacent stages is 136 ft. It can be clearly observed that no LGR in the vicinity of the hydraulic fractures is needed for the EDFM simulation.

The other main reservoir parameters used in the simulations are summarized in Table 6.5. The gas component of methane is considered, and only single-phase flow is simulated. The residual water saturation is 10%. A constant flowing BHP of 500 psi is assumed for the well. The measured gas

Table 6.5 Basic reservoir parameters used for the verification study between the EDFM and LGR method (Yu et al., 2018a).

Parameter	Value	Unit
Model dimension ($x \times y \times z$)	2,168 × 1,000 × 100	ft
Initial reservoir pressure	4,300	psi
Reservoir temperature	130	°F
Reservoir permeability	800	nd
Reservoir porosity	11.4%	–
Residual water saturation	10%	–
Total compressibility	3×10^{-6}	psi^{-1}
Reservoir thickness	100	ft
Reservoir depth	7,354	ft
Well length	1,968	ft
Number of stages	7	–
Clusters per stage	3	–
Cluster spacing	68	ft
Fracture half-length	350	ft
Fracture height	100	ft
Fracture conductivity	5	md-ft

content (scf/ton) vs. pressure from the Marcellus Shale is considered in the model, as presented in Fig. 6.21. The non-Darcy flow effect in fractures is taken into account using the non-Darcy Forchheimer coefficient proposed by Evans and Civan (1994). The Forchheimer equation is directly solved by the compositional reservoir simulator for both the LGR case and the EDFM case.

Fig. 6.22 compares the simulated production profiles including gas flow rate and cumulative gas production in a 30-year period between the LGR

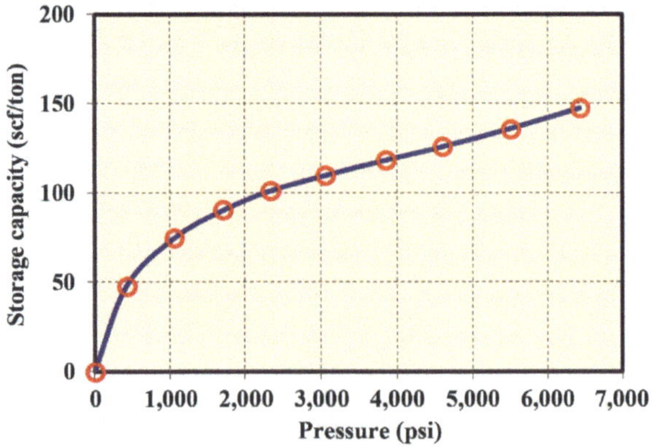

Figure 6.21 ***Measured gas content vs. pressure from the Marcellus Shale (Yu et al., 2018a).***

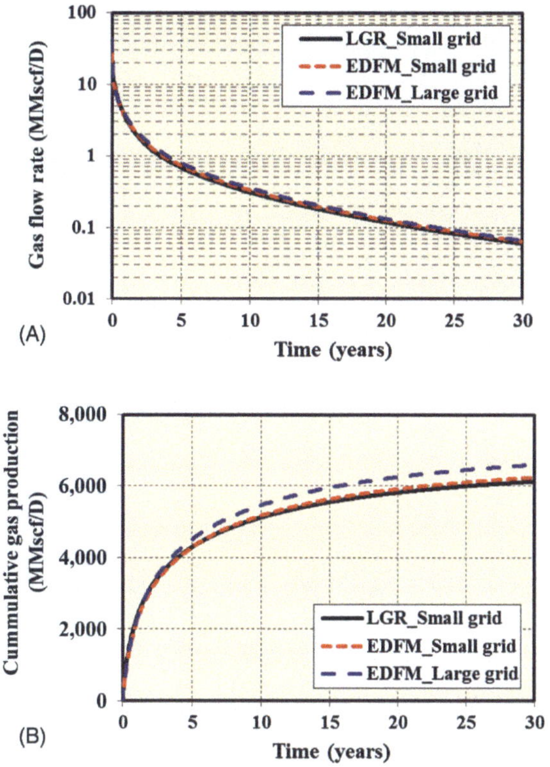

Figure 6.22 ***Comparison of gas flow rate and cumulative gas production between the LGR and the EDFM using small and large block sizes (Yu et al., 2018a).*** (A) Gas flow rate. (B) Cumulative gas production.

and the EDFM method with small and large block sizes. As can be seen, a good match is achieved between the LGR and EDFM method with small block size, indicating that the EDFM method in conjunction with the compositional reservoir simulator can accurately simulate shale-gas production with simple bi-wing fractures. In addition, the ratio of CPU time between the LGR and the EDFM method is about 1.3 (77 seconds compared with 60 seconds). Hence, the EDFM method is faster than the LGR method while maintaining similar results. When comparing the simulation results of the EDFM using small and large block sizes, a small difference in production profiles is observed. The relative difference in cumulative gas production after 30 years is about 5%. However, the ratio of CPU time between the two EDFM simulations is about 5.5 (60 seconds compared with 11 seconds). Hence, the EDFM becomes more efficient when increasing the matrix block size, but there is a tradeoff between its accuracy and efficiency.

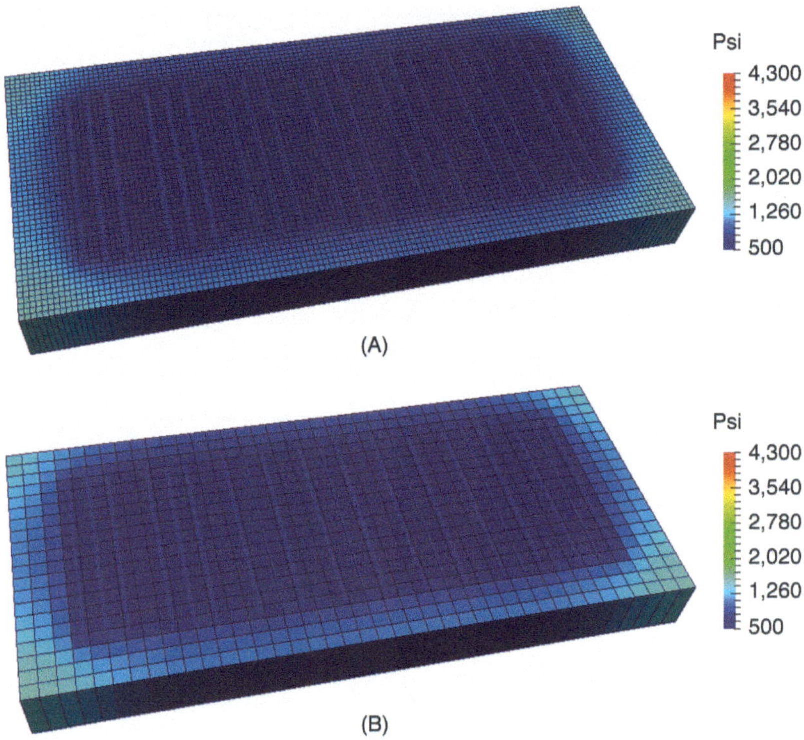

Figure 6.23 ***Comparison of pressure distribution after 30 years of production predicted by the EDFM using the small and large block sizes (Yu et al., 2018a).*** (A) The EDFM with small block size. (B) The EDFM with large block size.

Fig. 6.23 compares the pressure distribution after 30 years of production predicted by the EDFM using the small and large block sizes. Red indicates the initial reservoir pressure of 4,300 psi and blue represents the flowing BHP of 500 psi. It can be seen that a similar drainage area is obtained.

6.2.3 History matching and production forecasting

To investigate the impacts of complex fractures and gas transport mechanisms on well productivity in shale gas reservoirs, an actual well with available production data of 190 days from the Marcellus Shale was selected to perform simulations. The well was stimulated with seven perforation stages with three clusters per stage. The cluster spacing is 68 ft, and the separation distance between stages is 136 ft. A total of 53,814 bbls of water and up to 3,079,918 pounds of proppant were injected.

First, a complex hydraulic-fracture-propagation model developed by Wu and Olson (2015a, 2016) was used to simulate hydraulic fracture geometries in this well while considering the existence of natural fractures. The fracture model couples rock deformation and fluid flow in both the fractures and the wellbore. A simplified three-dimensional displacement discontinuity proposed by Wu and Olson (2015b) is used to determine fracture opening and shearing. The lubrication theory is used to handle fluid flow in the fracture, and the fluid is non-Newtonian with a power-law behavior. The partitioning of the total injection flow rate into different clusters is dynamically simulated. Several primary physical mechanisms controlling fracture growth are considered, including the stress shadow effects, flow rate distribution among multiple fractures, and the interaction between hydraulic and natural fractures. The interaction between hydraulic and natural fractures is simulated through the calculation of induced stresses at the fracture tip. The validation of the fracture model against the analytical solution and numerical model can be found in Olson and Wu (2012). In this study, the key parameters used for the fracture model include fluid injection rate of 60 bbl/minue, injection time of 128 minutes for each stage, Young's modulus of 3×10^6 psi, Poisson ratio of 0.23, and the difference between the minimum and maximum horizontal stress of 200 psi. Two sets of natural fractures are considered. The natural fracture spacing of each set is 32.8 ft. The natural fracture directions with respect to the horizontal wellbore are 30 and 150 degrees, respectively. Fig. 6.24 presents the simulated complex

(A)

Figure 6.24 ***Complex hydraulic fracture and natural fracture geometry predicted by the fracture-propagation model (Yu et al., 2018a).*** (A) Complex hydraulic fractures. (B) Complex hydraulic fractures (blue) and active natural fractures (green). (C) Complex hydraulic fractures (blue), active (green) and non-active (gray) natural fractures.

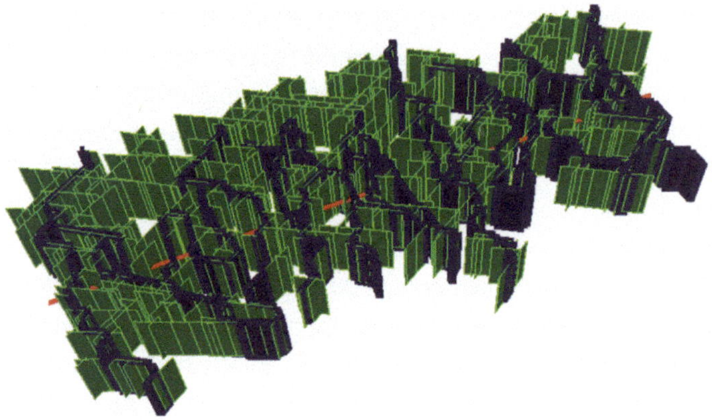

(B)

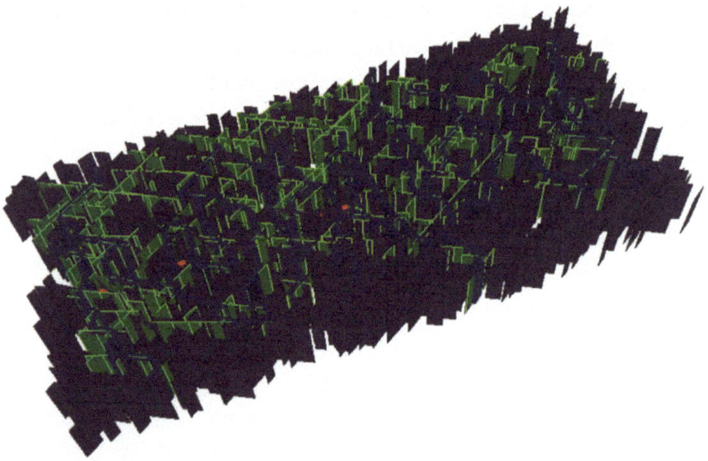

(C)

Figure 6.24 **(*Cont.*)**

fracture geometry, which consists of complex hydraulic fractures, active natural fractures, and non-active natural fractures. In this study, there are 793 active natural fractures and 178 non-active natural fractures.

Next, the complex fracture geometry was directly transferred and embedded into the reservoir model, as shown in Fig. 6.25A. The reservoir model has dimensions of 2,168 × 1,000 × 100 ft. The block size is 21.68 × 20 × 100 ft. Fracture half-length ranges from 86 to 652 ft. The total hydraulic fracture length is ~13,312 ft. In addition, a model with simple bi-wing fractures that have a similar total hydraulic fracture length was also created, as shown in Fig. 6.25B. The other reservoir parameters are kept the same as in Table 6.5.

During the history matching with actual production data, the measured flowing BHP was used as the simulation constraint, as shown in Fig. 6.26A. The measured gas flow rate and cumulative gas production were treated as the target history-matching variables. Matrix permeability,

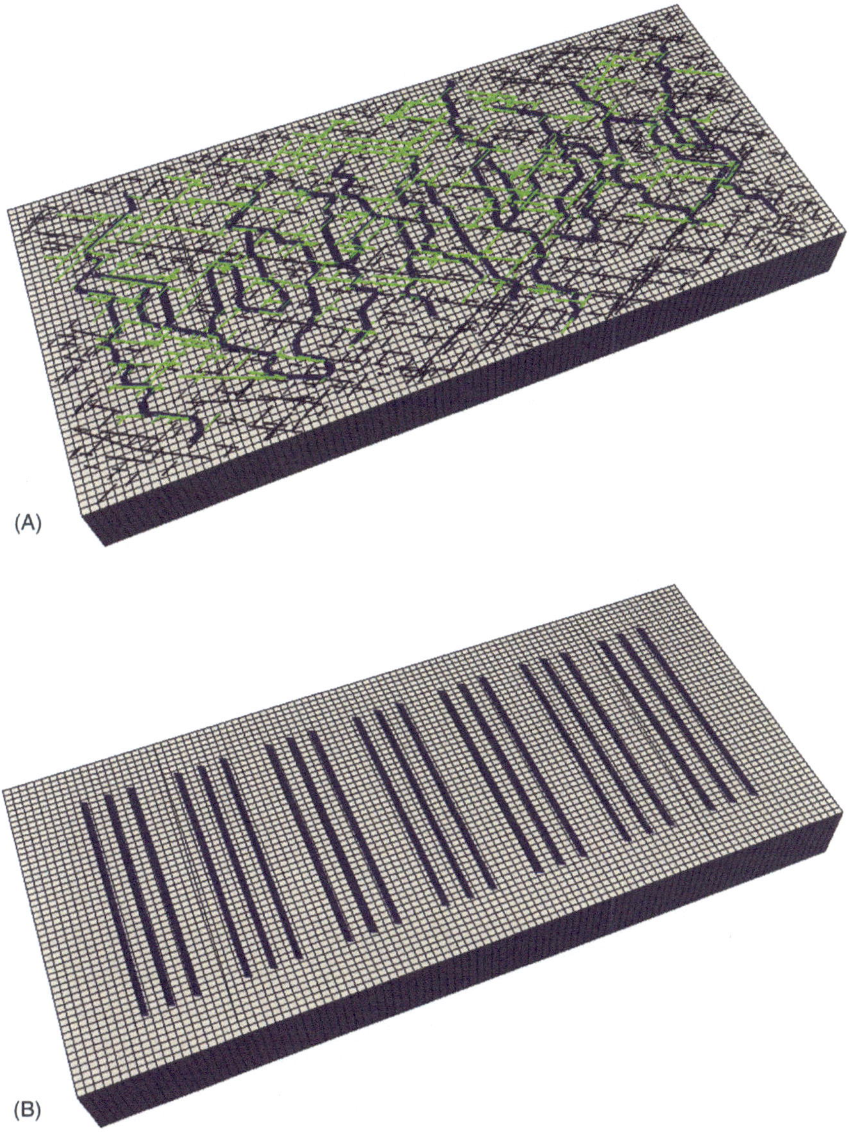

(A)

(B)

Figure 6.25 ***Comparison of simple and complex fracture geometries in the field case study (Yu et al., 2018a).*** (A) Complex fracture geometry directly embedded into the reservoir model. (B) Simple fracture geometry directly embedded into the reservoir model.

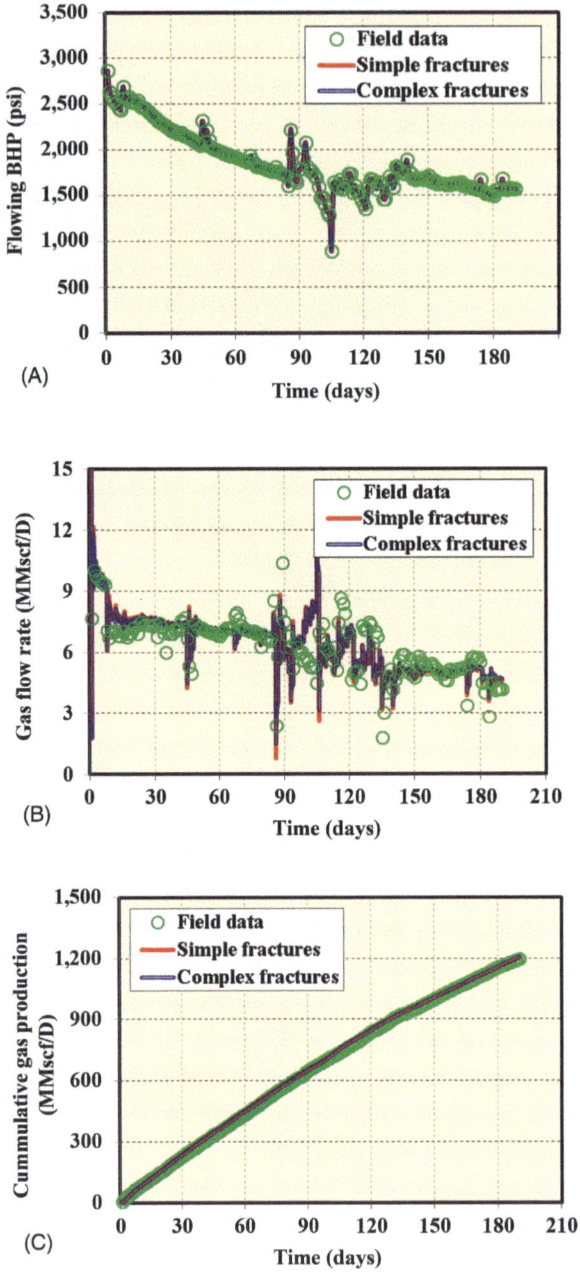

Figure 6.26 ***Comparison of BHP and well performance between actual data and simulation results using both simple and complex fractures (Yu et al., 2018a).*** (A) Flowing BHP. (B) Gas flow rate. (C) Cumulative gas production.

hydraulic-fracture conductivity, active natural fracture conductivity, and non-active natural fracture conductivity were the major tuning parameters to achieve a good history match. Also, fracture half-length is another tuning parameter for the case with simple fractures. It should be mentioned that the effects of pressure-dependent matrix permeability and fracture conductivity were not considered during the history-matching process.

The comparisons of gas flow rate and cumulative gas production between the actual data and the simulation results using both the simple and complex fracture geometries are displayed in Fig. 6.26B and C, illustrating that a good agreement was obtained for both cases with simple and complex fractures. For the case with simple fractures, the hydraulic-fracture conductivity is 5.5 md-ft, and the fracture half-length is 360 ft. For the case with complex fractures, the hydraulic-fracture conductivity is 5 md-ft, the active natural fracture conductivity is 0.03 md-ft, and the non-active natural fracture conductivity is 0.001 md-ft. The matrix permeability is 800 nd for both cases.

After history matching, we performed production forecasting for 30 years using a constant flowing BHP of 500 psi for both cases. Figs. 6.27 and 6.28 show comparisons of gas flow rate and cumulative gas production for 30 years between the two cases. There is a smaller difference during the early period of production (before 2 years). However, the difference increases with reservoir depletion later in the time frame. The case with complex fractures performs better than the case with simple fractures. The relative difference in cumulative gas production after 30 years is 36.4%. The reason is that the case with complex fractures has a larger conductive fracture surface area, resulting in a higher gas flow rate and larger ultimate gas

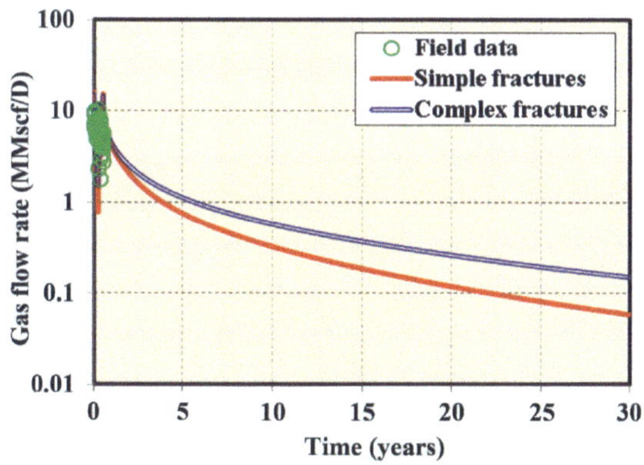

Figure 6.27 ***Comparison of 30-year gas flow rate between the cases with simple and complex fractures (Yu et al., 2018a).***

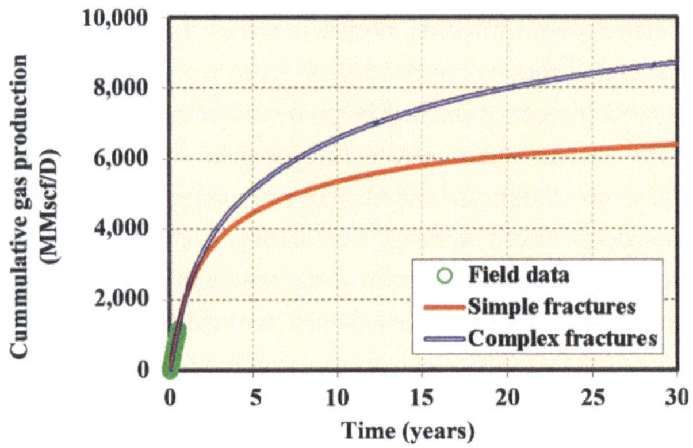

Figure 6.28 ***Comparison of 30-year cumulative gas production between the cases with simple and complex fractures (Yu et al., 2018a).***

recovery. It should be mentioned that the well production history in this study is short, compared with the forecasting period. Hence, the difference in long-term production forecasting between simple and complex fractures might become smaller given a longer well-production history.

Fig. 6.29 presents the comparison of pressure distribution after 190 days, 10 years, and 30 years of production for both cases. It can be clearly seen that the effective drainage area is larger for the case with complex fractures. In addition, at 190 days, there is only a little difference between the effective drainage area of the two cases. This is because both cases have similar hydraulic fracture length and the hydraulic fractures with higher conductivity are controlling gas production at the early stage.

The LBM is used to generate the relationship between the ratio of apparent permeability over original matrix permeability and pressure under the nanopore size of 10 nm in this study, as shown in Fig. 6.30. It can be seen that the ratio increases significantly when the pressure decreases, and it is about 4.6 at 500 psi. Accordingly, the gas slippage effect in nanopores should be considered in the model to obtain a more accurate production forecasting.

Fig. 6.31 shows the effects of gas adsorption and gas slippage on cumulative gas production. The case with complex fractures is used. It can be seen that the relative contribution of gas desorption on ultimate gas recovery after 30 years is ~35.6% compared with the case without considering gas desorption. In addition, the combined contribution of gas desorption and gas slippage effects to the ultimate gas recovery after 30 years is ~46.8%. Hence, both gas desorption and gas slippage effects significantly affect well

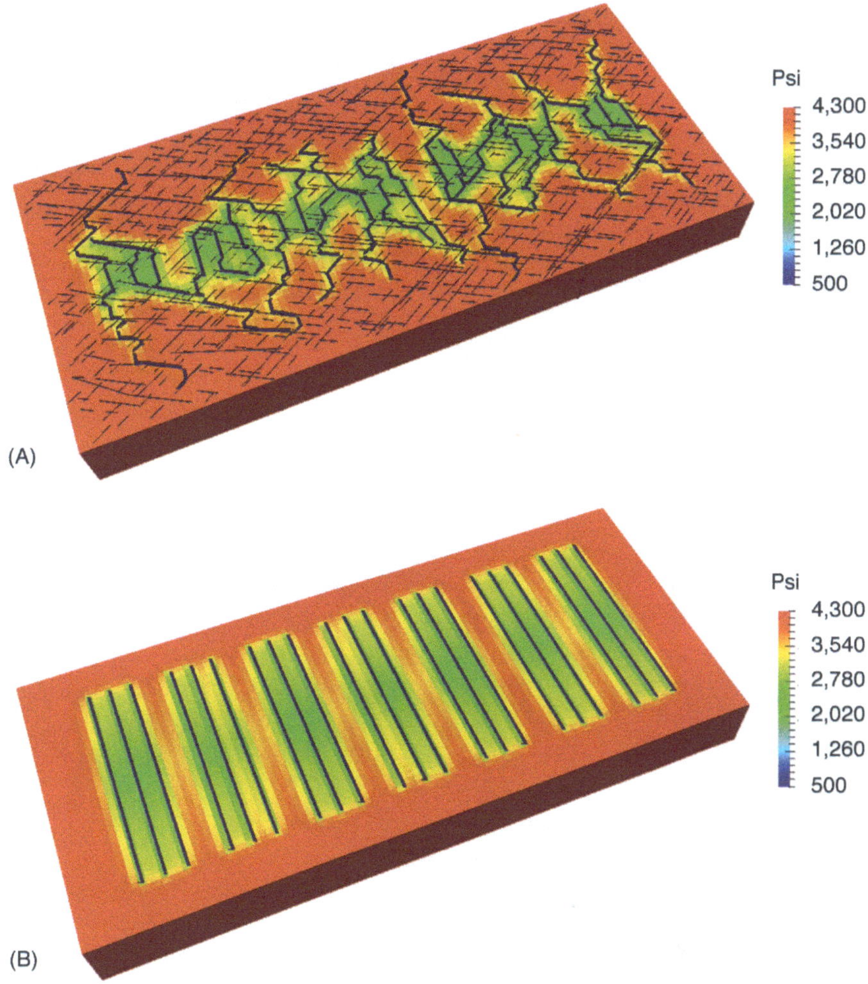

Figure 6.29 ***Comparison of pressure distribution at different times between the cases with simple and complex fractures (Yu et al., 2018a).*** (A) Pressure distribution of the case with complex fractures after the history matching period. (B) Pressure distribution of the case with simple fractures after the history matching period. (C) Pressure distribution of the case with complex fractures after 10 years of production. (D) Pressure distribution of the case with simplex fractures after 10 years of production. (E) Pressure distribution of the case with complex fractures after 30 years of production. (F) Pressure distribution of the case with simplex fractures after 30 years of production.

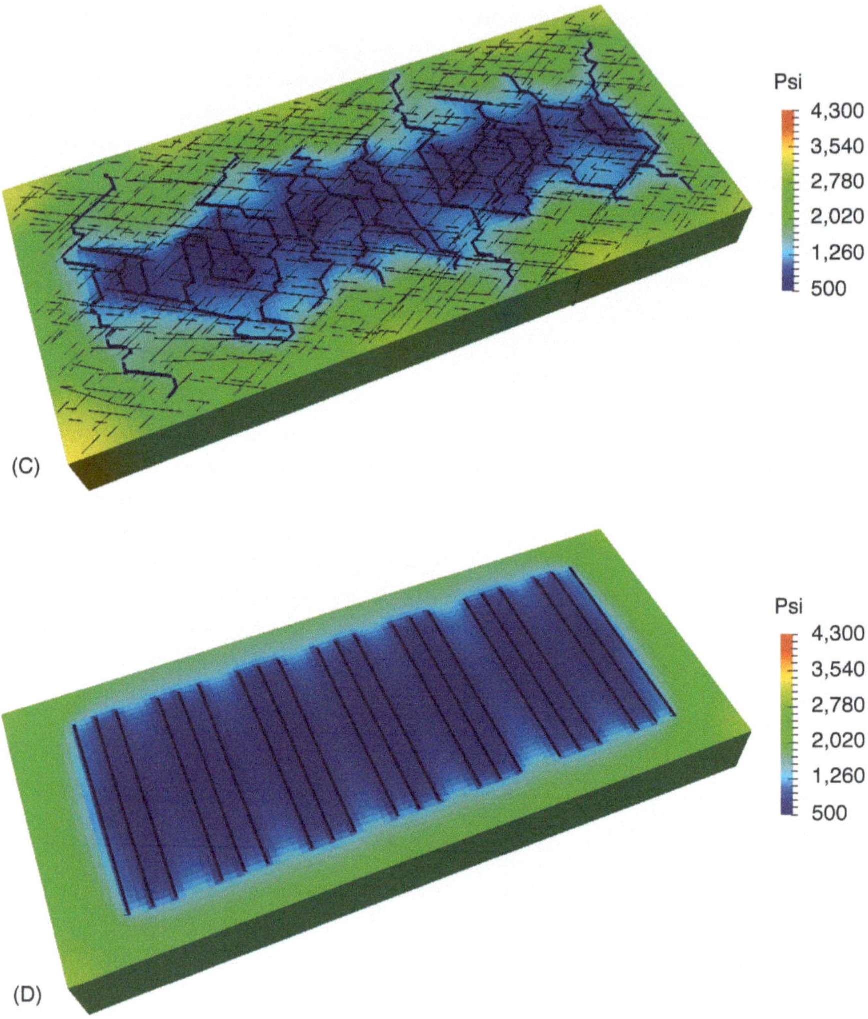

Figure 6.29 (*Cont.*)

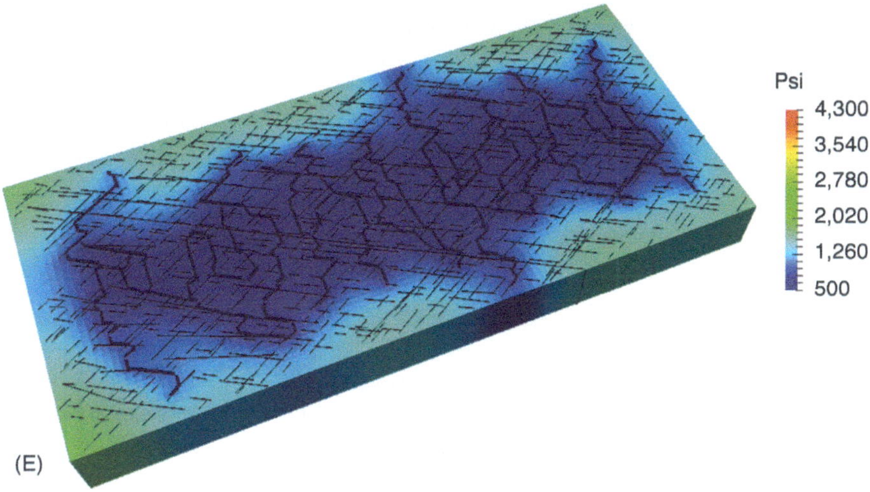

(E)

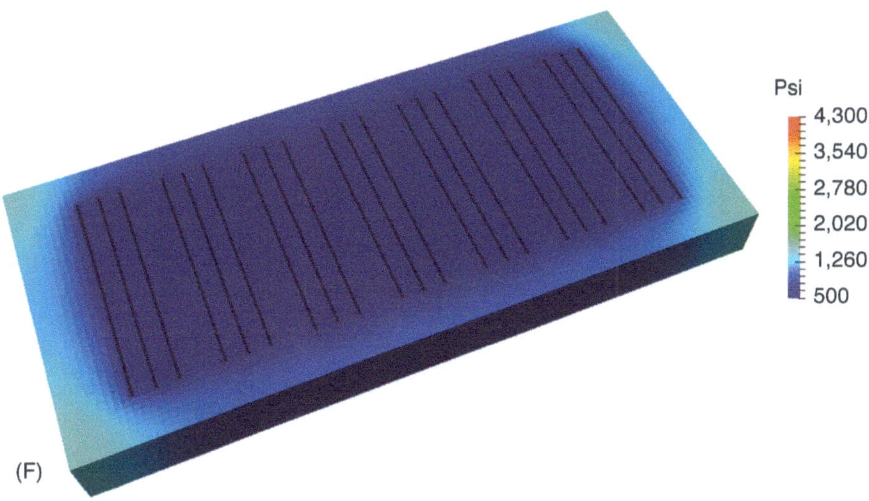

(F)

Figure 6.29 (*Cont.*)

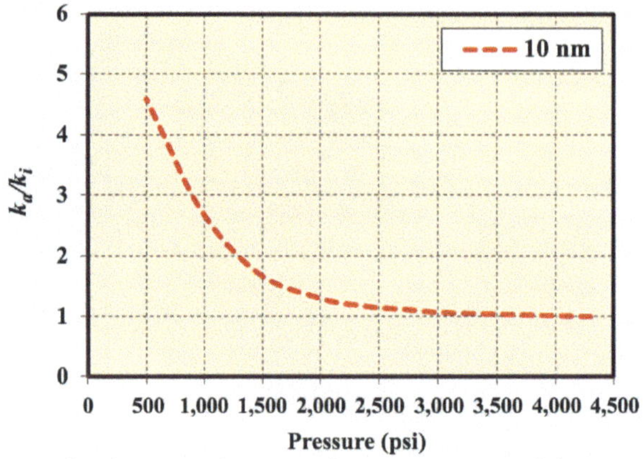

Figure 6.30 ***Relationship between the ratio of apparent permeability over original matrix permeability and pressure under the nanopore size of 10 nm (Yu et al., 2018a).***

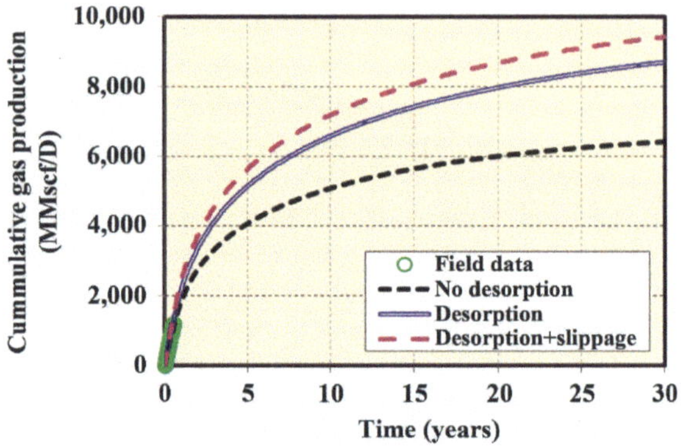

Figure 6.31 ***Effects of gas desorption and gas slippage on cumulative gas production using the complex fractures (Yu et al., 2018a).***

productivity, which should be considered in shale-gas reservoir simulation to obtain more accurate long-term production forecasting. Next, the effects of pressure-dependent matrix permeability and fracture conductivity on well performance are investigated. We used the following equation to generate pressure-dependent fracture conductivity based on the assumption that the Marcellus Shale is stiff (Yu et al., 2018b):

$$\log(FC_N) = -0.0004 \times \sigma_c + 0.2191, R^2 = 0.998 \tag{6.12}$$

where FC_N is the normalized fracture conductivity (fracture conductivity divided by the initial fracture conductivity), and σ_c is the closure pressure in psi.

We used the following empirical equation developed by Jones (1988) to generate pressure-dependent matrix permeability:

$$\frac{k}{k_i} = \exp\left\{a_k\left[\exp(-\sigma_e/\sigma^*)-1\right]\right\}\Big/(1+C\sigma_e), \tag{6.13}$$

where a_k is a constant, $\sigma^\star$ is a decay constant, which is 3,000 psi, C is a constant, which is set as 3×10^{-6} psi^{-1} (Jones, 1988), and σ_e is the net effective stress.

We use Eqs. (6.12) and (6.13) with a_k of 1.9 to generate the pressure-dependent fracture conductivity and matrix permeability curves, respectively, as shown in Fig. 6.32. The normalized reduction ratio means that the matrix permeability and fracture conductivity at different pressures are divided by initial matrix permeability and fracture conductivity, respectively. Correspondingly, these pressure-dependent permeability curves are transferred to different pressure-dependent compaction tables and given to the compositional reservoir simulator. As can be seen, the fracture conductivity and matrix permeability at BHP of 500 psi are ~20% and 31% of the initial values at a reservoir pressure of 4300 psi, respectively.

Fig. 6.33 presents the impacts of pressure-dependent matrix permeability and fracture conductivity on well performance individually and simultaneously. Here, all cases for comparison use the complex fractures, and they consider the gas desorption effect. It can be clearly observed that considering the pressure-dependent effect significantly reduces the ultimate gas recovery. The cumulative gas production after 30 years decreases by 5%, 10%, and 14% when considering pressure-dependent matrix permeability,

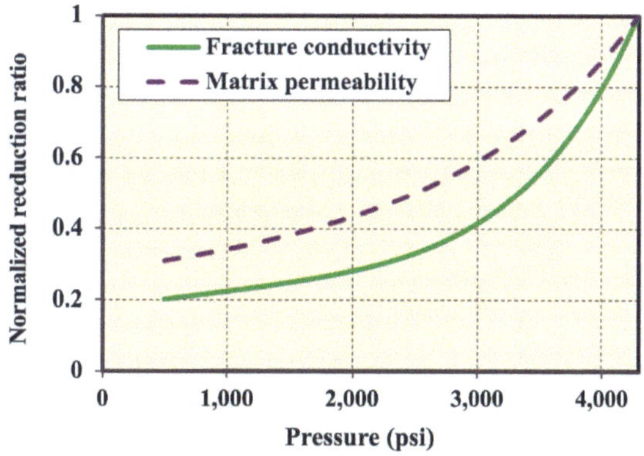

Figure 6.32 ***Pressure-dependent matrix permeability and fracture conductivity curves used in this study (Yu et al., 2018a).***

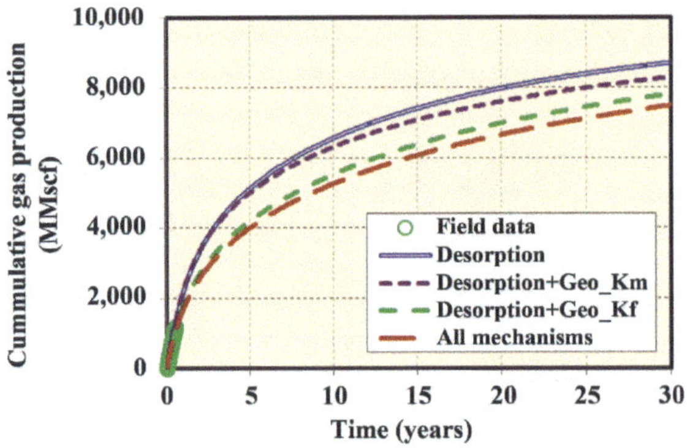

Figure 6.33 ***Effects of pressure-dependent matrix permeability and fracture conductivity on 30-year cumulative gas production (Yu et al., 2018a).***

pressure-dependent fracture conductivity, and both, respectively, when compared with the case without considering these effects.

6.2.4 Discussions

This study shows the application of the EDFM in conjunction with a compositional reservoir simulator to simulate shale gas production. The complex gas transport mechanisms such as gas desorption, gas slippage, and pressure-dependent matrix permeability and fracture conductivity were considered.

The comparison between simple bi-wing fractures and complex fractures generated from a fracture propagation model shows that both can obtain similar history matching results. However, the complex fractures produce more cumulative gas production over the long term than the simple fractures. These results indicate the importance of considering the complexity of fracture geometries.

The simulation results also show that gas desorption and gas slippage significantly affect ultimate gas recovery. Pressure-dependent matrix permeability and fracture conductivity play a negative impact on well productivity. Hence, complex fracture geometries and gas transport mechanisms should be properly accounted for in the numerical model to achieve accurate long-term production forecasting for shale-gas reservoirs.

Nomenclature

f_i = distribution function
$\mathbf{e}_i$ = discretization velocities
τ = relaxation factor

K_n = Knudsen number
λ = molecular mean free path
k_B = Boltzmann constant
d = molecular diameter
P = fluid pressure
T = ambient temperature
H = characteristic length of flow
Ω = relaxation time
Q = flow flux
L = flow length
A = cross-section area
μ = kinematic viscosity
k_a = apparent permeability
k_i = intrinsic permeability
R = radius of nanopore size
γ = ratio of apparent permeability over intrinsic permeability
FC_N = normalized fracture conductivity divided by the initial fracture conductivity
σ_c = closure pressure
a_k = constant value
$\sigma^\star$ = decay constant value
C = constant value
a_e = net effective stress

References

Akkutlu, I.Y., Fathi, E., 2012. Multiscale gas transport in shales with local kerogen heterogeneities. SPE J. 17 (4), 1002–1011.

Awada, A., Santo, M., Lougheed, D., Xu, D., Virues, C., 2016. Is that interference? A work flow for identifying and analyzing communication through hydraulic fractures in a multiwell pad. SPE J. 21 (5), 1554–1566.

Chen, L., Zhang, L., Kang, Q., Viswanathan, H.S., Yao, J., Tao, W., 2015. Nanoscale simulation of shale transport properties using the Lattice Boltzmann method: permeability and diffusivity. Sci. Rep. 8089, 1–8.

Cipolla, C.L., Lolon, E.P., Erdle, J.C., Rubin, B., 2010. Reservoir modeling in shale-gas reservoirs. SPE Reservoir Eval. Eng. 13 (4), 638–653.

Civan, F., Rai, C.S., Sondergeld, C.H., 2011. Shale-gas permeability and diffusivity inferred by improved formulation of relevant retention and transport mechanisms. Transp. Porous Media 86 (3), 925–944.

CMG, 2017a. GEM User's Guide. Computer Modeling Group.

CMG, 2017b. IMEX User Guide. Computer Modeling Group.

Evans, R.D., Civan, F., 1994. Characterization of non-Darcy multiphase flow in petroleum bearing formations. Final Report. Report, U.S. DOE Contract No. DE-AC22-90BC14659, School of Petroleum and Geological Engineering. University of Oklahoma.

Florence, F.A., Rushing, J.A., Newsham, K.E., Blasingame, T.A., 2007. Improved permeability prediction relations for low permeability sands. Paper SPE 107954, SPE Rocky Mountain Oil & Gas Technology Symposium. Denver, Colorado.
Fung, L.S.K., Du, S., 2016. Parallel-simulator framework for multipermeability modeling with discrete fractures for unconventional and tight gas reservoirs. SPE J. 21 (4), 1370–1385.
Javadpour, F., Fisher, D., Unsworth, M., 2007. Nanoscale gas flow in shale gas sediments. J. Can. Pet. Technol. 46 (10), 55–61.
Jones, S.C., 1988. Two-point determinations of permeability and PV vs. net confining stress. SPE Form. Eval. 3 (1), 235–241.
King, G.E., Valencia, R.L., 2016. Well integrity for fracturing and re-fracturing: what is needed and why? Paper SPE 179120, SPE Hydraulic Fracturing Technology Conference. The Woodlands, Texas.
Lawal, H., Jackson, G., Abolo, N., Flores, C., 2013. A novel approach to modeling and forecasting frac hits in shale gas wells. Paper SPE 164898, EAGE Annual Conference & Exhibition. London, United Kingdom.
Li, J., Sultan, A.S., 2017. Klinkenberg slippage effect in the permeability computations of shale gas by the pore-scale simulations. J. Nat. Gas Sci. Eng. 48, 197–202.
Li, N., Lolon, E., Mayerhofer, M., Cordts, Y., White, R., Childers, A., 2017. Optimizing well spacing and well performance in the Piceance Basin Niobrara formation. Paper SPE 184848, SPE Hydraulic Fracturing Technology Conference and Exhibition. The Woodlands, Texas.
Li, N., Mayerhofer, M., Childers, A., Weitzel, B., White, R., Lolon, E., Melcher, H., 2016. Optimizing well placement and completion strategies in the Piceance Basin Niobrara formation. Paper URTeC 2435473, Pages 16, Unconventional Resources Technology Conference. San Antonio, Texas, http://archives.datapages.com/data/urtec/2016/2435473.htm.
Li, X., Zhang, D., Li, S., 2015. A multi-continuum multiple flow mechanism simulator for unconventional oil and gas recovery. J. Nat. Gas Sci. Eng. 26, 652–669.
Liang, B., Khan, S., Tang, Y., 2017. Fracture hit monitoring and its mitigation through integrated 3D modeling in the Wolfcamp stacked pay in the Midland Basin. Paper URTeC 2671336, Unconventional Resources Technology Conference. Austin, Texas.
Liu, M., Yu, Z., Wang, T., Wang, J., Fan, L.S., 2010. A modified pseudopotential for a Lattice Boltzmann simulation of bubbly flow. Chem. Eng. Sci. 65 (20), 5615–5623.
Malpani, R., Sinha, S., Charry, L., Sinosic, B., Clark, B., Gakhar, K., 2015. Improving hydrocarbon recovery of horizontal shale wells through refracturing. Paper SPE 175920, SPE/CSUR Unconventional Resources Conference. Calgary, Alberta, Canada.
Marongiu-Porcu, M., Lee, D., Shan, D., Morales, A., 2016. Advanced modeling of interwell-fracturing interference: an Eagle Ford shale-oil study. SPE J. 21 (5), 1567–1582.
Molina, O., Zeidouni, M., 2017. Analytical model to estimate the fraction of fracture hits in a multi-well pad. Paper SPE 187501, SPE Liquids-Rich Basins Conference-North America. Midland, Texas.
Olson, J.E., Wu, K., 2012. Sequential versus simultaneous multi-zone fracturing in horizontal wells: insights from a non-planar, multi-frac numerical model. Paper SPE 152602, SPE Hydraulic Fracturing Technology Conference. The Woodlands, Texas.
Portis, D.H., Bello, H., Murray, M., Barzola, G., Clarke, P., Canan, K., 2013. Searching for the optimal well spacing in the Eagle Ford Shale: a practical tool-kit. Paper URTeC 1581750, Unconventional Resources Technology Conference. Denver, Colorado.
Sakhaee-Pour, A., Bryant, S.L., 2012. Gas permeability of shale. SPE Reservoir Eval. Eng. 15 (4), 401–409.
Sani, A.M., Podhoretz, S.B., Chambers, B.D., 2015. The use of completion diagnostics in Haynesville Shale horizontal wells to monitor fracture propagation, well communication,

and production impact. Paper SPE 175917, SPE/CSUR Unconventional Resources Conference. Calgary, Alberta, Canada.
Sardinha, C., Petr, C., Lehmann, J., Pyecroft, J.F., Merkle, S., 2014. Determining interwell connectivity and reservoir complexity through frac pressure hits and production interference analysis. Paper SPE 171628, SPE/CSUR Unconventional Resources Conference. Calgary, Alberta, Canada.
Scott, K.D., Chu, W.C., Flumerfelt, R.W., 2015. Application of real-time bottom-hole pressure to improve field development strategies in the Midland Basin Wolfcamp Shale. Paper URTeC 2154675, Unconventional Resources Technology Conference. San Antonio, Texas.
Sun, H., Chawathe, A., Hoteit, H., Shi, X., Li, L., 2015. Understanding shale gas flow behavior using numerical simulation. SPE J. 20 (1), 142–154.
Tang, G.H., Zhang, Y.H., Gu, X.J., Emerson, D.R., 2008. Lattice Boltzmann modelling Knudsen layer effect in non-equilibrium flows. EPL 83 (4), 40008.
Tang, H., Chai, Z., Yan, B., Killough, J., 2017. Application of multi-segment well modeling to simulate well interference. Paper URTeC 2668100, SPE/AAPG/SEG Unconventional Resources Technology Conference. Austin, Texas.
Vanorsdale, C.R., Boring, P.A., 1987. Evaluation of initial reservoir data to estimate Devonian Shale gas reserves. Paper SPE 16862, SPE Annual Technical Conference and Exhibition. Dallas, Texas.
Wang, F.P., Reed, R.M., 2009. Pore networks and fluid flow in gas shales. Paper SPE 124253, SPE Annual Technical Conference and Exhibition. New Orleans, Louisiana.
Wang, Z., Wang, M., Chen, S., 2018. Coupling of high Knudsen number and non-ideal gas effects in microporous media. J. Fluid Mech. 840, 56–73.
Wu, K., Olson, J.E., 2015a. Simultaneous multifracture treatments: fully coupled fluid flow and fracture mechanics for horizontal wells. SPE J. 20 (2), 337–346.
Wu, K., Olson, J.E., 2015b. A simplified three-dimensional displacement discontinuity method for multiple fracture simulations. Int. J. Fract. 193 (2), 191–204.
Wu, K., Olson, J.E., 2016. Numerical investigation of complex hydraulic-fracture development in naturally fractured reservoirs. SPE Prod. Oper. 31 (4), 300–309.
Wu, Y.S., Li, J., Ding, D., Wang, C., Di, Y., 2014. A generalized framework model for the simulation of gas production in unconventional gas reservoirs. SPE J. 19 (5), 845–857.
Xu, Y., Yu, W., Li, N., Lolon, E., Sepehrnoori, K., 2018. Modeling well performance in Piceance Basin Niobrara Formation using embedded discrete fracture model. Paper URTeC 2901327, Unconventional Resources Technology Conference. Houston, Texas.
Yaich, E., Diaz de Souza, O.C., Foster, R.A., Abou-sayed, I., 2014. A methodology to quantify the impact of well interference and optimize well spacing in the Marcellus Shale. Paper SPE 171578, SPE/CSUR Unconventional Resources Conference. Calgary, Alberta, Canada, 30.
Yu, W., Sepehrnoori, K., Patzek, T.W., 2016a. Modeling gas adsorption in Marcellus Shale with Langmuir and BET isotherms. SPE J. 21 (2), 589–600.
Yu, W., Wu, K., Zuo, L., Tan, X., Weijermars, R., 2016b. Physical models for inter-well interference in shale reservoirs: relative impacts of fracture hits and matrix permeability. Paper URTeC 2457663, Unconventional Resources Technology Conference. San Antonio, Texas.
Yu, W., Wu, K., Liu, M., Sepehrnoori, K., Miao, J., 2018a. Production forecasting for shale gas reservoirs with nanopores and complex fracture geometries using an innovative non-intrusive EDFM method. Paper SPE 191666. SPE Annual Technical Conference and Exhibition. Dallas, Texas.
Yu, W., Xu, Y., Liu, M., Wu, K., Sepehrnoori, K., 2018b. Simulation of shale gas transport and production with complex fractures using embedded discrete fracture model. AIChE J. 64 (6), 2251–2264.

Yu, W., Xu, Y., Weijermars, R., Wu, K., Sepehrnoori, K., 2018c. A numerical model for simulating pressure response of well interference and well performance in tight oil reservoirs with complex-fracture geometries using the fast embedded-discrete-fracture-model method. SPE Reservoir Eval. Eng. 21 (2), 489–502.

Zhang, Y., Qin, R., Emerson, D.R., 2005. Lattice Boltzmann simulation of rarefied gas flows in microchannels. Phys. Rev. E 71 (4), 047702.

CHAPTER SEVEN

EDFM for field-scale reservoir simulation with complex corner-point grids

7.1 Introduction

Corner-point grids are widely used in reservoir simulation to represent realistic reservoir geometry. In corner-point grids, a series of coordinate lines and depths are given to construct the reservoir model (Wadsley, 1980; Ponting, 1989). For a single gridblock, the positions of its eight corner points are specified individually. Compared with Cartesian grids, corner-point grids have more flexible definition of block geometries, which provides the capability to represent complex reservoir boundaries and certain geological features, such as sloping faults and cross-stratified beds (Ding and Lemonnier, 1995). Most commercial reservoir simulators support this type of grid.

Similar to faults, fractures are also an important geological feature in hydrocarbon reservoirs. A large number of fractures, especially natural fractures, might exist in certain types of reservoirs. When corner-point grids are used for such reservoirs, although coordinate lines might be adjusted to represent the orientations of a limited number of fractures, it is difficult and sometimes unrealistic to explicitly grid all fractures in such a manner. Therefore, an effective and efficient approach needs to be adopted to simulate fractures in corner-point grids.

As mentioned in previous chapters, an important advantage of the embedded discrete fracture model (EDFM) is its compatibility with traditional reservoir simulators. Because the corner-point grid is a standard and one of the most frequently used grid types in traditional simulators, it is a natural idea to extend the EDFM to corner-point grids to provide a general solution to simulate fractures in geologically complex reservoirs. However, the flexibility in the corner-point grid definition presents challenges for the implementation of the EDFM. Several researchers have reported their work on using the EDFM in corner-point grids (Panfili and

Embedded Discrete Fracture Modeling and Application in Reservoir Simulation.
http://dx.doi.org/10.1016/B978-0-12-821872-3.00007-1

Cominelli, 2014; Panfili et al., 2015; Fumagalli et al., 2016; Shah et al., 2016; Fumagalli et al., 2017). For example, the EDFM was implemented in the "hfm" module of the open-source MATLAB Reservoir Simulation Toolbox (MRST) (Lie et al., 2012; Lie, 2016), and this module was used in several studies (Shah et al., 2016; Wong et al., 2018). Also, Fumagalli et al. (2016) developed a numerical upscaling technique based on the EDFM in corner-point grids. However, because the handling of complex grid geometries was not the focus of their research, the challenges and difficulties related to corner-point grids were not comprehensively illustrated. It is necessary to further develop the methodology and formulations of the EDFM for an effective extension of this approach to general corner-point grids with complex geometries.

The focus of this chapter is to address the challenges in the application of the EDFM in general corner-point grids, and to investigate the properties of the EDFM approach in such applications. Both the challenges related to corner-point grid geometries and the algorithms and formulations to solve these challenges are discussed. Subsequently, a verification of the EDFM against local grid refinement (LGR) models and a grid sensitivity study with different matrix gridding are presented to illustrate the accuracy of the model. Finally, two three-dimensional (3D) full-field studies are presented to show the applicability of the EDFM in geologically complex reservoirs. The case studies also demonstrate the convenience of using the EDFM to place fractures in complex reservoir models.

7.2 EDFM in corner-point grids

There are several differences in the definition of corner-point grids compared with that of Cartesian grids. The main aspect of these differences concerns block geometry, and they greatly extend the type of geological features that corner-point grids can represent. However, as the transmissibility calculations of the EDFM relies on parameters obtained from geometrical calculations, the special features of corner-point grids also greatly complicate the calculation of transmissibilities in the EDFM. In this section, the unique challenges and solutions pertaining to corner-point grids are discussed in detail.

7.2.1 Non-cuboid block geometry

A unique type of connection in the EDFM is the connection between a fracture segment and the matrix block it penetrates. Eq. (4.7) is used for the calculation of transmissibility factor for this type of connection, and the calculation of fracture segment area and average normal distance in the equation requires accurate geometrical calculations, whose complexity heavily depends on block geometries. Different from cases of Cartesian grids where all vertice coordinates of a block can be defined by six numbers (two for x, two for y, and two for z), in corner-point grids, 24 floating-point numbers are required to completely define the geometry of a single block (eight for x, eight for y, and eight for z). This results in irregular matrix block shapes, in which the angles between neighboring faces might not be 90 degrees, or the block faces might not have a rectangular shape. As shown in Fig. 7.1, the gridblock is a hexahedron instead of a cuboid. New geometrical algorithms to calculate the intersection between a hexahedron (matrix gridblock) and a polygon (fracture) need to be designed because of these complexities. As an example of the difference between a cuboid and a hexahedron in terms of geometrical calculation, the basic idea to check whether a point is inside a cuboid is not suitable for hexahedron, while the latter is more complicated to implement and will be discussed later.

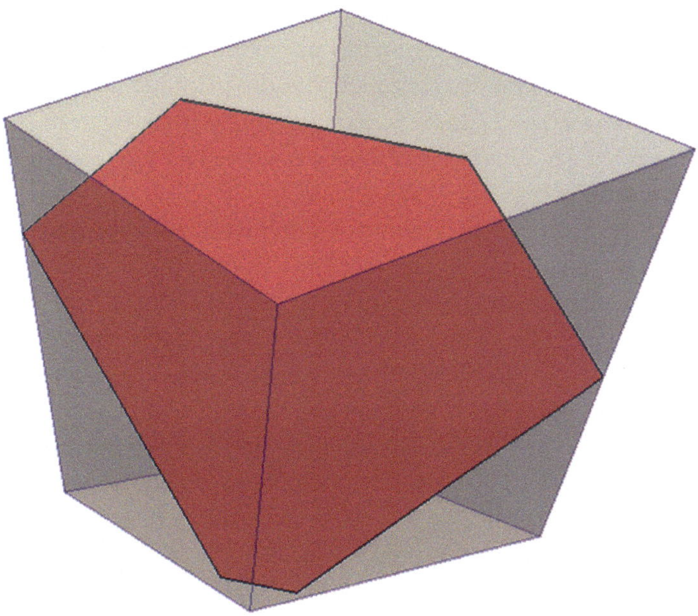

Figure 7.1 ***An example of matrix gridblock geometry in corner-point grids (Xu and Sepehrnoori, 2019).*** The gridblock is a hexahedron. The red polygon shows a fracture segment inside this gridblock.

7.2.2 Definition of block faces

Each face of a hexahedron is defined by four points. In corner-point grids, because the coordinates of the four points in a block face are independently defined (using coordinate lines and z-coordinates of vertices), these four points may not be coplanar. Such faces are not suitable for geometrical calculation. Several possible methods might be applied to overcome this issue. In Fumagalli et al. (2016), both the matrix block surfaces and the fractures were treated as bilinear surfaces. Their intersections were approximated as line segments. However, the bilinear-surface representation of fractures makes it difficult for the EDFM to handle certain fracture geometries, such as nonplanar fracture geometry and irregular fracture shape. In addition, it is difficult to extend this geometrical algorithm to other types of grids, for example, unstructured grids using element types such as triangular prism, hexagonal prism, or tetrahedron. Therefore, a new method is developed in an effort to solve this problem. For each face with non-coplanar points, a line is drawn to split it into two triangles. It should be noted that the lines should be drawn in a way that, for the same face belonging to different gridblocks, the splitting lines are the same. After the splitting, each face is divided into two triangular faces, and the total number of actual faces of the gridblock will increase. In general, this number can increase from 6 to 12, as shown in Fig. 7.2. The final geometry of the gridblock can be described as a polyhedron with 6–12 faces. Correspondingly, if the fracture plane fully penetrates such a gridblock, the resulting fracture segment can also have up to 12 edges, as shown in Fig. 7.2D.

It is highly probable that concave polyhedra will be generated after splitting the faces defined by non-coplanar points. In fact, if a split face is shared by two neighboring gridblocks, a concave polyhedron will always be created on one side of the face. When a polygon intersects a concave polyhedron, there is a possibility that it might produce a concave polygon, as shown in Fig. 7.2D. Most geometrical algorithms for polygons assume convexity to make use of the special properties of convex polygons and simplify the calculation. However, to solve the problem here, a general intersection algorithm applicable to both convex and concave polygons needs to be developed. A unique feature of concave polygons is the connectivity between vertices, which is illustrated in Fig. 7.3. For a convex polygon, once the vertices of the polygon v_1, v_2, ..., and v_n are found, an easy way to arrange the vertices is to find a central point c, connect each vertex v_i with c, and arrange the sequence of the vertices on the basis of the angle between

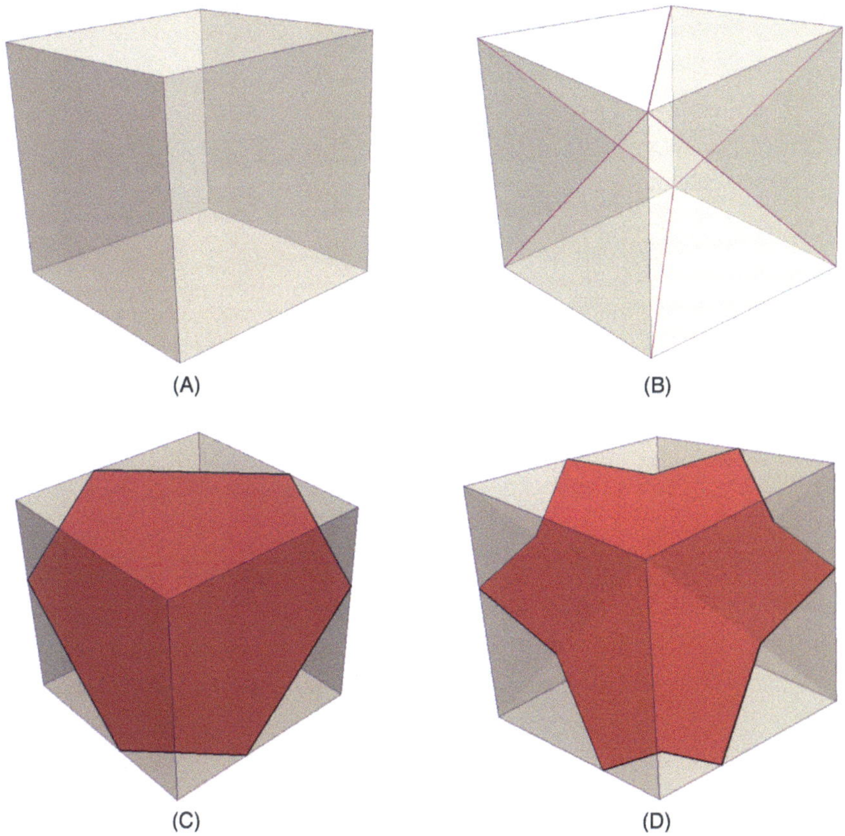

Figure 7.2 ***Illustration of the splitting of faces when the four points defining a gridblock face are non-coplanar (Xu and Sepehrnoori, 2019).*** (A) A gridblock in which all the points in a single face are coplanar. (B) A gridblock in which all faces are defined by four non-coplanar points. The purple lines represent the additional lines drawn to split the faces. (C) The created fracture segment (colored in red) when a fracture fully penetrates the gridblock in (A). (D) The created fracture segment (colored in red) when a fracture fully penetrates the gridblock in (B). When a fracture fully penetrates a regular gridblock as shown in (A), the fracture segment can have up to six vertices, as shown in (C). However, for the gridblock geometry in (B), the fracture segment can have up to 12 vertices, as shown in (D). It should also be noted that the fracture segment in (D) is a concave polygon.

v_i-c and any axis on the same plane. However, for a concave polygon, even if all vertices have been found, more than one simple polygon can be formed by connecting the vertices, as shown in Fig. 7.3B. A key to solve this issue is that while calculating the intersection points between a matrix block and a fracture, the connectivity between the points needs to be recorded. Thus,

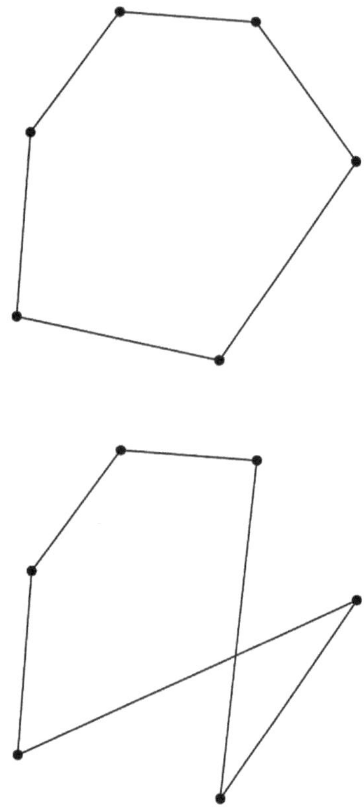

(A)

Figure 7.3 ***Illustration of the difference between convex and concave polygons in terms of vertex connectivity (Xu and Sepehrnoori, 2019).*** (A) For convex polygons, once the vertices have been found, the connectivity between vertices can be easily determined as there is only one way to connect the vertices such that the final polygon is a simple polygon (top row). The other connecting sequences will result in complex polygons (bottom row), in which the edges cross over each other. The fracture segments in the EDFM do not have such geometry. (B) For concave polygons, with a set of vertices, more than one simple polygon can be formed by connecting the vertices in different sequences. The correct sequence to connect the vertices can only be determined with the help of other information.

after obtaining the coordinates of all vertices, the recorded connectivity can be used to determine the sequence of the points to connect them and form a polygon. Because in later steps of the EDFM, the calculation of connections between fracture segments is completely dependent on the geometry of each fracture segment, this approach is necessary to find each fracture-segment polygon accurately.

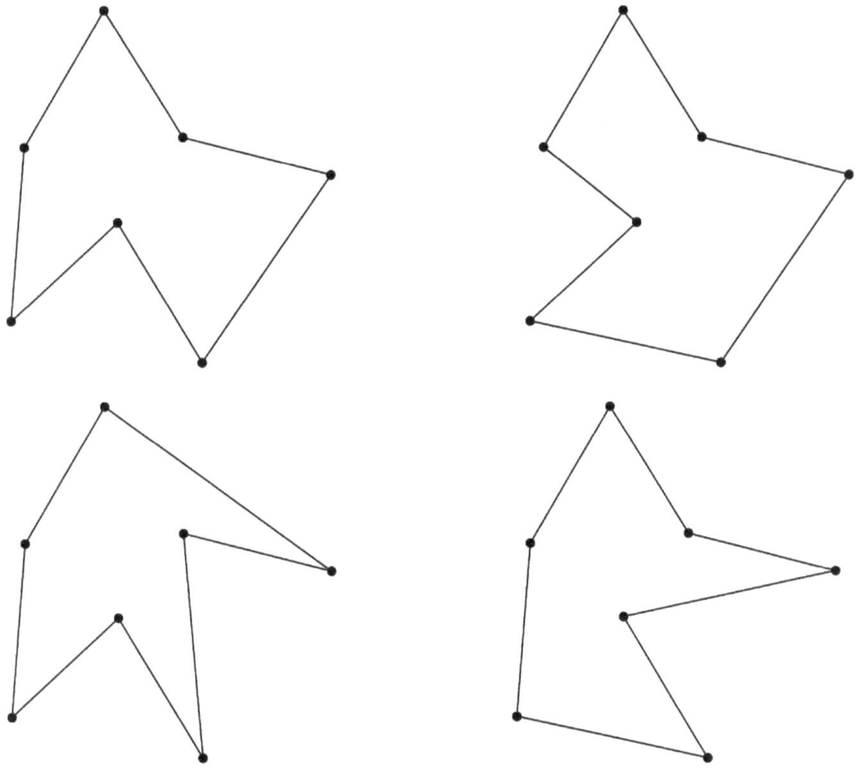

(B)

Figure 7.3 (*Cont.*)

7.2.3 Degeneracy of block geometry

Degenerate block geometry is a unique feature of corner-point grids. Degeneracy occurs when two or more vertices of a gridblock have the same coordinates. It is typically used in the modeling of erosion surfaces. Compared with an ordinary gridblock, a degenerate gridblock has fewer than eight vertices. Fig. 7.4 shows the possible variations of a cube because of degeneracy, in which the shape of the gridblock ranges from a hexahedron with eight vertices to a tetrahedron with four vertices or a polygon with zero volume. Considering all these possibilities, the corner-point block geometry can be even more challenging than that of unstructured grids for the calculation of matrix–fracture intersection. In fact, several geometries in Fig. 7.4 are basic types of elements in unstructured grids, including

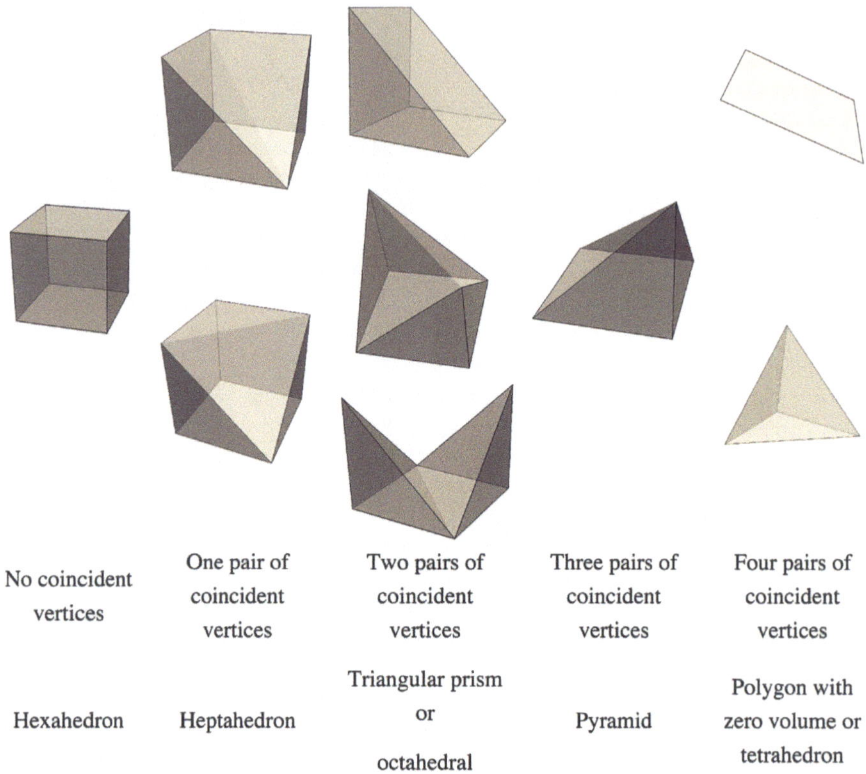

Figure 7.4 ***Possible geometries of gridblocks in corner-point grids due to degeneracy (Xu and Sepehrnoori, 2019).*** **All these geometries are variations from a cube, but from the second column on, one or more pairs of vertices have the same coordinates.**

tetrahedron, pyramid, and triangular prism. Before calculating the intersection, an additional step for corner-point grids is to detect the degeneracy of gridblocks and merge vertices with the same coordinates.

Another issue brought by degeneracy is that a gridblock might have zero or close-to-zero volume. When a fracture intersects such gridblocks, the intersection should be ignored, and the fracture segment must not be generated. Otherwise, the small pore volume of the fracture segment will severely affect the simulation time step. In addition, similar to how matrix–matrix connections are handled in many simulators, the connections between fracture segments across such gridblocks need to be checked to maintain fracture connectivity.

7.2.4 Algorithm for matrix–fracture intersection

A new matrix–fracture intersection algorithm is developed to meet the requirements for practical applications of the EDFM in general corner-point grids. The development of this algorithm considers the challenges discussed previously, including non-cuboid block geometry, non-coplanar face points, and degeneracy. The new algorithm has the capability of handling the intersection between a general polyhedron and a general polygon (convex or concave). Details of the algorithm are provided in Appendix A. Compared with previous studies on the EDFM in corner-point grids, using this new algorithm makes it possible to find the geometries of fracture segments accurately, even along complex geological features such as faults and pinch-out layers. This algorithm could also be used for different types of complex fracture geometries, such as nonplanar fracture geometry and irregular fracture shape represented by arbitrary polygons.

The implemented algorithm has been tested on BedModel2, an open dataset in the MRST (https://www.sintef.no/projectweb/mrst/modules/mrst-core/data-sets). Shah et al. (2016) also used this dataset in their study on the EDFM with a multiscale method. The corner-point grid of this model is shown in Fig. 7.5A. This grid has a large portion of gridblocks with degenerate geometry. Two fractures are placed inside the reservoir, and the intersections between the matrix gridblocks and fractures are shown in Fig. 7.5B. The developed algorithm successfully calculates the geometries of the fracture segments.

Another test case uses a 300 × 600 × 20 corner-point grid ("Flat UP1") from the Impact of Realistic Geologic Models on Simulation of CO_2 Storage (IGEMS) open dataset (https://www.nr.no/en/IGEMS), which is a synthetic dataset from the IGEMS project (Nilsen et al., 2012). Similar to BedModel2, this dataset is also one of the standard datasets in the MRST. The reservoir has a very complex geometry, as shown in Fig. 7.6, and the total number of matrix gridblocks is 3.6 million. A nonplanar fracture with a complex fracture geometry (as shown in Fig. 7.7A) is placed inside the reservoir, and the fracture is represented as 473 interconnected small fractures. In the vertical direction, the fracture fully penetrates the formation. After the geometrical calculation, 38,642 fracture segments are obtained in this example. The fracture segments are shown in Fig. 7.7B. This example demonstrates the generality of the algorithm

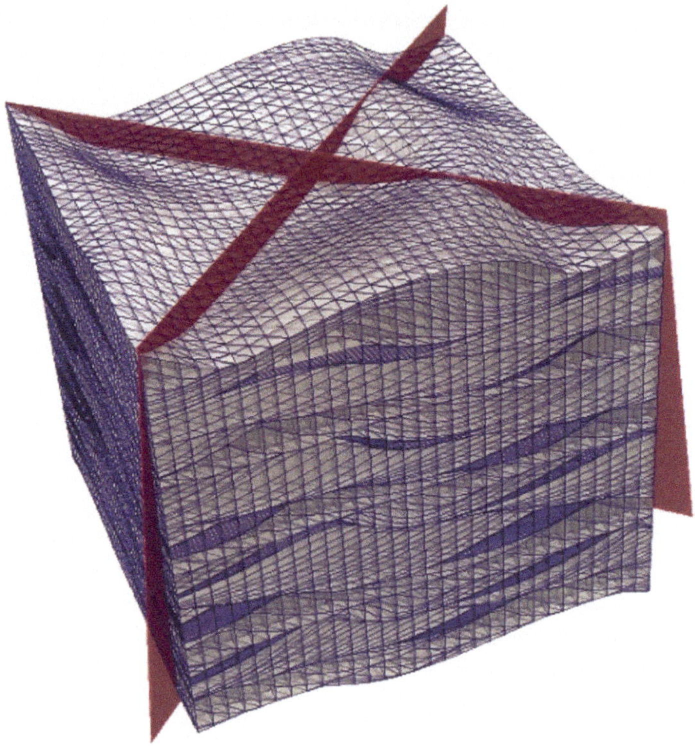

(A)

Figure 7.5 ***An example with two fractures intersecting the BedModel2 model (Xu and Sepehrnoori, 2019).*** In (A), the red polygons represent two fractures. In (B), the fracture segments are colored by the number of vertices. The total number of matrix gridblocks is 299,700. The number of fracture segments is 10,705. The scale in the *z*-direction is 5.

for handling large-scale corner-point grids representing reservoirs with complex structures.

7.2.5 Connections between fracture segments

The discretization of fractures in the EDFM is dependent on the discretization of the matrix. Correspondingly, the connections between fracture segments mimic the connections between the corresponding matrix gridblocks. In a regular Cartesian grid, for a given fracture segment, to find its connections with other fracture segments in the same fracture, only those fracture segments in neighboring matrix gridblocks need to be searched. In 3D problems, this means that, at most, six other fracture segments need

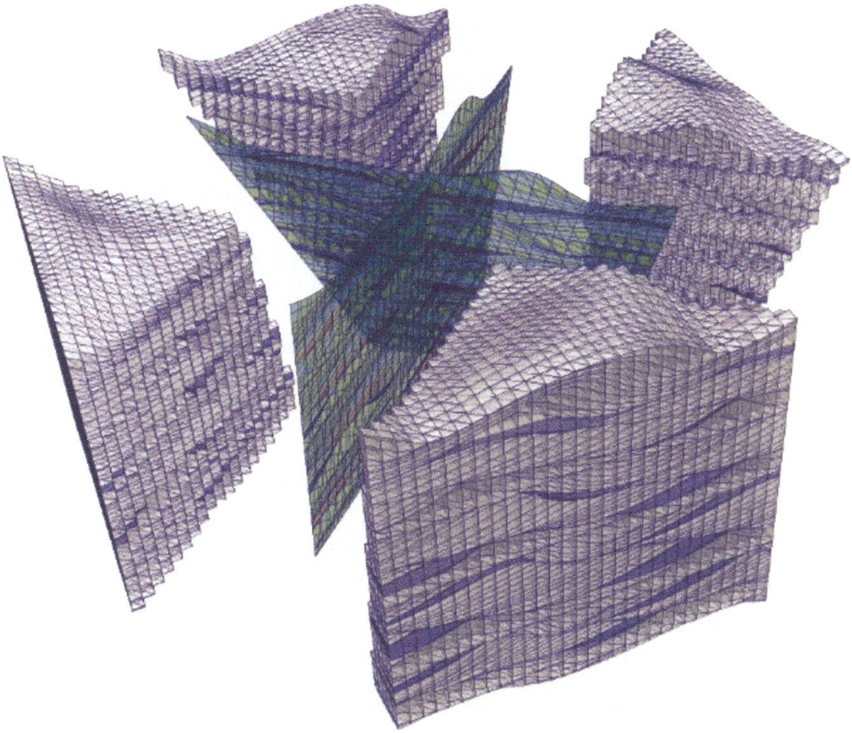

(B)

Figure 7.5 (*Cont.*)

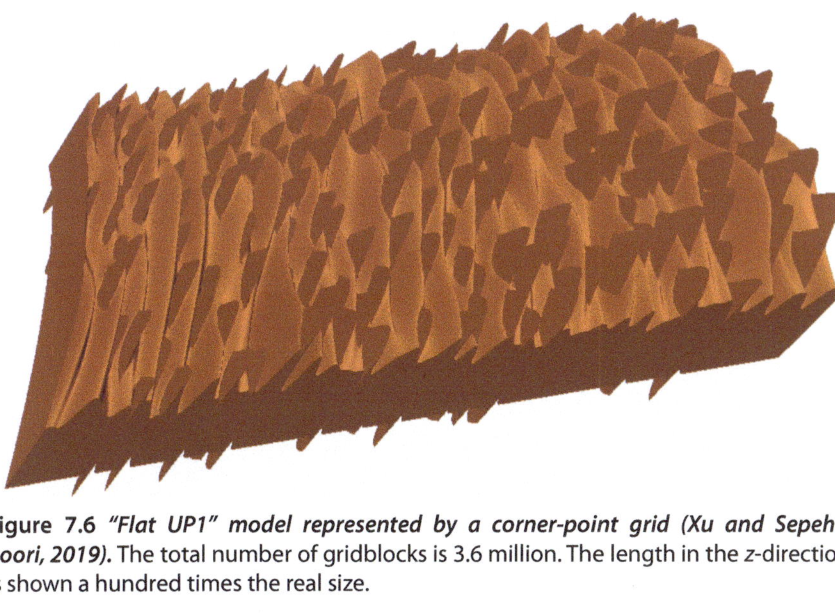

Figure 7.6 ***"Flat UP1" model represented by a corner-point grid (Xu and Sepehrnoori, 2019).*** The total number of gridblocks is 3.6 million. The length in the *z*-direction is shown a hundred times the real size.

(A)

(B)

Figure 7.7 ***Geometry of the fracture and the fracture segments in the testing case using the "Flat UP1" model (Xu and Sepehrnoori, 2019).*** (A) The nonplanar fracture geometry. (B) Fracture segments obtained from geometrical calculation. The length in the *z*-directions is shown a hundred times the real size.

to be considered (if the fracture is not crossing through the edges of the gridblock). However, in corner-point grids, because of the existence of faults or pinched-out layers, the fracture segments in two gridblocks that are not neighbors in the (I, J, K) index space might be connected, which greatly increases the complexity of connections between fracture segments.

In addition, in corner-point grids, the corresponding faces between neighboring gridblocks might not overlap. In such cases, the common area between the two gridblocks is the overlapping area of two quadrilaterals (Ponting, 1989; Sammon, 2000). Correspondingly, for the EDFM, the common edge between the fracture segments is the overlapping edge between the polygons representing the fracture segments. Thus, the required geometrical calculation to find the common edge between two fracture segments is more complicated than that in Cartesian grids, as a collinearity test and an overlapping calculation need to be performed between the corresponding edges of the two polygons. Fig. 7.8 illustrates several possibilities of the connections between fracture segments in neighboring gridblocks, where the edges might completely overlap, partially overlap, or not overlap. The formulation to calculate the transmissibility factor between fracture segments also needs to be modified accordingly. Here, the idea for calculating matrix–matrix connection transmissibility in two-dimensional (2D) reservoir simulation is borrowed. For the connection between the two fracture segments (*A* and *B*) in Fig. 7.9, the transmissibility factor can be calculated as

$$
\begin{aligned}
T_{A-B} &= \frac{T_A T_B}{T_A + T_B} \\
T_A &= \frac{k_{fA} A_c d_A}{r_A^2}, T_B = \frac{k_{fB} A_c d_B}{r_B^2}.
\end{aligned}
\tag{7.1}
$$

In Eq. (7.1), k_{fA} and k_{fB} are the permeabilities of the fracture segments, A_c is the contact area between the two fracture segments, which can be calculated as the product of fracture width and the overlapping length between edges, r_A is the distance from the centroid of Fracture Segment *A* to the midpoint of the corresponding edge of *A*, d_A is the normal distance from the centroid of Fracture Segment *A* to the corresponding edge of *A*, and r_B and d_B have similar definitions as r_A and d_A, respectively. Eq. (7.1) uses a standard two-point flux approximation to evaluate the transmissibility, and the ratio of the overlapping-edge length to the full-edge-length serves as a multiplier in the calculation of both T_A and T_B (Cordazzo et al., 2002). An accurate calculation of the overlapping-edge length is crucial for the treatment of connections between fracture segments along faults.

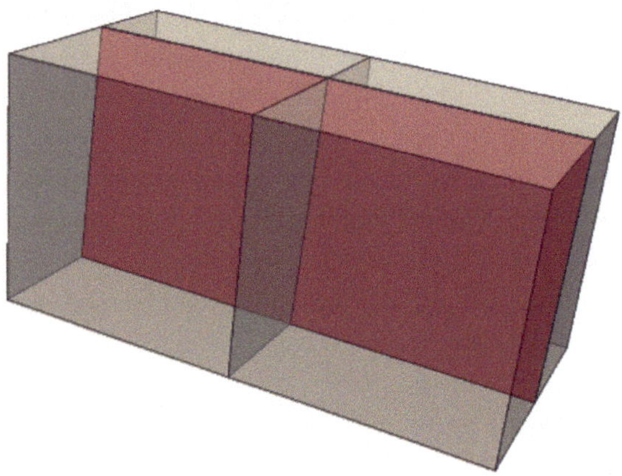

(A)

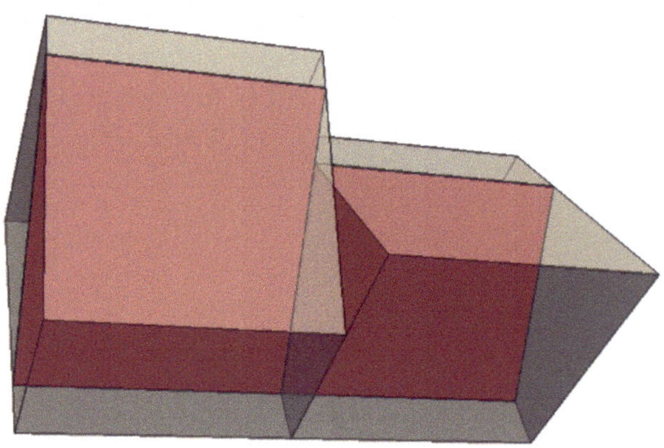

(B)

Figure 7.8 ***Connections between fracture segments (Xu and Sepehrnoori, 2019).*** (A) The simple scenario in which the two corresponding gridblocks that contain the fracture segments have a matching face. (B)–(D) Complex scenarios in which the corresponding gridblocks have non-matching faces. The fracture segments are shown in pink. In (A), the two fracture segments have a common edge; in (B), the edges of the two fracture segments have a common vertex; in (C), the edges of the two fracture segments have no common vertex; in (D), the two fracture segments might or might not have an overlapping edge, depending on the location of the fracture.

(C)

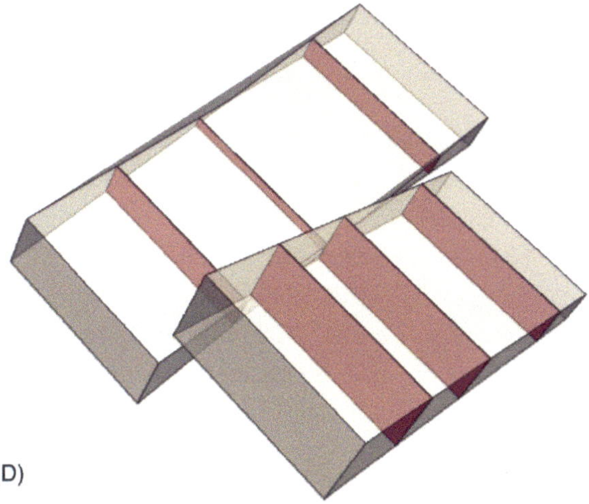
(D)

Figure 7.8 *(Cont.)*

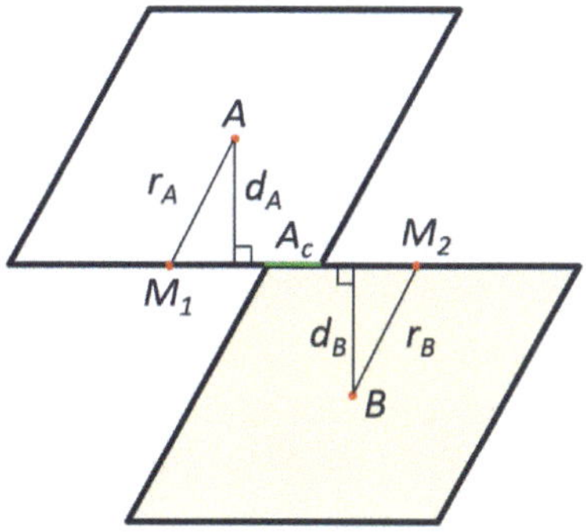

Figure 7.9 Calculation of transmissibility factor between two fracture segments (*A* and *B*) with a partially overlapping edge (Xu and Sepehrnoori, 2019). The two fracture segments are shown in white and yellow, respectively. The overlapping edge is shown in green. M_1 and M_2 are the midpoints of the corresponding edges of Fracture Segments *A* and *B*, respectively.

7.3 Model verification

In the following sections, several case studies will be presented for verification and application of the methodology presented in this chapter. These studies include 2D and 3D black-oil and compositional simulations. Three commercial reservoir simulators are used to illustrate the generality of the method.

The first study is a verification of the EDFM approach in a 2D corner-point grid. A commonly used method to simulate hydraulic fractures is LGR, where the fractures are represented as high-permeability sub-grid-blocks. Therefore, in this study, the EDFM is compared with LGR to verify its accuracy. A black-oil commercial simulator CMG–IMEX (CMG, 2017b) was used for the simulations.

7.3.1 Case 1: Comparison of EDFM with LGR in a shale-gas reservoir

The matrix grid is shown in Fig. 7.10. The reservoir dimensions are 1,600 × 1,380 × 10 ft, respectively, and the reservoir is discretized using a 75 × 69 × 1 matrix grid. All gridblocks have similar dimensions. The key simulation parameters are summarized in Table 7.1. For this shale-gas reservoir, the gas desorption mechanism is considered using the method by Seldle and Arri (1990). Three hydraulic fractures are placed inside the reservoir, with a fracture width of 0.01 ft and a fracture permeability of 10,000 md. The length of each fracture is 729 ft, and the fracture planes are not parallel to the coordinate axes. The angles between the fracture planes and the y-axis are around 16 degrees. In order to simulate these fractures using LGR, the grid is distorted around the fractures in a way that the x- and x+ faces of the gridblocks are parallel to the fracture planes. Therefore, in the LGR model, the gridblocks intersected by the fractures can be refined in the x-direction to easily represent the fractures. Fig. 7.11 shows the LGR models for the simulation. Two LGR models, LGR1 and LGR2, are used to simulate the same case. Compared with LGR1, LGR2 has a higher level of refinement for the gridblocks intersected by fractures. Therefore, if LGR1 and LGR2 obtain the same results, it can be asserted that these models are giving reliable solutions that can be used as the reference solution. The EDFM is applied in the same matrix grid (without refinement) to simulate these fractures.

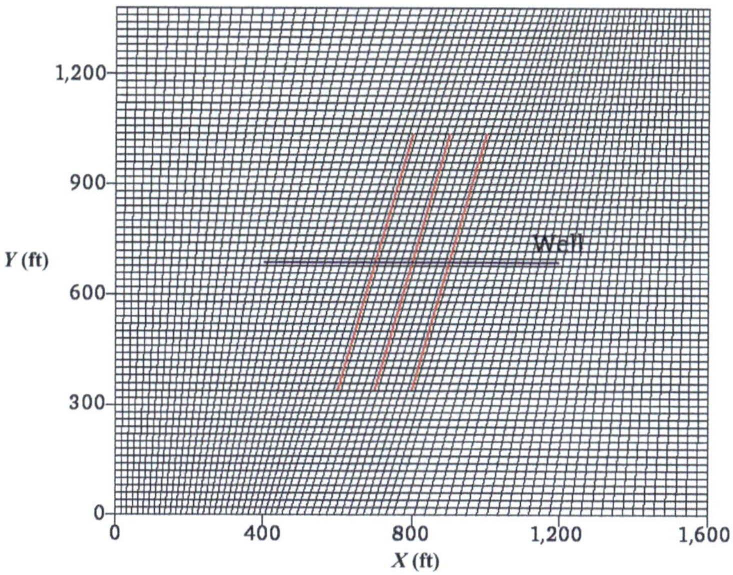

Figure 7.10 ***Simulation grid in Case 1 (Xu and Sepehrnoori, 2019).*** The red lines represent hydraulic fractures. In the LGR models, the gridblocks intersected by fractures are refined.

Table 7.1 Key reservoir and operation parameters for Case 1 (Xu and Sepehrnoori, 2019).

Parameter	Value	Unit
Reservoir permeability	0.0008	md
Reservoir porosity	10%	–
Reservoir pressure	4,500	psi
Reservoir thickness	10	ft
Rock compressibility	1×10^{-6}	psi^{-1}
Initial water saturation	10%	–
Initial gas saturation	90%	–
Langmuir pressure	530	psi
Langmuir volume	190	scf/ton
Bottomhole pressure	1,000	psi
Production time	1,000	day

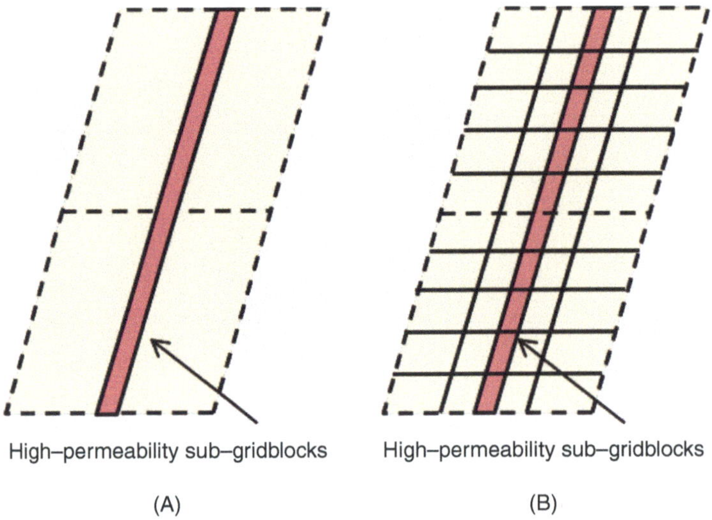

Figure 7.11 ***Illustration of the LGR settings in Case 1 (Xu and Sepehrnoori, 2019).*** In both (A) and (B), two gridblocks are shown, and the dashed lines indicate the boundary of each gridblock. The pink blocks are the high-permeability sub-gridblocks representing the hydraulic fractures. Their widths are enlarged for illustration purposes. In the simulation, their widths are the same as the fracture width. In (A), each gridblock is refined into 3 × 1 sub-gridblocks (LGR1); In (B), each gridblock is refined into 5 × 5 sub-gridblocks (LGR2).

Fig. 7.12 presents the predicted gas production rate by the three models: LGR1, LGR2, and EDFM. The three models give almost the same results. This demonstrates that the EDFM provides an accurate solution for this simulation, showing its applicability in matrix grids with corner-point geometry. Previous chapters have also shown the high accuracy of the EDFM approach in Cartesian grids. Overall, the EDFM is a reliable method for simulating fractures in structured grids.

7.4 Grid sensitivity study of the EDFM

An important advantage of the EDFM approach is that it does not require detailed gridding of the matrix around complex fractures to perform the simulation, making it convenient to use in different reservoir models. However, different gridding of the matrix might affect the accuracy of the EDFM, and this has seldom been studied. To study the influence of matrix gridding, we present a 2D case study in which the same reservoir is

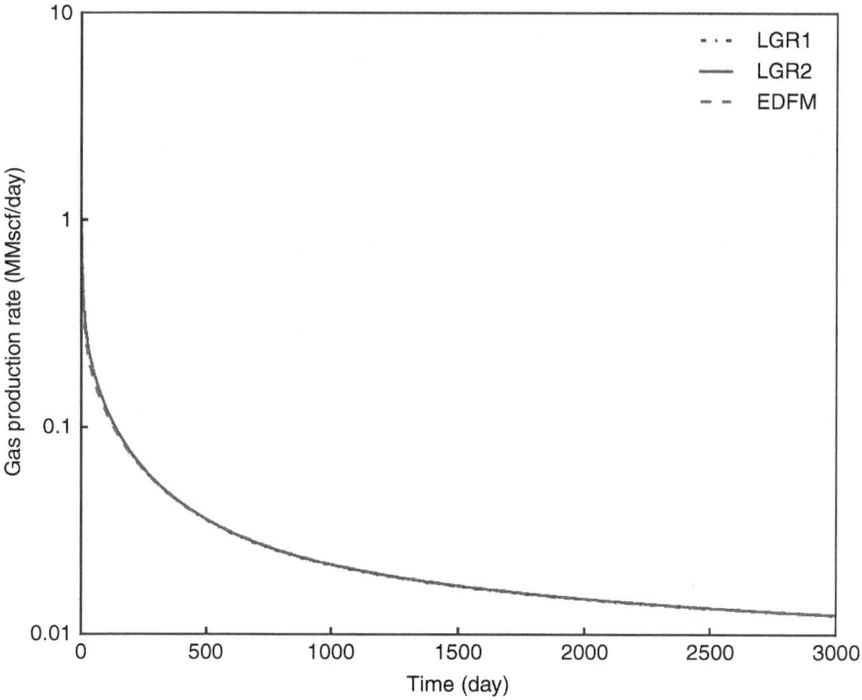

Figure 7.12 ***Comparison of gas production rate (surface condition) predicted by LGR models and the EDFM (Xu and Sepehrnoori, 2019).***

discretized in different ways, and the same set of fractures is simulated using the EDFM. The same reservoir simulator as in Case 1 is used.

7.4.1 Case 2: EDFM with different matrix gridding

Four different cases (Cases 2a through 2d) were created using different simulation grids, as shown in Fig. 7.13. The reservoir dimensions are 1,000 × 1000 × 10 ft, respectively, and the simulation parameters are the same as in Case 1. For each case, the reservoir is discretized using a 100 × 100 matrix grid in the horizontal direction. In Case 2a, a uniform Cartesian grid is used; in Case 2b, a corner-point grid with some distortion is used; in Cases 2c and 2d, the reservoir is discretized using randomly generated corner-point grids. Six fractures and a horizontal well are placed inside the reservoir, as shown in Fig. 7.14. The same set of fractures is simulated using the EDFM approach on the different matrix grids in the four cases. The fracture width is 0.01 ft, and the fracture permeability is 10,000 md,

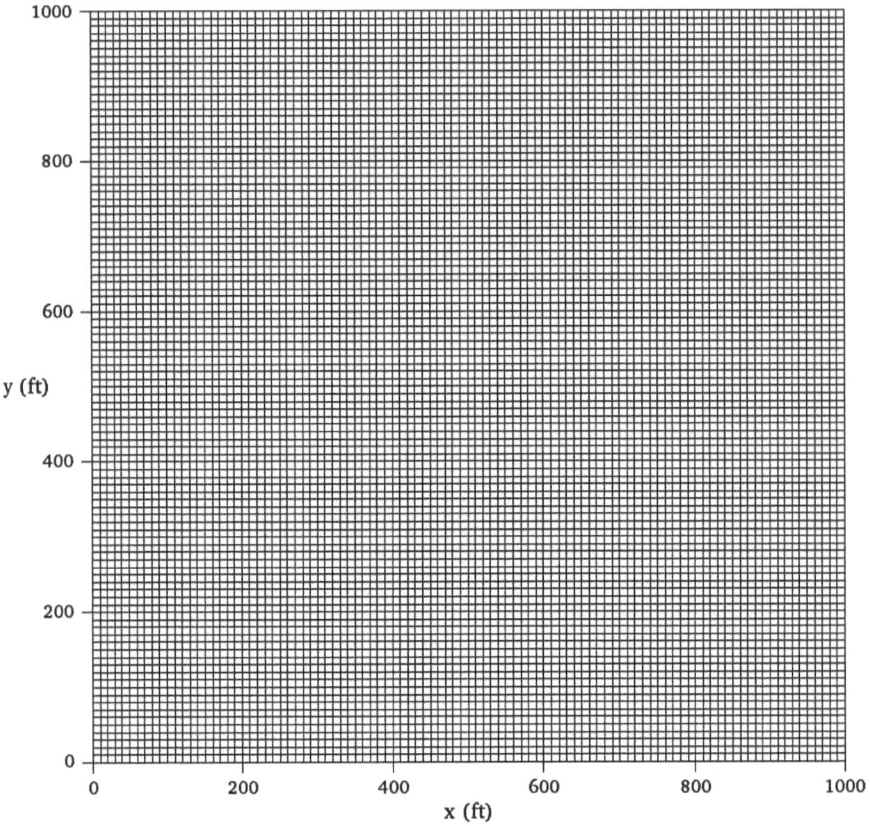

(A)

Figure 7.13 ***Grids A, B, C, and D used for the four different simulations in Case 2 (Xu and Sepehrnoori, 2019).*** (A) A uniform Cartesian grid for Case 2a; (B) a corner-point grid with some distortion for Case 2b; (C)–(D) two randomly generated corner-point grids for Cases 2c and 2d. The overall dimensions of the four grids are the same.

which gives a fracture conductivity of 100 md-ft. Because this value is high enough, the fractures can be regarded as infinitely conductive.

Fig. 7.15 shows the pressure profiles at 500 days for the four different cases as predicted by the EDFM. As can be observed, the pressure profiles match well with the cases. Fig. 7.16A shows the gas production rate for the four cases. The EDFM obtains similar results with different grids. As the fractures are infinitely conductive in these cases, the matrix gridding mainly affects the fluid-transport calculation between the matrix and the fractures. Therefore, these results show that the EDFM formulations can accurately capture the fluid flow between the matrix and the fractures in this case.

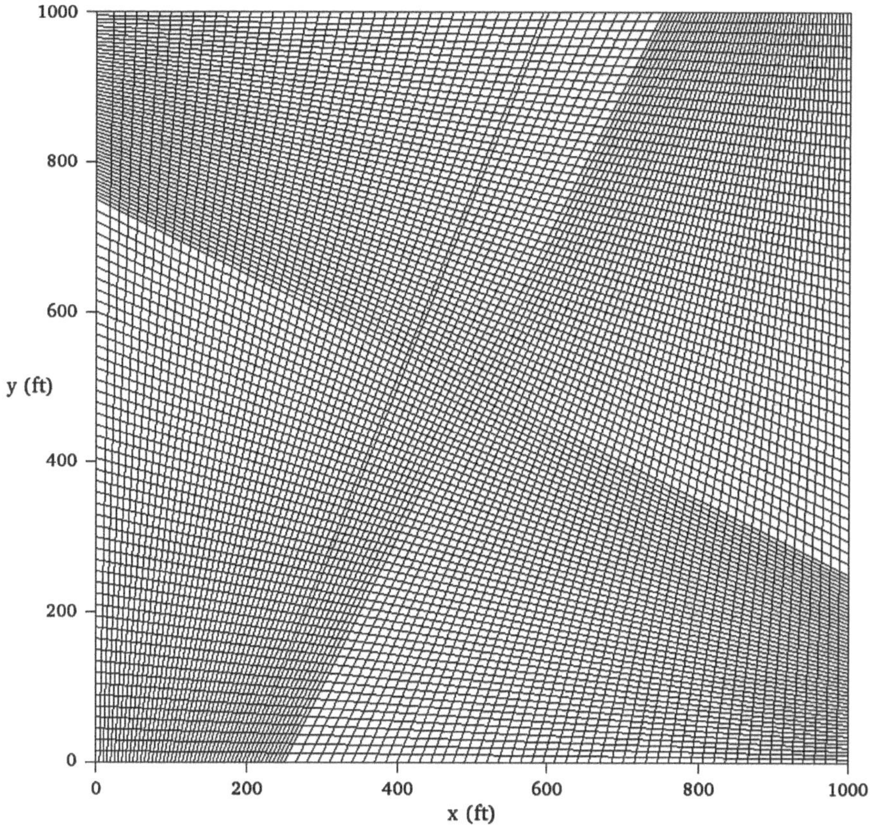

(B)

Figure 7.13 *(Cont.)*

In the EDFM, the gridding of fractures is also affected by the gridding of the matrix, because the fracture segments are created on the basis of the intersections between the matrix blocks and the fractures. When the fracture conductivity is not infinitely high, the fracture gridding might also influence the simulation results. Therefore, another set of simulations is performed to investigate the influence of fracture gridding on the accuracy of the EDFM. For the previous four cases, the conductivities of fractures are changed from 100 to 1 md-ft, and the fractures become finitely conductive. Fig. 7.16B presents the gas production rate for the new cases. The EDFM captures the finite conductivity of the fractures, as the gas flow rate decreases in all cases. Furthermore, the four different cases obtain very similar

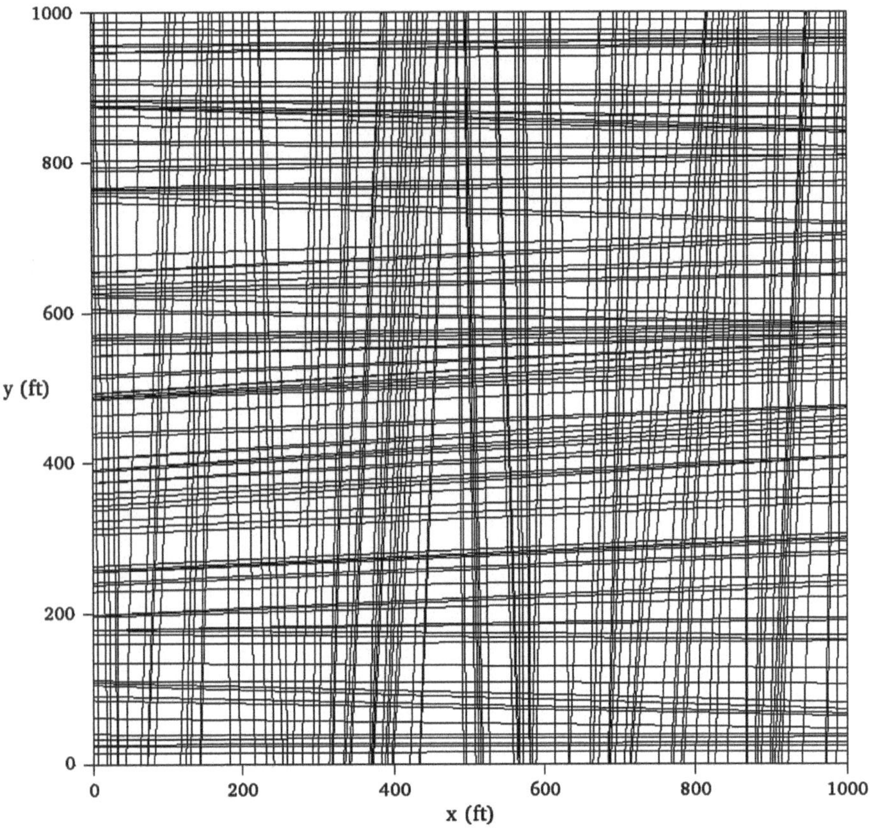

(C)

Figure 7.13 *(Cont.)*

results, showing that the EDFM can accurately model the connections between fracture segments.

As the four different grids have the same total number of matrix gridblocks, the results of this study indicate that the accuracy of the EDFM is not significantly affected by matrix gridding when the average gridblock size is similar. This shows an important feature of the EDFM: during the usage of this approach, with a moderate matrix gridblock size, no detailed gridding around the fractures is required for the prediction of long-term reservoir production. This feature confirms the convenience of using the EDFM, because it indicates that for long-term simulations with moderate gridblock sizes, there is no need to adjust the matrix grid after placing the fractures using the EDFM. For simulations of short-term near-fracture flow or simulations with

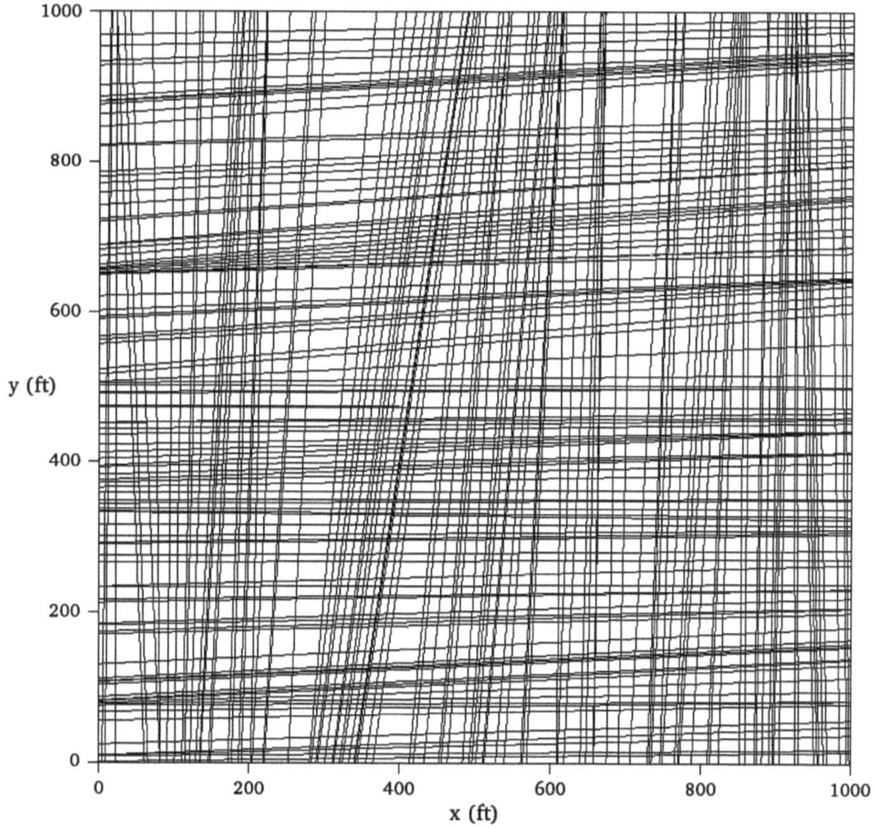

(D)

Figure 7.13 (*Cont.*)

large gridblock sizes, the EDFM could be combined with LGR to improve its accuracy (Panfili et al., 2015). It should also be noted that although different grids are used for the matrix, the fracture inputs in the EDFM preprocessor, which are the coordinates of fracture polygons, do not change. The type of input required by the EDFM makes it easy and convenient to prepare in real-life studies.

7.5 Model application

In previous sections, the accuracy of the EDFM in corner-point grids has been confirmed through case studies. In this section, we present two applications of the EDFM in 3D corner-point grids. The first case uses a

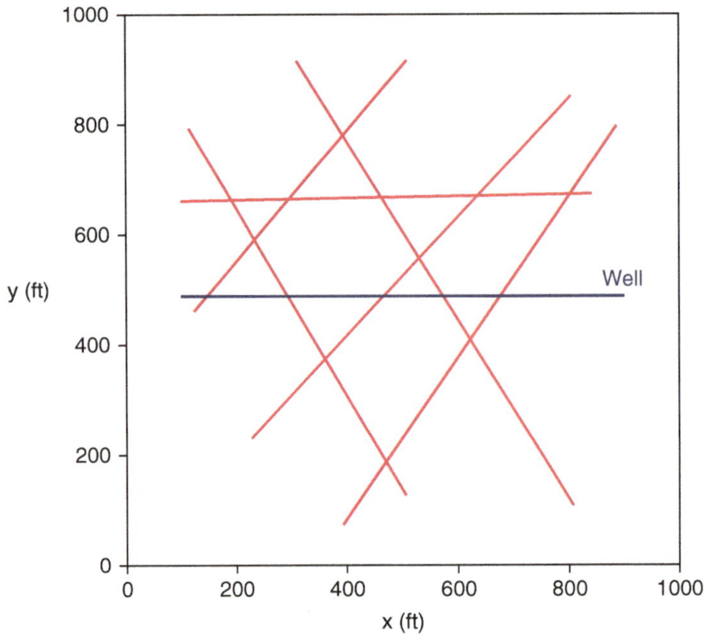

Figure 7.14 ***Locations of the fractures and the horizontal wellbore in Case 2 (Xu and Sepehrnoori, 2019).*** The fractures are shown in red, and the wellbore is shown in blue.

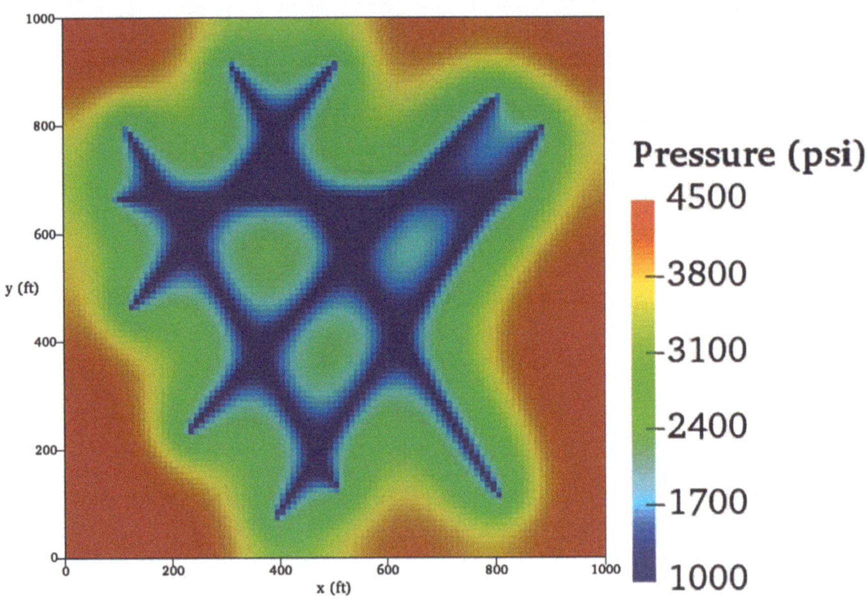

(A)

Figure 7.15 ***Pressure profiles after 500 days of production predicted by the EDFM for the four cases with a fracture conductivity of 100 md-ft (Xu and Sepehrnoori, 2019).*** (A) through (D) show the pressure profiles for Cases 2a through 2d, respectively.

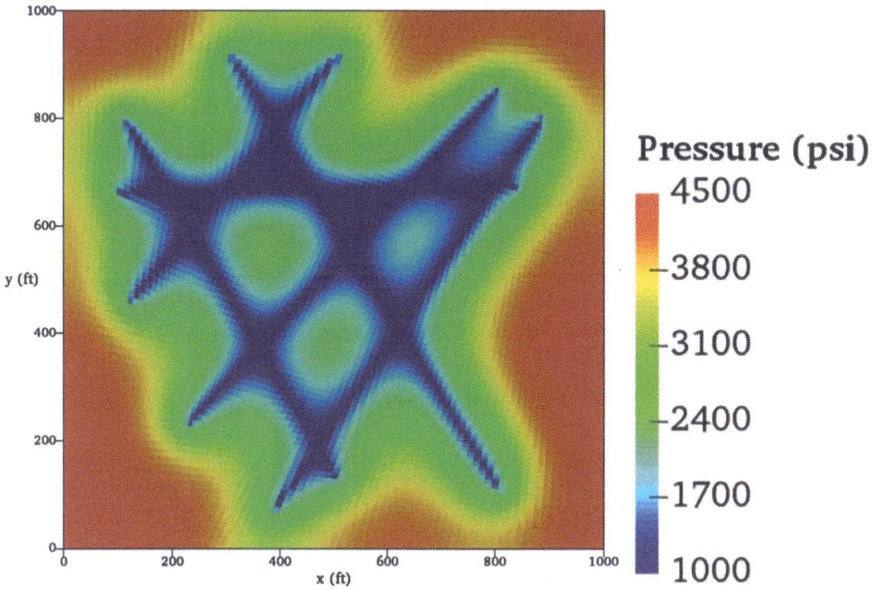

(B)

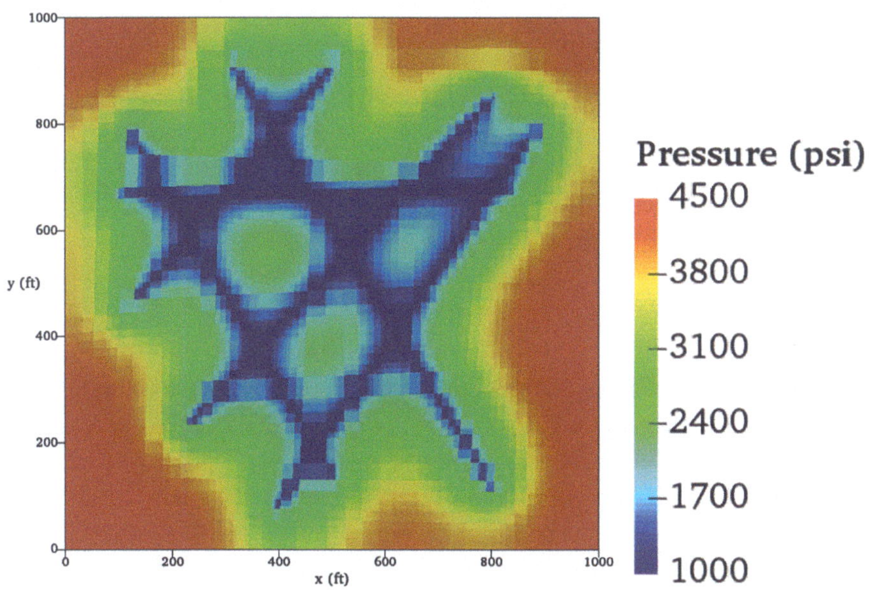

(C)

Figure 7.15 *(Cont.)*

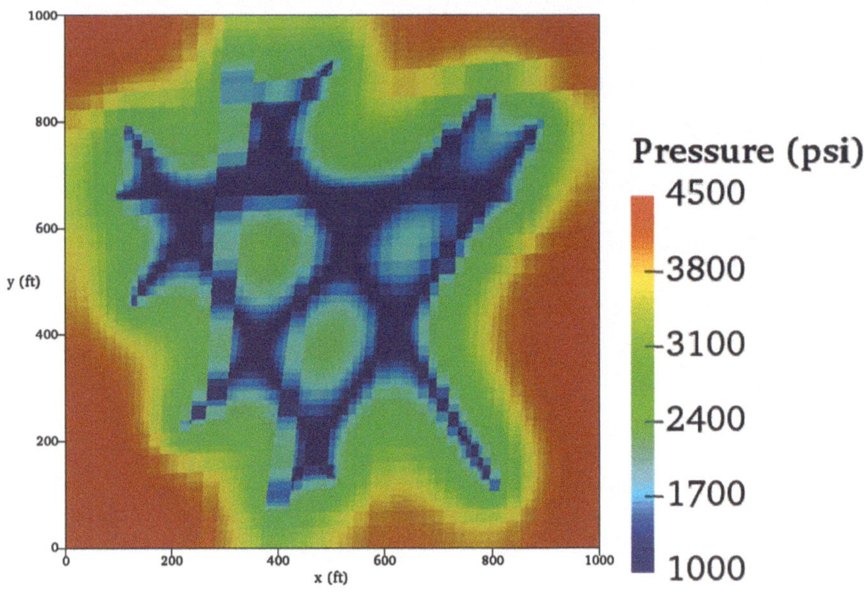

(D)

Figure 7.15 *(Cont.)*

synthetic reservoir model, and the second case uses a real reservoir model. Two commercial reservoir simulators, CMG-GEM (CMG, 2017a) and Eclipse 100 (Schlumberger, 2016), were used for the simulations.

7.5.1 Case 3: Three-phase-flow simulation

In this example, a compositional simulator, CMG–GEM, is used to simulate a waterflooding process in a light oil reservoir. A reservoir model with irregular shape and variable thickness (as shown in Figs. 7.17 and 7.18) is discretized using a 50 × 50 × 5 corner-point grid. Some basic simulation parameters are summarized in Table 7.2. A water injector and a producer are placed at the corners of the reservoir, as shown in Fig. 7.17. Four hydrocarbon components are considered, and their properties are listed in Table 7.3. The Peng–Robinson equation of state is used for the calculation of hydrocarbon phase behavior. The initial fluid phases in the reservoir include water and oil, and water is at irreducible saturation. The gas phase appears during reservoir depletion. For the matrix, Corey-type relative permeability curves are used for water and gas phases, and Stone's second model is used for

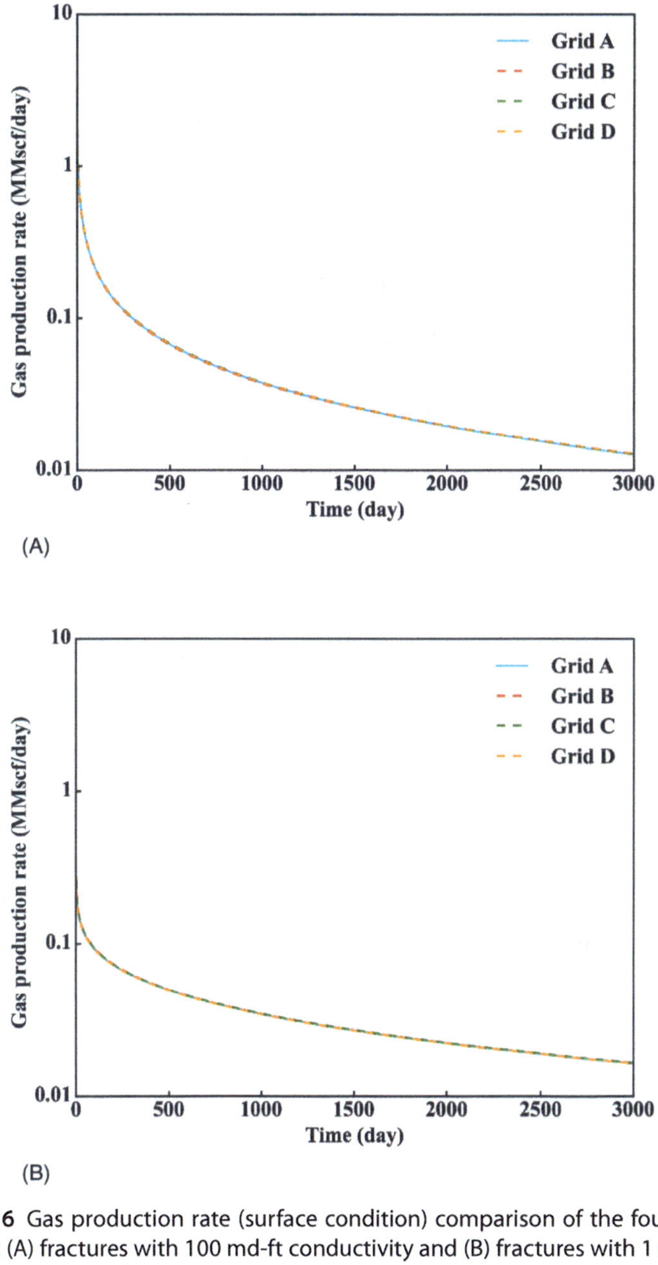

Figure 7.16 Gas production rate (surface condition) comparison of the four different cases with (A) fractures with 100 md-ft conductivity and (B) fractures with 1 md-ft conductivity (Xu and Sepehrnoori, 2019).

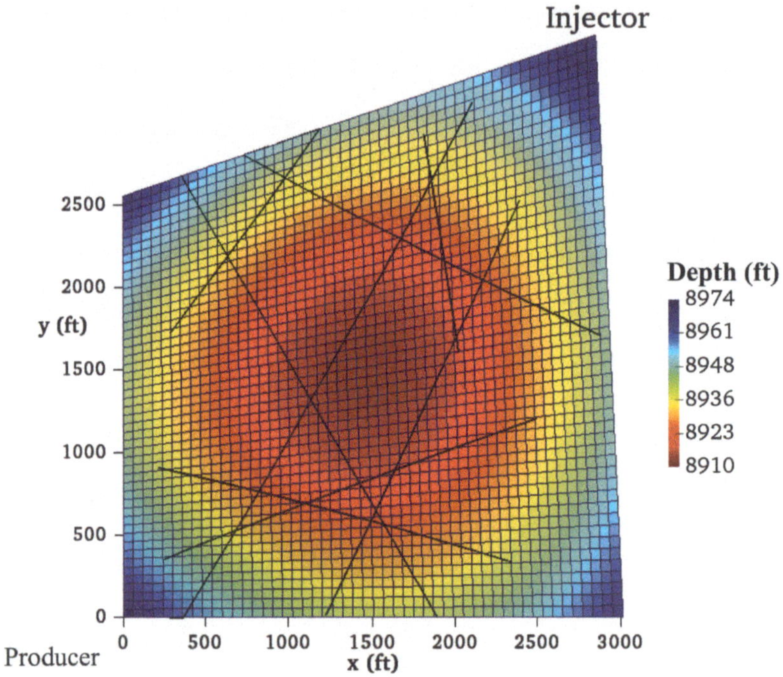

Figure 7.17 ***Top view of the reservoir model in Case 3 (Xu and Sepehrnoori, 2019).*** An injector and a producer are placed at two opposite corners of the reservoir. The eight black lines represent locations of fractures. The dip angle of the fractures ranges from 60 to 90 degrees.

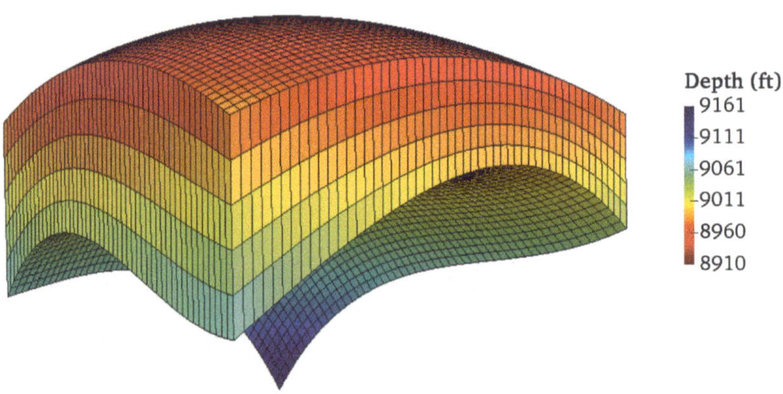

Figure 7.18 ***Side view of the reservoir model in Case 3 (Xu and Sepehrnoori, 2019).*** The scale in the *z*-direction is 10.

Table 7.2 Key simulation parameters in Case 3 (Xu and Sepehrnoori, 2019).

Parameter	Value	Unit
Permeability in the *x*-direction	150	md
Permeability in the *y*-direction	250	md
Permeability in the *z*-direction	20	md
Reservoir porosity	13%	–
Initial reservoir pressure	3,600	psi
Reservoir temperature	100	°F
Rock compressibility	4×10^{-6}	psi^{-1}
Initial water saturation	15.4%	–
Maximum oil rate (producer)	5,000	STB/day
Minimum bottomhole pressure (producer)	500	psi
Water injection rate (injector)	5,000	STB/day
Simulation time	10	year

Table 7.3 Properties of the hydrocarbon components used in the simulations (Xu and Sepehrnoori, 2019).

Component	CH_4	C_2H_6	C_3H_8	FC20
Molar fraction	0.30	0.22	0.12	0.36
Molecular weight (lb/lb-mol)	16.0	30.1	44.1	275.0
Critical pressure (psi)	667.2	708.3	615.8	211.0
Critical volume (ft^3/lb-mol)	1.586	2.371	3.252	16.450
Critical temperature (K)	190.6	305.4	369.8	782.9
Acentric factor	0.008	0.098	0.152	0.816
Parachor	77.0	108.0	150.3	990.3

Table 7.4 Parameters for water, oil, and gas relative permeability curves (Xu and Sepehrnoori, 2019).

Parameter	Value
Residual water saturation	15.4%
Residual oil saturation	10%
Residual gas saturation	0%
Water relative permeability endpoint	0.5
Oil relative permeability endpoint	0.8
Gas relative permeability endpoint	0.8
Water relative permeability exponent	2.0
Oil relative permeability exponent	2.0
Gas relative permeability exponent	2.0

the oil relative permeability. Parameters of the matrix relative permeability curves are summarized in Table 7.4. In fractures, the straight-line relative permeability model is used. The capillary pressure effect is ignored.

Eight large-scale fractures are placed inside the reservoir, as shown in Fig. 7.17. All fractures fully penetrate the reservoir in the vertical direction. The conductivity of the fractures is 10,000 md-ft, and the dip angle of the fractures ranges from 60 to 90 degrees. After the preprocessing calculation, the fractures are discretized into a total number of 2,509 fracture segments. These fracture segments are represented as additional gridblocks in the computational domain.

To study the impact of fractures on waterflooding efficiency, the case without considering the existence of fractures was also simulated. Fig. 7.19

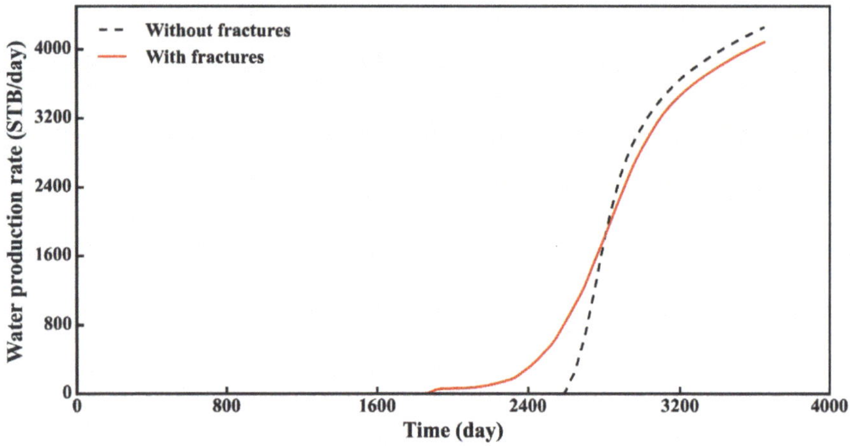

(A)

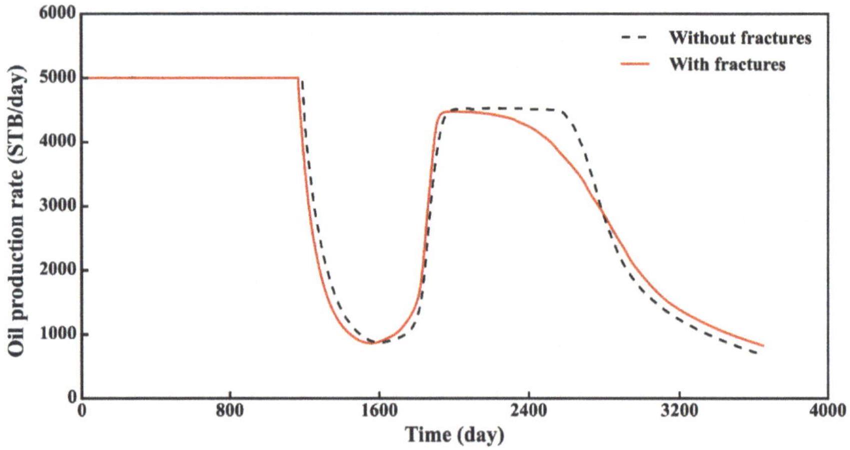

(B)

Figure 7.19 Predicted (A) water, (B) oil, and (C) gas production rates (surface condition) for the cases with and without fractures (Xu and Sepehrnoori, 2019).

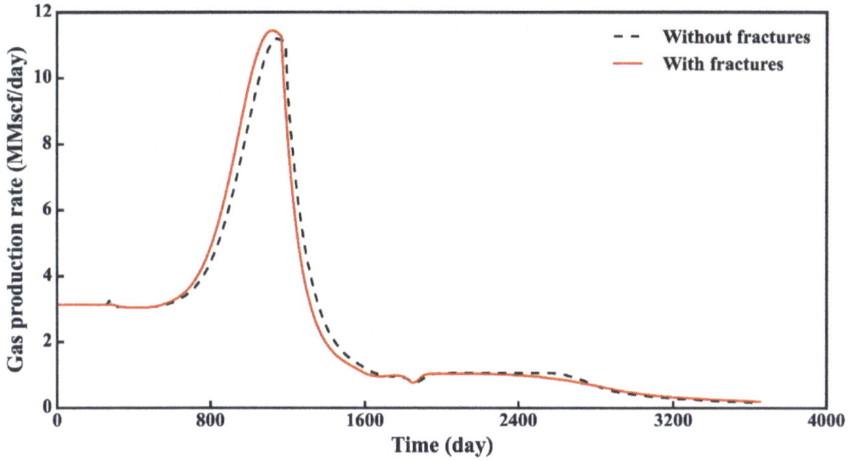

(C)

Figure 7.19 (*Cont.*)

compares the predicted water, oil, and gas production for the cases with and without fractures. For both cases, at the early stage, the flowing phases are water and oil, and oil is produced at the maximum oil rate. Subsequently, because the injector is not providing enough support, the reservoir pressure decreases, and the light hydrocarbon components form the gas phase, increasing gas production. As the waterfront moves, the reservoir pressure near the producer increases, resulting in an increase in oil saturation in this region and an increase in oil production rate. Finally, the water breaks through at ~2,550 days for the case without fractures. For the case with fractures, the water breakthrough occurs much earlier. This is mainly because of the large-scale fractures that "connect" the injector with the producer. Fig. 7.20 shows the water-saturation profiles at 1,500 days, in both the matrix and the fractures, to illustrate the advancing of the waterfront in the reservoir. In Fig. 7.20B, several "connecting" fractures can be clearly seen from the matrix water-saturation profiles, as these fractures serve as channels for the waterfront to move and they reduce the "resistance" between the injector and the producer. This results in a lower sweep efficiency in the horizontal direction. The gravitational effect can also be easily observed in the water-saturation profiles. As seen in Fig. 7.20, the bottom layer has a higher water saturation than the top layer in both the matrix and the fractures. In addition, the waterfront moves much faster in the bottom layers. As a result, water breaks through much earlier in the bottom layers compared with the top layers.

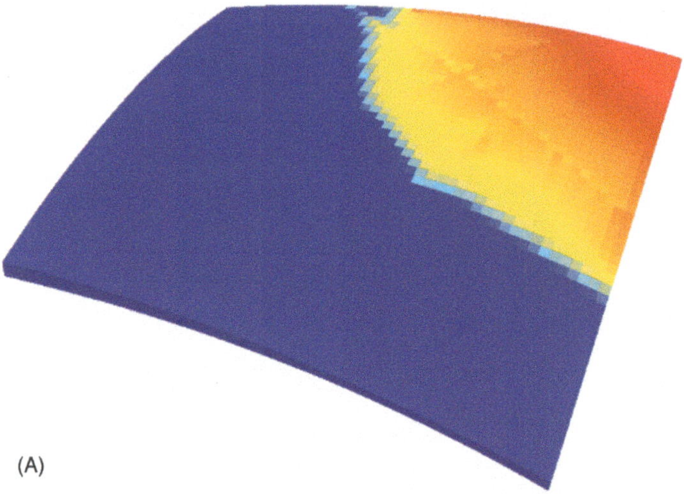

(A)

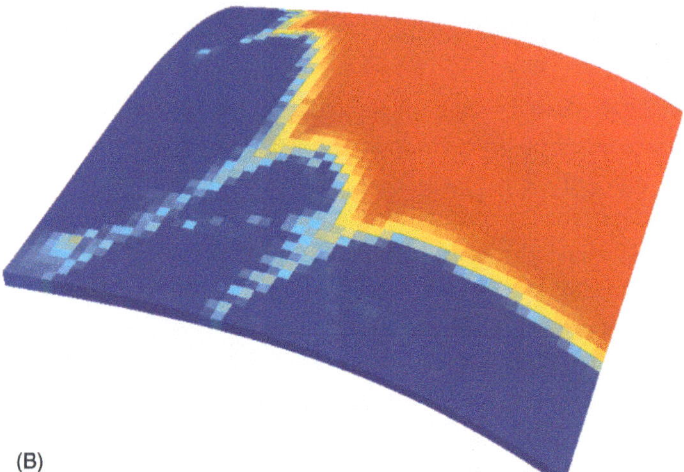

(B)

Figure 7.20 ***Predicted water-saturation profiles in the matrix and the fractures at 1,500 days (Xu and Sepehrnoori, 2019).*** (A) Water saturation in the top layer of the matrix; (B) water saturation in the bottom layer of the matrix; (C) water saturation in the fractures. In the fracture water-saturation profile, the black lines represent the connections between fracture segments. The scale in the *z*-direction is 5.

Because spontaneous imbibition is an important oil-recovery mechanism during waterflooding processes in fractured reservoirs (Karimi-Fard and Firoozabadi, 2003), the influence of water–oil capillary pressure is also investigated. For the case with fractures, the water–oil capillary pressure is considered assuming that the rock is water-wet. Fig. 7.21A presents

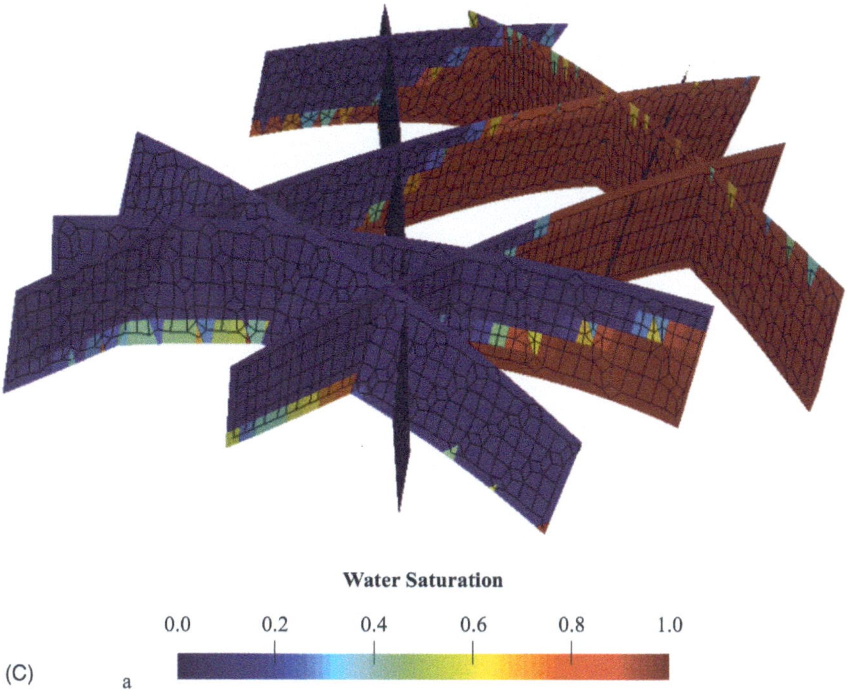

(C)

Figure 7.20 *(Cont.)*

the capillary pressure curve used in the simulation and Fig. 7.21B shows the comparison of water rate for the cases with and without capillary pressure. A much later water breakthrough is predicted when the capillary pressure effect is taken into consideration, owing to the spontaneous imbibition of water from the fracture network to the matrix. To confirm this, Fig. 7.22 shows the water saturation in the top and bottom layers of the reservoir at 1,500 days. Compared with the case without capillary pressure, the waterfront moves faster in the matrix, giving rise to a better sweep efficiency. In addition, in different layers of the matrix, the waterfront moves at a similar speed. Therefore, the top layers are better swept by the injected water.

This example illustrates the capability of the EDFM to simulate fractures in geologically complex reservoirs. Different recovery mechanisms involved in waterflooding processes in fractured reservoirs can be effectively simulated. The results of the simulations also demonstrate the importance of large-scale fractures and the capillary pressure effect during waterflooding.

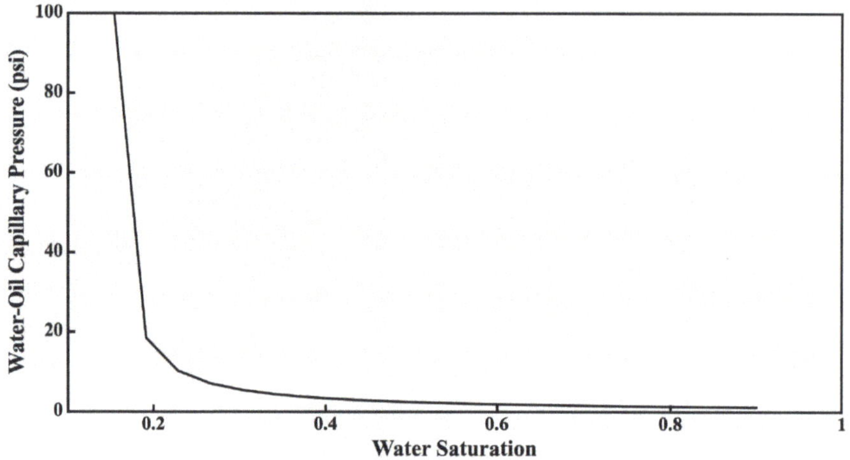

(A)

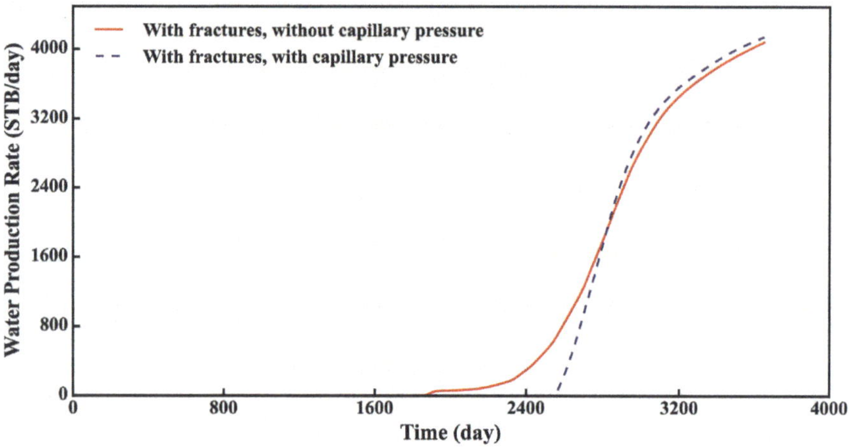

(B)

Figure 7.21 (A) Water–oil capillary pressure curve (water-wet) used in the simulation. (B) Comparison of water production rate for the cases with and without capillary pressure (Xu and Sepehrnoori, 2019).

7.5.2 Case 4: Application of the EDFM in the Norne Field

Here, we present case studies using the reservoir model of the Norne Field in the Norwegian Sea (Verlo and Hetland, 2008). The Norne Field was discovered in 1991, and the development started in 1997. The reservoir permeability ranges from 0.1 to 4,000 md, and porosity ranges from 9% to 35%. An Eclipse 100 reservoir simulation model using a corner-point

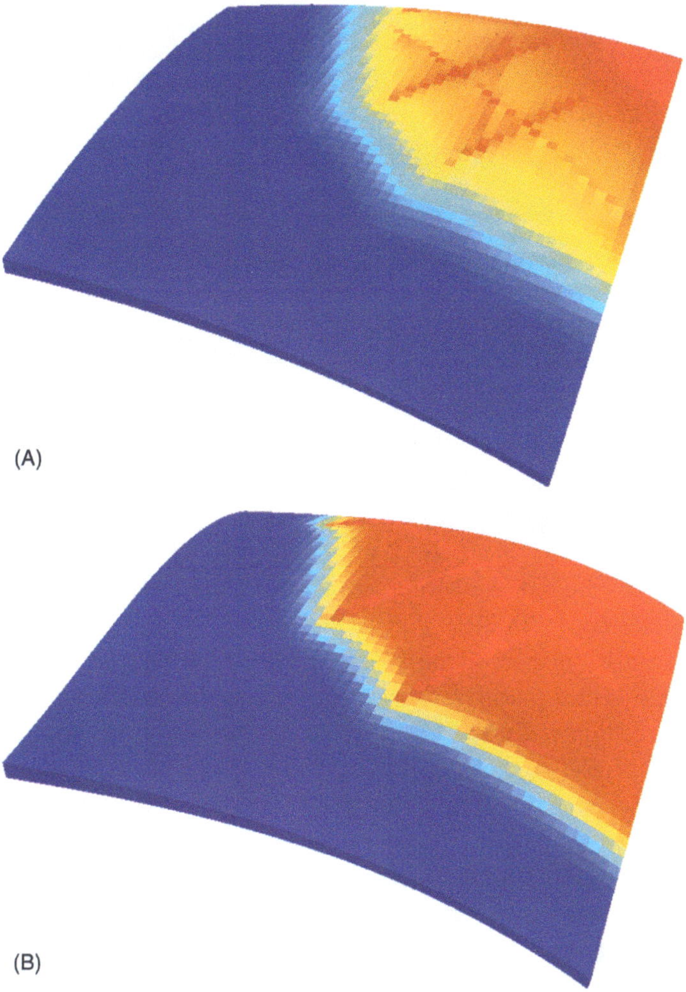

Figure 7.22 ***Predicted water-saturation profiles in the matrix at 1,500 days for the case with capillary pressure (Xu and Sepehrnoori, 2019).*** (A) Water saturation in the top layer of the matrix; (B) water saturation in the bottom layer of the matrix.

grid (as shown in Fig. 7.23) was made available by the operator through the IO center at the Norwegian University of Science and Technology as an open dataset (http://www.ipt.ntnu.no/~norne/wiki/doku.php?id=english:nornebenchmarkcase2). It is also available in the Open Porous Media dataset (https://github.com/OPM/opm-data/tree/master/norne) under the open database license. This model consists of 113,604 (46 × 112 × 22) gridblocks, of which 45,187 are active. The gridblock

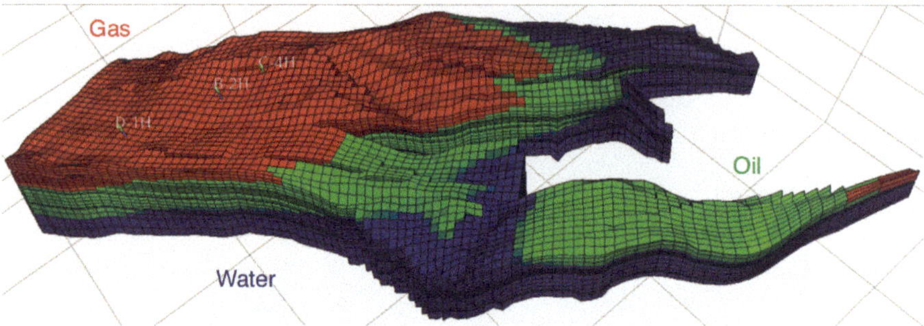

Figure 7.23 ***Reservoir simulation grid and initial phase saturation of the Norne Field (Xu and Sepehrnoori, 2019).*** The scale in the *z*-direction is 5.

dimensions are roughly 150 × 150 ft in the horizontal direction. The initial three-phase saturation is shown in Fig. 7.23, and the water, oil, and gas phases are shown in blue, green, and red, respectively. Corey-type relative permeability curves are used, and the capillary pressure curves are also provided with the dataset. More details about the model can be found in Verlo and Hetland (2008).

The simulation grid of the Norne Field is complex in several aspects. More than 60 faults in different orientations are present in this model, making the grid highly distorted. Inactive cells, non-coplanar points in gridblock faces, partially matched gridblock faces, eroded layers, and pinch-outs can all be found in this simulation grid. This presents great challenges for the robustness of the geometrical calculation in the EDFM.

The field production history of oil, water, and gas phases from 1997 to 2006 is available together with the reservoir-simulation model to help researchers perform history-matching studies. No production data are available from 2007 on, and production prediction or field-development-plan optimization on the basis of reservoir simulation can be performed.

Using this reservoir model, we performed two case studies. In the first study, we added three hydraulic fractures using the LGR method and the EDFM, respectively, to verify the accuracy of the EDFM in a 3D realistic case study. In the second study, two sets of natural fractures are placed inside the reservoir to simulate the influence of fractures on oil, water, and gas production. For both case studies, during 1997–2006, the well-control parameters were kept the same as in the dataset, with injectors controlled by injection rate and producers controlled by liquid rate; after 2006, the well-control parameters were set to be the same as the values at the end of the history period.

7.5.2.1 Case 4a: Verification of EDFM with LGR for simulating hydraulic fractures

In this case study, we assume that a deviated well, B-3AH, was drilled and hydraulically fractured in the oil zone at the beginning of 2007. Three hydraulic fractures were created, as shown in Fig. 7.24. The shape of the hydraulic fractures is nearly rectangular. Some matrix gridblocks are partially penetrated by the fractures. The fracture length and height are around 600 and 100 ft, respectively, and the fracture conductivity is ~5,000 md-ft. The rock permeability near the well is ~100 md in the horizontal direction and 15 md in the vertical direction.

An LGR model is created as a reference solution. In the LGR model, each gridblock containing fractures is refined into three sub-gridblocks, and a high permeability value is assigned to the sub-gridblock in the middle to represent the fractures. In the EDFM, each fracture is defined as a polygon with four vertices. Water, oil, and gas production rates from 2007 to 2009 are predicted by both models. The maximum liquid rate for well B-3AH is set to 3,774 STB/day.

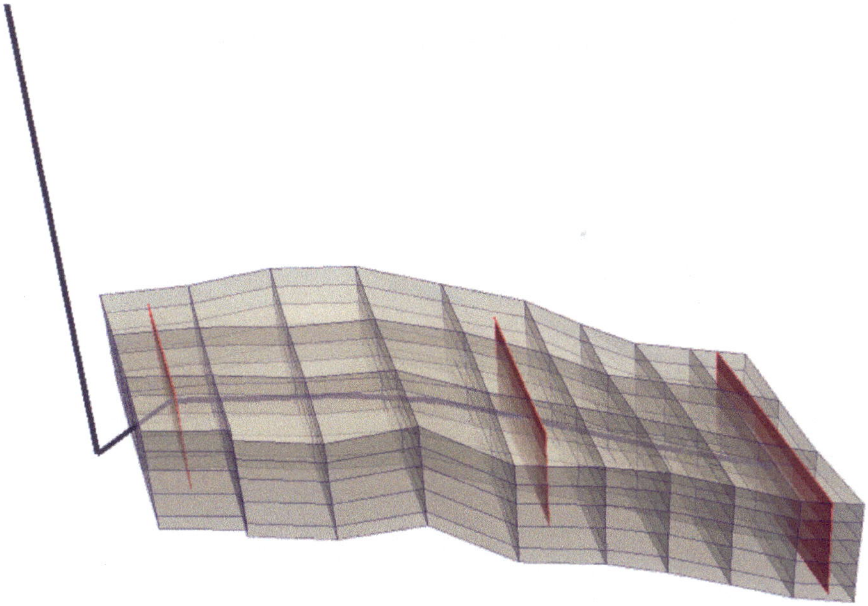

Figure 7.24 ***Locations of the three hydraulic fractures and the trajectory of Well B-3AH in Case 4a (Xu and Sepehrnoori, 2019).*** The blue line represents the well trajectory, and the three red rectangles represent three hydraulic fractures. The gridblocks around the hydraulic fractures are also shown. A scale of 5 in the *z*-direction is used for visualization.

Fig. 7.25 shows the field water, oil, and gas production rates from 1997 to 2009 predicted by both models. Production curves for the case without hydraulic fractures are also shown for comparison. Before 2007, all three curves in each figure match, indicating that the modified simulator input file based on the EDFM keeps all of the original information. From 2007 (3,400 days), it can be observed that the added fractures do not have a

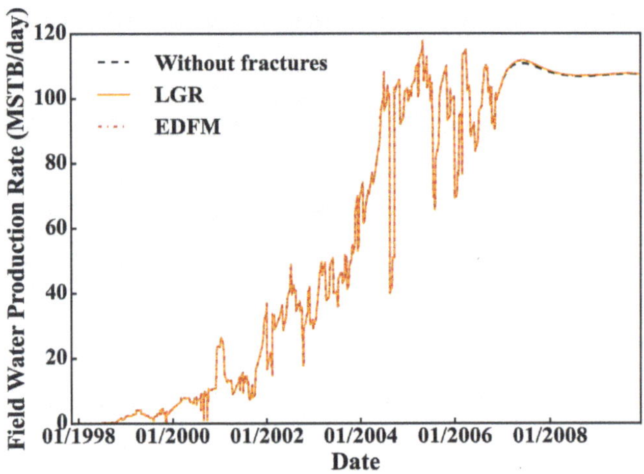

(A)

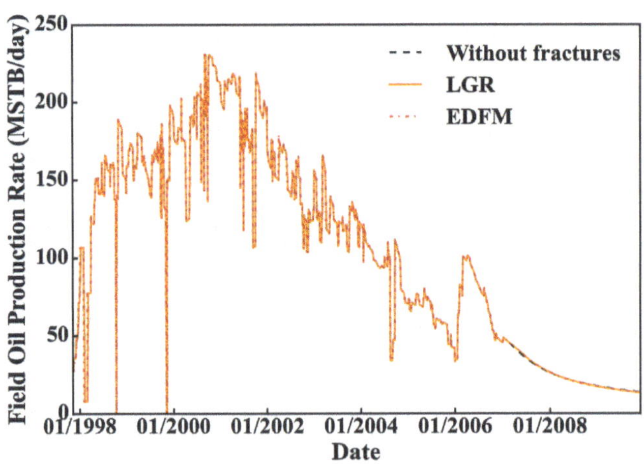

(B)

Figure 7.25 ***Field production rates of (A) water, (B) oil, and (C) gas predicted by the LGR model and the EDFM (Xu and Sepehrnoori, 2019).*** The production curves for the case without fractures are also shown as references.

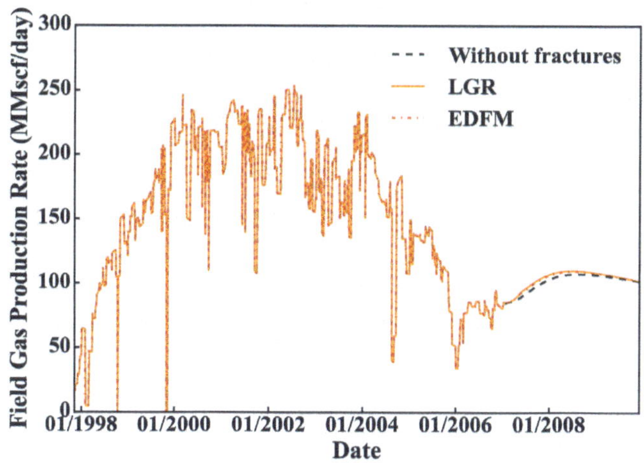

(C)

Figure 7.25 *(Cont.)*

significant impact on oil production. For water and gas production, the impact of fractures can be noticed in Fig. 7.25A and C. Overall, the EDFM obtains a good match with the LGR model for production rates of all three phases, showing its good accuracy in 3D corner-point grids.

7.5.2.2 Case 4b: Simulation of natural fractures

In this case, on the basis of the original reservoir model, two sets of randomly generated fractures are placed into the reservoir using the EDFM to create a possible "scenario" of the reservoir, as shown in Fig. 7.26. The total number of fractures is 400. All fracture planes are perpendicular to the z-axis. The fracture width is 0.0328 ft, and the fracture permeability ranges from 1,000 to 10,000 md. After the preprocessing calculation, 120 fractures penetrate the reservoir, and the total number of fracture segments in the EDFM is 15,312, as shown in Fig. 7.27. These fracture segments are represented by additional layers in the x-direction in the simulation grid, as shown in Fig. 7.28. The added layers align with the existing layers in the y and z directions. It is worth mentioning that adding new layers in a corner-point grid is nontrivial compared with doing so in a Cartesian grid. Attention needs to be paid to avoid adding gridblocks with zero volume or gridblocks that overlap with existing gridblocks. The total number of connections between fracture segments is 26,495, of which 26,040 connections are within individual fractures and 455 connections are from fracture intersections.

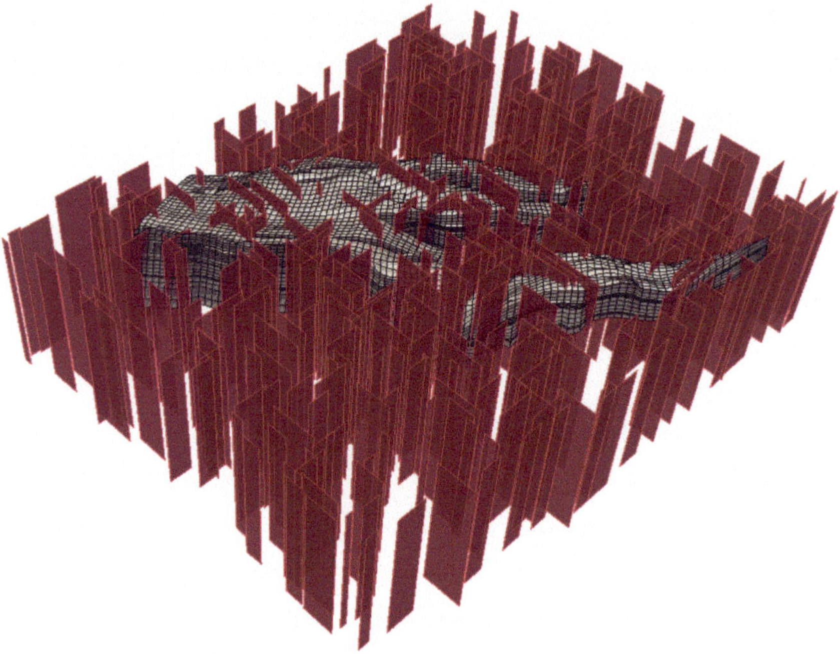

Figure 7.26 ***Two sets of fractures in Case 4b (Xu and Sepehrnoori, 2019).*** The red polygons represent fractures, and the total number of fractures is 400. The scale in the *z*-direction is 5.

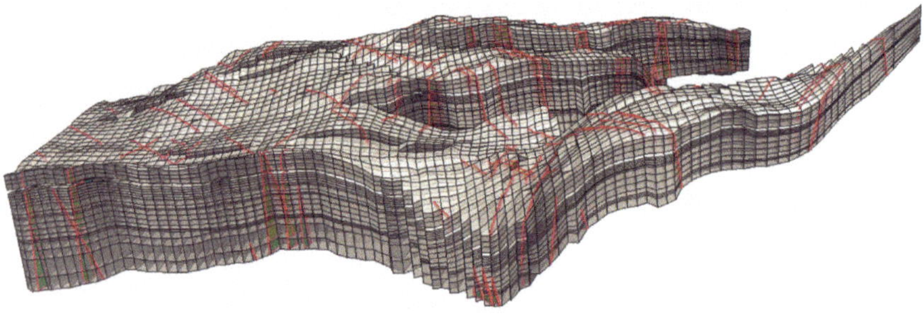

Figure 7.27 ***Fracture segments calculated from the EDFM preprocessing (Xu and Sepehrnoori, 2019).*** The red polygons represent fracture segments. The scale in the *z*-direction is 5.

The objective of this case study is to evaluate the influence of fractures on the field-development plan. Fig. 7.29 presents the field water, oil, and gas production rates with and without fractures. The results show that the existence of fractures decreases the oil and gas production and increases the water production. The water-saturation profiles in Fig. 7.30 provide an

Figure 7.28 ***Visualization of the grid domain after adding fracture gridblocks in the EDFM (Xu and Sepehrnoori, 2019).*** The blocks representing fracture segments are shown in red, and the original matrix grid is shown in blue. For the matrix, both the active and non-active gridblocks are shown in the figure. The scale in the *z*-direction is 5.

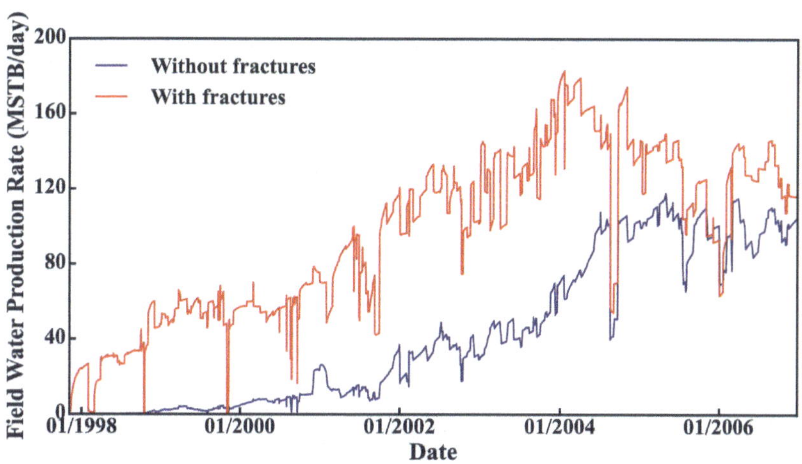

(A)

Figure 7.29 Comparison of (A) field water production rate, (B) field oil production rate, and (C) field gas production rate for the cases with and without fractures (Xu and Sepehrnoori, 2019).

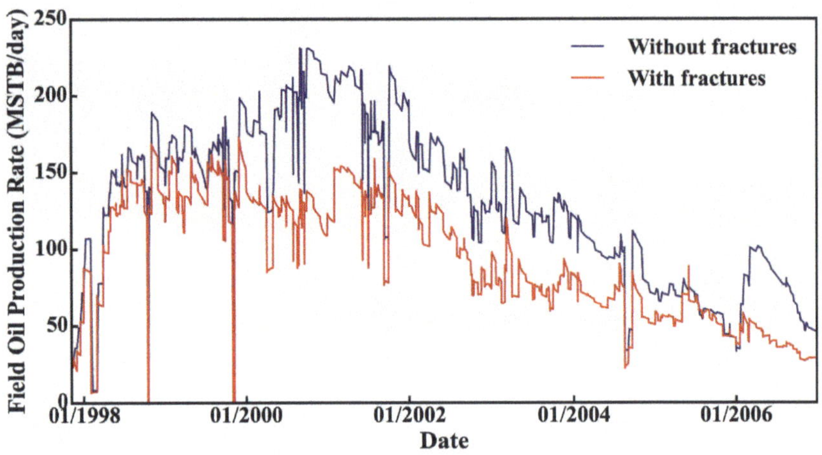

(B)

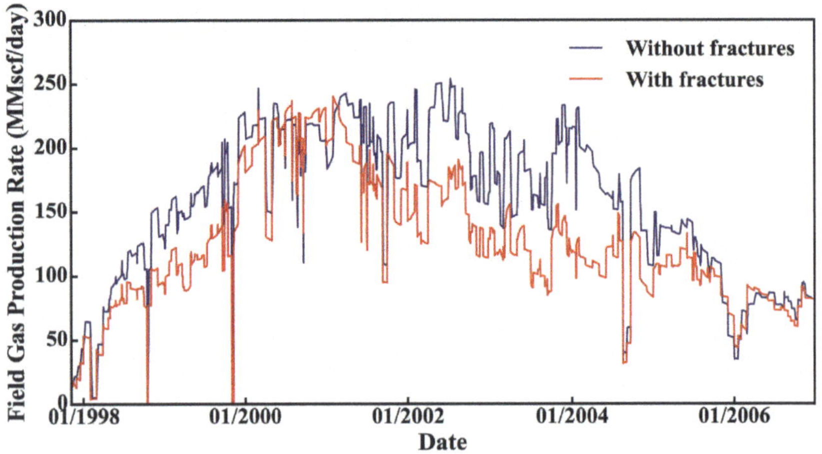

(C)

Figure 7.29 (*Cont.*)

explanation. In Fig. 7.30A, the water saturation in the fracture segments after 7 years of production is presented. Correspondingly, Fig. 7.30B shows the water saturation in the matrix blocks at the same locations as these fractures for the case without fractures. It can be observed that for the case with fractures, the water saturation in the middle layers of the reservoir is higher. The reason is that the fractures improve the vertical connectivity of the reservoir, and they provide an additional pathway for water in the bottom layers (water zone) to enter the middle layers (oil zone), which increases the water saturation in the oil zone. Therefore, the original field-development plan will not work properly for this scenario. For reservoirs

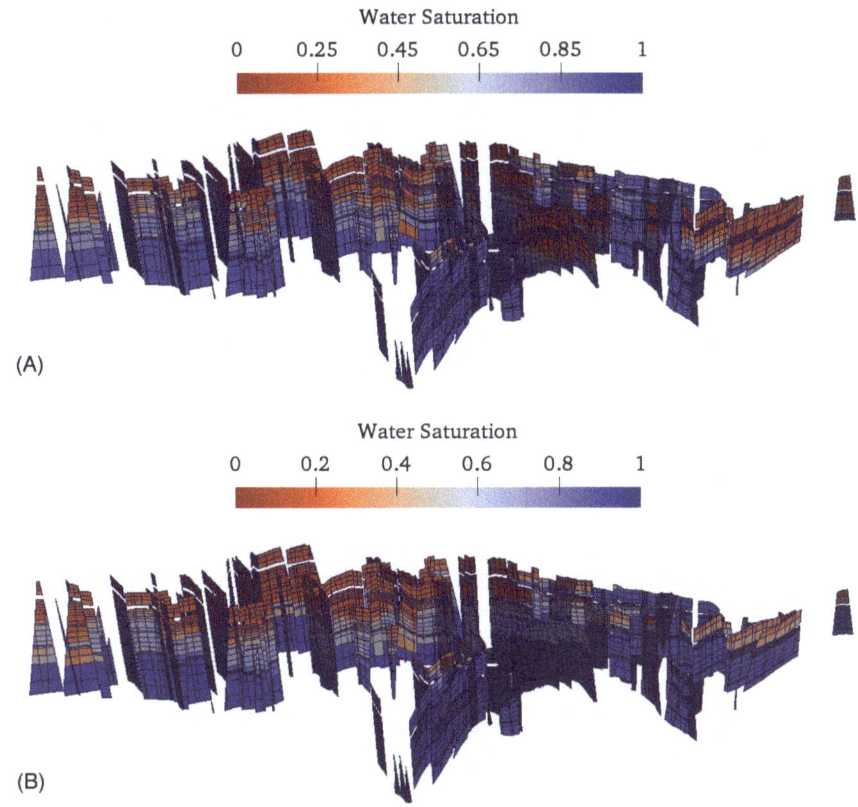

Figure 7.30 ***Comparison of water saturation after 7 years of production for the cases with and without fractures (Xu and Sepehrnoori, 2019).*** (A) Water saturation in fracture segments for the case with fractures. (B) Water saturation in the matrix at the same location as the fracture segments for the case without fractures. The scale in the *z*-direction is 5.

with fully penetrating fractures in the vertical direction, the increase in vertical connectivity, owing to the existence of fractures, needs to be taken into account while making the field-development plan, as shown in this study.

This example illustrates the possible significance of fractures in a realistic reservoir-simulation model, and it also shows the convenience of using the EDFM to place fractures in reservoirs with complex structures. It will be very challenging to simulate this case using other methods. In addition, it should be noted that the simulation of fractured reservoirs is, in general, more computationally expensive than the simulation of non-fractured reservoirs, especially for cases with complex fractures. In this example, with a single processor, the central processing unit time for the case with fractures is ~30 times greater than that for the case without fractures, mainly because of the small pore volume of fractures compared with the matrix and the large number of NNCs introduced by the EDFM.

It should also be emphasized that this study uses an existing reservoir model. With the EDFM, a traditional simulator can be used to simulate complex fractures, and previously prepared input data can be used with small modifications. The procedure of using the EDFM does not require rebuilding of the reservoir model, which makes it convenient to place fractures in existing reservoir models for sensitivity study and history matching.

Nomenclature

A = area, ft
d = distance, ft
k_f = fracture permeability, md
r = distance

Appendix A: Algorithm to calculate the intersection between a polyhedron (representing matrix gridblock) and a polygon (representing fracture)

Given a gridblock G and a fracture polygon P, the steps to calculate the intersection between G and P are as follows:

1. Find a bounding box B_1 for G. The bounding box can be found by calculating the minimum and maximum x-, y-, and z-coordinates of the vertices of G.
2. Find a bounding box B_2 for P.
3. Check whether the two bounding boxes B_1 and B_2 intersect. If not, stop the algorithm, and there is no intersection between G and P.
4. Check whether all vertices of G are on one side of the plane containing P (denoted as P'). If yes, stop the algorithm, and there is no intersection between G and P.
5. Find the edge list $\{E\}$ of G.
6. For each edge e in $\{E\}$, check whether the two vertices of e have the same coordinates (degeneracy). If yes, remove e from $\{E\}$, and merge the two vertices.
7. For each edge e in $\{E\}$, check whether the two vertices belonging to e have already been merged. If yes, remove e from $\{E\}$.
8. Find the original face list $\{F\}$ of G. Each element f in $\{F\}$ should contain four vertices.
9. For each face f in $\{F\}$, list the vertices and check whether any pair of the vertices has already been merged. If yes, remove one vertice in this pair from f.

10. For each face f in $\{F\}$, if f contains less than three vertices, remove f from $\{F\}$. If f contains four vertices, check whether the four vertices are coplanar. If not, split f into two faces, f_1 and f_2, with each containing three vertices, remove f from $\{F\}$, and add f_1 and f_2 into $\{F\}$.
11. Prepare three empty point lists $\{P_1\}$, $\{P_2\}$, and $\{P_3\}$ and an empty connection list $\{C\}$.
12. For each edge e in $\{E\}$, check the intersection between e and P'. If e and P' intersect, add the intersection point i into $\{P_1\}$. Record the gridblock faces in $\{F\}$ that e belongs to. Notice that e might belong to multiple faces in $\{F\}$.
13. If $\{P_1\}$ contains less than three points, stop the algorithm, and there is no intersection between G and P.
14. For each point p in $\{P_1\}$, check whether p is inside the fracture polygon P. If not, remove p from $\{P_1\}$. If all points in $\{P_1\}$ are inside P and P is convex, go directly to Step 17.
15. Find the edge list $\{E'\}$ of the fracture polygon P. For each edge e' in $\{E'\}$ and each face f in $\{F\}$, check the intersection between e' and f. If e' and f intersect, add the intersection point i' into $\{P_2\}$. Record the gridblock faces in $\{F\}$ and the fracture edges in $\{E'\}$ that i' belongs to.
16. For each vertex v of the fracture polygon P, check whether v is inside G. If yes, add v into $\{P_3\}$. Record the fracture edges in $\{E'\}$ that v belongs to.
17. Let $\{P_{12}\}=\{P_1\}\cup\{P_2\}$. For any pair of points p_1 and p_2 in $\{P_{12}\}$, if both p_1 and p_2 belong to a face in $\{F\}$, no other point in $\{P_{12}\}$ is on line segment p_1–p_2, and the midpoint of line segment p_1–p_2 is inside P, add (p_1, p_2) into $\{C\}$.
18. Let $\{P_{23}\}=\{P_2\}\cup\{P_3\}$. For any pair of points p_1 and p_2 in $\{P_{23}\}$, if both p_1 and p_2 belong to an edge e in $\{E'\}$, no other point in $\{P_{23}\}$ is on line segment p_1-p_2, and the midpoint of line segment p_1-p_2 is inside G, add (p_1, p_2) into $\{C\}$.
19. Let $\{P_{123}\}=\{P_1\}\cup\{P_2\}\cup\{P_3\}$. Check if any two points p_1 and p_2 in $\{P_{123}\}$ have the same coordinates. If yes, remove p_2 from $\{P_{123}\}$ and replace p_2 with p_1 in each element of $\{C\}$. Repeat this step until there are no coincident points in $\{P_{123}\}$.
20. Check the number of points in $\{P_{123}\}$. If $\{P_{123}\}$ contains less than three points, stop the algorithm, and there is no intersection between G and P. Otherwise, there is an intersection between G and P.
21. For each element c in $\{C\}$, check if the two points in c are the same. If yes, remove c from $\{C\}$.
22. Use the connection list $\{C\}$ to arrange the sequence S of the points in $\{P_{123}\}$. The points should be arranged such that all pairs of the neigh-

boring points (p_1, p_2) in S, including the pair containing the first and the last points in S, appear in the connection list $\{C\}$.

23. Generate a polygon with the points in $\{P_{123}\}$ using sequence S. This is the intersection polygon between G and P.

This algorithm finds whether the fracture intersects the gridblock. If there is an intersection, it generates the polygon representing the fracture segment. The input of the algorithm is a general polyhedron and a general polygon, which indicates that the application of this algorithm is not limited to corner-point grids. It is also valid for Cartesian grids and unstructured grids. Several simpler problems are embedded in this algorithm. For example, Step 15 finds the intersection between a polygon and a line segment, which can be divided into three problems: the intersection between a line and a plane, checking whether a point is inside a polygon, and checking whether a point belongs to a line segment. Steps 14 and 16 involve checking whether a point is inside a polygon or a polyhedron, which can be solved by first checking whether the point is inside the bounding box of the polygon or the polyhedron, and then using the ray-casting algorithm (Shimrat, 1962). Several steps are added to this algorithm to handle special situations. For instance, Step 7 is necessary when more than two vertices of the gridblock have the same coordinates. Step 19 also plays an important role if the fracture polygon crosses through the edges or the vertices of the gridblock. Steps 1 through 4 are added to speed up the calculation; in most cases, the algorithm will stop after Step 3 or Step 4. When there is an intersection between G and P, the total number of required basic computations (such as line-plane intersection and line–line intersection calculation) is a function of the number of edges of G, the number of vertices of P, and the number of vertices of the intersection polygon.

Because in a computer, several steps of the algorithm are working with floating-point numbers, it is vital to set appropriate tolerances for the geometrical calculations. For example, tolerances need to be set when checking whether two points have the same coordinates or whether three points are collinear. In addition, floating-point errors can have a significant influence on the results of special cases. Therefore, much effort might be required on floating-point number calculation and special case treatment to obtain a robust code.

References

CMG, 2017a. GEM User Guide. Computer Modeling Group.

CMG, 2017b. IMEX User Guide. Computer Modeling Group.

Cordazzo, J., Maliska, C.R., da Silva, A.F.C., 2002. Interblock transmissibility calculation analysis for petroleum reservoir simulation. 2nd Meeting on Reservoir Simulation. Buenos Aires, Argentina.

Ding, Y., Lemonnier, P., 1995. Use of corner point geometry in reservoir simulation. Paper SPE 29933, International Meeting on Petroleum Engineering. Beijing, China.

Fumagalli, A., Pasquale, L., Zonca, S., Micheletti, S., 2016. An upscaling procedure for fractured reservoirs with embedded grids. Water Resour. Res. 52 (8), 6506–6525.

Fumagalli, A., Zonca, S., Formaggia, L., 2017. Advances in computation of local problems for a flow-based upscaling in fractured reservoirs. Math. Comput. Simul. 137, 299–324.

Karimi-Fard, M., Firoozabadi, A., 2003. Numerical simulation of water injection in fractured media using the discrete-fracture model and the Galerkin method. SPE Reservoir Eval. Eng. 6 (2), 117–126.

Lie, K.-A., 2016. An Introduction to Reservoir Simulation Using MATLAB: User Guide for the Matlab Reservoir Simulation Toolbox (MRST), Dec 2016. SINTEF ICT, Oslo, Norway.

Lie, K.-A., Krogstad, S., Ligaarden, I.S., Natvig, J.R., Nilsen, H.M., Skaflestad, B., 2012. Open-source MATLAB implementation of consistent discretisations on complex grids. Comput. Geosci. 16 (2), 297–322.

Nilsen, H.M., Syversveen, A.R., Lie, K.-A., Tveranger, J., Nordbotten, J.M., 2012. Impact of top-surface morphology on CO_2 storage capacity. Int. J. Greenhouse Gas Control 11, 221–235.

Panfili, P., Cominelli, A., 2014. Simulation of miscible gas injection in a fractured carbonate reservoir using an embedded discrete fracture model. Paper SPE 171830, Abu Dhabi International Petroleum Exhibition and Conference. Abu Dhabi.

Panfili, P., Colin, R., Cominelli, A., Giamminonni, D., Guerra, L., 2015. Efficient and effective field scale simulation of hydraulic fractured wells: methodology and application. Paper SPE 175542, SPE Reservoir Characterisation and Simulation Conference and Exhibition. Abu Dhabi, UAE.

Ponting, D.K., 1989. Corner point geometry in reservoir simulation. 1st European Conference on the Mathematics of Oil Recovery. Cambridge, UK.

Sammon, P.H., 2000. Calculation of convective and dispersive flows for complex corner point grids. Paper SPE 62929, SPE Annual Technical Conference and Exhibition. Dallas, Texas.

Schlumberger, 2016,. Eclipse Technical Description. Schlumberger.

Seldle, J.P., Arri, L.E., 1990. Use of conventional reservoir models for coalbed methane simulation. Paper PETSOC 90-118, CIM/SPE International Technical Meeting. Calgary, Alberta.

Shah, S., Møyner, O., Tene, M., Lie, K.-A., Hajibeygi, H., 2016. The multiscale restriction smoothed basis method for fractured porous media (F-MsRSB). J. Comput. Phys. 318, 36–57.

Shimrat, M., 1962. Algorithm 112: position of point relative to polygon. Commun. ACM 5 (8), 434–440.

Verlo, S.B., Hetland, M., 2008. Development of a field case with real production and 4D data from the Norne Field as a benchmark case for future reservoir simulation model testing. Master Thesis. Norwegian University of Science and Technology.

Wadsley, W.A., 1980. Modelling reservoir geometry with non-rectangular coordinate grids. Paper SPE 9369, SPE Annual Technical Conference and Exhibition. Dallas, Texas.

Wong, D.L.Y., Doster, F., Geiger, S., Kamp, A., 2018. Systematic hybrid modelling using fracture subset upscaling. 80th EAGE Conference and Exhibition. Copenhagen, Denmark.

Xu, Y., Sepehrnoori, K., 2019. Development of an embedded discrete fracture model for field-scale reservoir simulation with complex corner-point grids. SPE J. 24 (4), 1552–1575.

CHAPTER EIGHT

Advanced EDFM approach using unstructured grids

8.1 Introduction

Unstructured grids have been used in reservoir simulation to represent geologically complex reservoirs (Fung et al., 1992; Durlofsky and Chien, 1993; Verma and Aziz, 1997; Prévost et al., 2002; Edwards, 2002; Karimi-Fard et al., 2004; Prévost et al., 2005; Hoteit and Firoozabadi, 2006; Vitel and Souche, 2007; Marcondes and Sepehrnoori, 2007; Marcondes and Sepehrnoori, 2010; Araújo et al., 2016; Fernandes et al., 2016, 2017; Samier and Masson, 2017). In comparison with structured grids, unstructured grids are more capable of representing complex reservoir structures and irregular reservoir boundaries, as they provide more flexibility regarding the geometry of gridblocks and their discretization.

There are several types of numerical approximation schemes used for unstructured grids, including the finite-volume method (Fung et al., 1992; Edwards, 2002; Marcondes et al., 2013; Fernandes et al., 2017), finite-element method, and mixed finite-element method (Durlofsky and Chien, 1993; Hoteit and Firoozabadi, 2006). The control-volume finite-element method (CVFEM) (Baliga and Patankar, 1980; Forsyth, 1990; Fung et al., 1992) is a widely used numerical approximation scheme for unstructured grids. It was initially proposed for solving Navier-Stokes flow in computational fluid dynamics (Baliga and Patankar, 1980). This method borrows the concepts of elements and shape functions from the finite-element method. The computational grids used in this method are defined as a series of elements. Physical properties such as pressure, phase saturations, and number of moles of each component are evaluated at the vertices of the elements, and porosity and permeability are stored in the elements. An advantage of this method is that it can be easily implemented in simulators with the capability to construct arbitrary connections between cells (Forsyth, 1990). As CVFEM still uses a finite-volume formulation, it is also called the element-based finite-volume method (EbFVM) (Maliska, 2004; Cordazzo et al., 2005; Marcondes and Sepehrnoori, 2007), which better reflects its nature. Therefore, in the remainder of this chapter, this method will be referred to as EbFVM.

Embedded Discrete Fracture Modeling and Application in Reservoir Simulation.
http://dx.doi.org/10.1016/B978-0-12-821872-3.00008-3

Similar to corner-point grids, unstructured grids have the capability to explicitly represent certain geological features, such as fractures, by creating conforming meshes honoring the geological features. However, as the number and complexity of fractures increase, it can be time-consuming and sometimes extremely challenging to generate the desired mesh. Other issues involve the large number of gridblocks due to the discretization around fractures and the large linear system of equations, as have been discussed in previous chapters. These issues can be solved to some extent by using nonconforming meshes to simulate fractures, which can be achieved by combining the usage of unstructured grids with the embedded discrete fracture model (EDFM).

The use of the EDFM in unstructured grids has not been studied extensively. Fang et al. (2016) combined the EDFM in a two-dimensional (2D) triangular mesh with the dual-continuum model to simulate the transient flow of horizontal wells with complex fracture networks. Norbeck et al. (2016) developed a numerical framework to calculate the coupled processes of fluid flow and geomechanics, where the fluid flow problem was solved with the EDFM in 2D triangular meshes. The use of the EDFM in 2D unstructured grids with mixed elements or in three-dimensional (3D) unstructured grids has not been studied by previous researchers. Furthermore, to the best of our knowledge, there has not been research on the application of the EDFM in unstructured grids using the EbFVM.

Here, the EDFM is applied to 2D and 3D unstructured grids (with mixed elements) using the EbFVM to combine the flexibility provided by unstructured gridding with the EDFM and demonstrate the extension of the EDFM to unstructured grids. First, the numerical formulations of the EbFVM are presented. Subsequently, methodology and formulations are developed to apply the EDFM in unstructured grids using the EbFVM, with a focus on the discretization of fractures and the transmissibility calculation between matrix blocks and fractures. Thereafter, case studies are presented to demonstrate the effectiveness of the methodology.

8.2 Unstructured grids using the EbFVM

The EbFVM is a general approach that can be applied to both 2D and 3D unstructured grids. In this section, the basic ideas of the EbFVM are introduced briefly. The types of elements, the subdivision of elements, and the evaluation of advective terms between matrix blocks are discussed.

8.2.1 Two-dimensional grids

For 2D simulations using the EbFVM, linear triangular and bi-linear quadrilateral elements can be used to construct the grids. The porosity and absolute permeability tensor are defined for each element, and other physical properties are evaluated at the vertices. Each element in the grid is divided into several sub-elements, and the conservation equations are integrated for each sub-element. For this reason, sub-elements are usually termed sub-control-volumes (SCVs). The material balances are built around the vertices of each element, giving rise to a cell-vertex approach. Herein, it is important to stress that heterogeneous and anisotropic reservoirs can easily be handled with the EbFVM, as shown by Marcondes and Sepehrnoori (2010), Marcondes et al. (2013), and Santos et al. (2013), because all fluxes are evaluated on the interfaces of each SCV. In Fig. 8.1A, a 2D grid is shown. This grid contains thirteen vertices and nine elements. In Fig. 8.1B, each element in the grid is divided into several parts by connecting the centroid of the element to the middle points of the element edges. Therefore, each triangular element is divided into three parts, and each quadrilateral element is divided into four parts. Each part of the element is called an SCV. In the EbFVM, the control volume (CV) around each vertex in the grid is created through the contribution of all SCVs that share that vertex. In Fig. 8.1B, the CV around Vertex 4 (shown in red) is made up of SCVs from Elements 1, 4, 7, and 8. Using this approach, the total number of CVs is always the same as the number of vertices in the grid. Therefore, in Fig. 8.1B, a total number of thirteen CVs are created. In the simulator, physical properties such as pressure, phase saturation, and number of moles of each component are calculated for each CV. More details can be found in Marcondes and Sepehrnoori (2010) and Santos et al. (2013).

8.2.2 Three-dimensional grids

The basic concepts used in 3D grids are similar to those in 2D grids. However, 3D grids are typically much more complicated than 2D grids. Four types of elements can be used in 3D grids: tetrahedron, prism, hexahedron, and pyramid. Each element is discretized into several SCVs following a similar process as in 2D grids. Figs. 8.2 through 8.5 show the splitting of the tetrahedron, prism, hexahedron, and pyramid elements into SCVs, respectively. The splitting points are the middle points of element edges and the centroids of element faces. More details can be found in Marcondes et al. (2013).

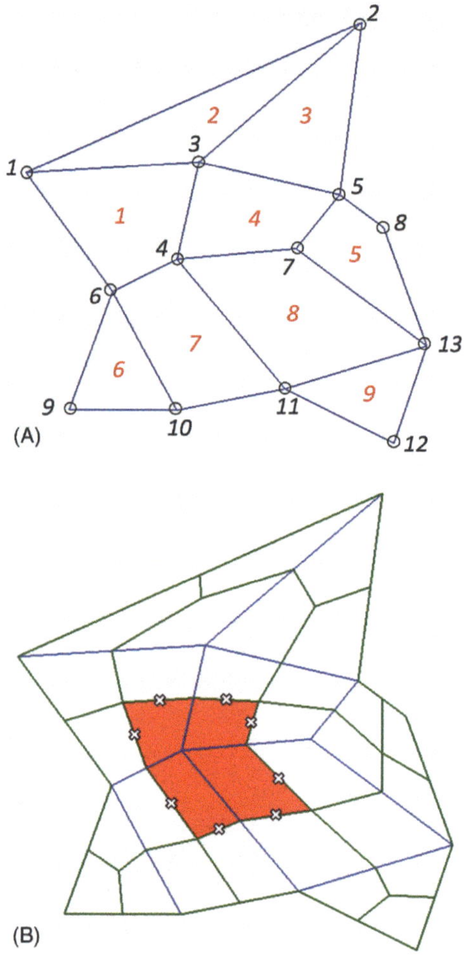

Figure 8.1 ***Illustration of the elements and the creation of CVs in a 2D unstructured grid (Xu et al., 2020).*** (A) Elements. (B) Elements and CVs. The blue lines represent the boundaries of the elements, and the green lines represent the boundaries of the CVs. In (A), the thirteen vertices and nine elements are labeled in black and red, respectively. In (B), a CV containing four SCVs (around Vertex 4) is shown in red. The integration points for the CV are shown as well.

After the discretization of the elements, the SCVs that share the same vertex will form a CV. Fig. 8.6A shows a 3D grid with three types of elements. These elements are split into SCVs, and the discretization is shown in Fig. 8.6B. In Fig 8.6B, the SCVs that belong to one CV are shown in blue. It can be seen that each CV could have a very irregular geometry. Fig. 8.6C shows all CVs formed by SCVs.

Figure 8.2 *Splitting of a tetrahedron element (Xu et al., 2020). The SCVs are shown in different colors.*

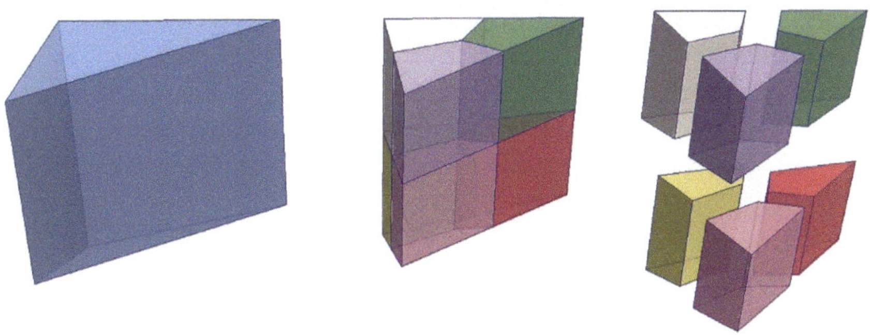

Figure 8.3 *Splitting of a prism element (Xu et al., 2020). The SCVs are shown in different colors.*

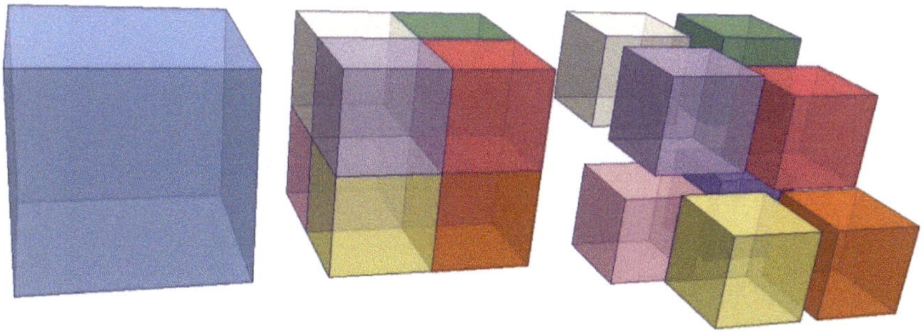

Figure 8.4 *Splitting of a hexahedron element (Xu et al., 2020). The SCVs are shown in different colors.*

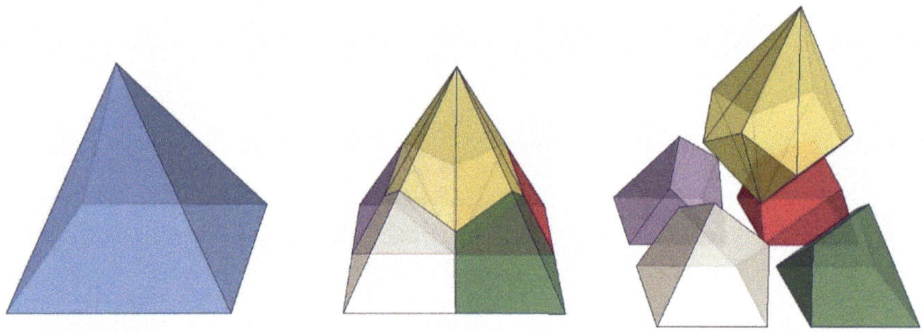

Figure 8.5 ***Splitting of a pyramid element (Xu et al., 2020). The SCVs are shown in different colors.***

8.2.3 Evaluation of flux

The reason elements are subdivided into SCVs in the EbFVM is to make it convenient to evaluate the flux between CVs. In this section, the approach to evaluating the flux in the EbFVM is discussed briefly.

As mentioned before, in the EbFVM, physical properties such as fluid pressure are evaluated at the vertices (or CVs). The coordinates and physical properties at any point inside an element can be approximated using the coordinates and properties at the vertices of the element. For 2D elements,

$$
\begin{aligned}
x(\xi,\eta) &= \sum_{i=1}^{N_v} N_i x_i; \\
y(\xi,\eta) &= \sum_{i=1}^{N_v} N_i y_i; \\
\Phi(\xi,\eta) &= \sum_{i=1}^{N_v} N_i \Phi_i.
\end{aligned}
\tag{8.1}
$$

For 3D elements,

$$
\begin{aligned}
x(\xi,\eta,\gamma) &= \sum_{i=1}^{N_v} N_i x_i; \\
y(\xi,\eta,\gamma) &= \sum_{i=1}^{N_v} N_i y_i; \\
z(\xi,\eta,\gamma) &= \sum_{i=1}^{N_v} N_i z_i; \\
\Phi(\xi,\eta,\gamma) &= \sum_{i=1}^{N_v} N_i \Phi_i.
\end{aligned}
\tag{8.2}
$$

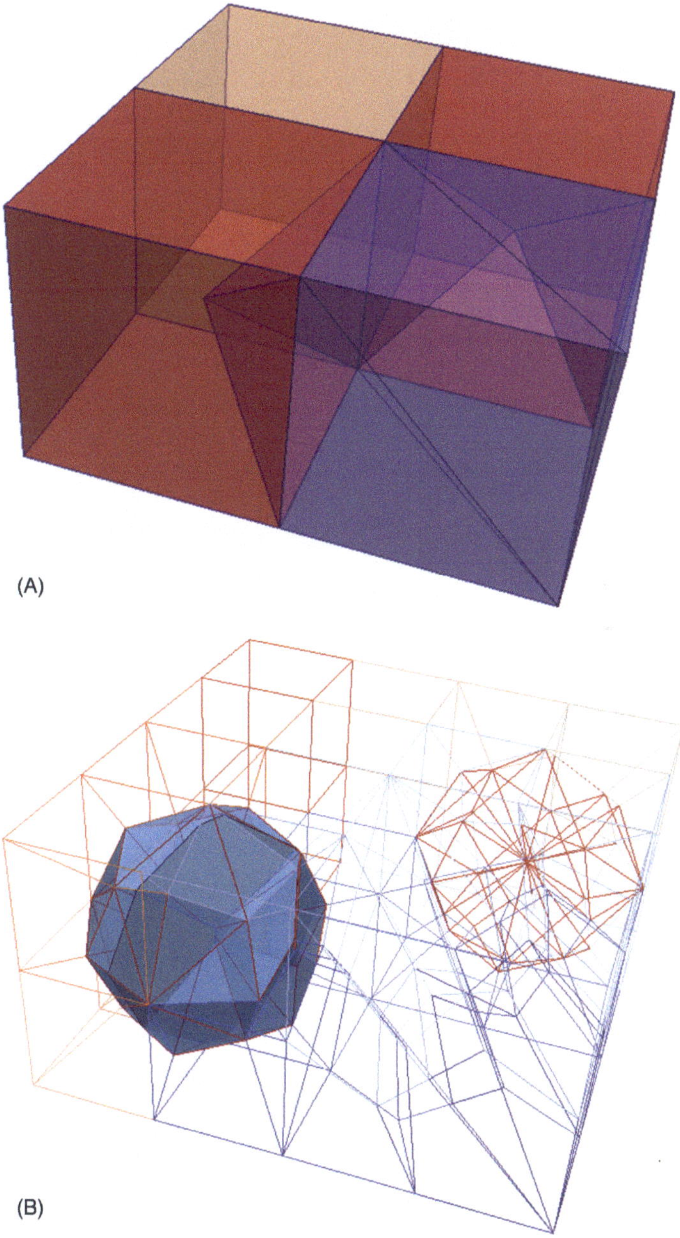

Figure 8.6 ***Illustration of the construction of CVs in 3D grids (Xu et al., 2020).*** (A) A grid made up of ten tetrahedron elements (purple), one hexahedron element (orange), and ten pyramid elements (red). (B) Discretization of elements. A CV in the grid is shown in blue. (C) All CVs shown in different colors.

(C)

Figure 8.6 (*Cont.*)

In Eqs. (8.1) and (8.2), x, y, and z are the Cartesian coordinates of a point in the element, ξ, η, and γ are the local coordinates of the point in the computational plane, Φ is the physical property value at the point, N_v is the number of vertices of the element, N_i is the i-th shape function, x_i, y_i, and z_i are the Cartesian coordinates of Vertex i, and Φ_i is the physical property value at Vertex i. The shape functions for 2D and 3D elements in the computational plane are presented in Appendix B.

Using Eqs. (8.1) and (8.2), the gradient of physical property Φ can be evaluated as

$$\begin{aligned} \frac{\partial \Phi}{\partial x} &= \sum_{i=1}^{N_v} \frac{\partial N_i}{\partial x} \Phi_i; \\ \frac{\partial \Phi}{\partial y} &= \sum_{i=1}^{N_v} \frac{\partial N_i}{\partial y} \Phi_i; \\ \frac{\partial \Phi}{\partial z} &= \sum_{i=1}^{N_v} \frac{\partial N_i}{\partial z} \Phi_i. \text{(for 3D elements)}. \end{aligned} \tag{8.3}$$

For 2D grids, $\frac{\partial N_i}{\partial x}$ and $\frac{\partial N_i}{\partial y}$ can be obtained by solving the following linear system:

$$\frac{\partial N_i}{\partial \xi} = \frac{\partial N_i}{\partial x}\frac{\partial x}{\partial \xi} + \frac{\partial N_i}{\partial y}\frac{\partial y}{\partial \xi};$$
$$\frac{\partial N_i}{\partial \eta} = \frac{\partial N_i}{\partial x}\frac{\partial x}{\partial \eta} + \frac{\partial N_i}{\partial y}\frac{\partial y}{\partial \eta}. \tag{8.4}$$

For 3D grids, the following system should be solved to obtain $\frac{\partial N_i}{\partial x}$, $\frac{\partial N_i}{\partial y}$, and $\frac{\partial N_i}{\partial x}$:

$$\frac{\partial N_i}{\partial \xi} = \frac{\partial N_i}{\partial x}\frac{\partial x}{\partial \xi} + \frac{\partial N_i}{\partial y}\frac{\partial y}{\partial \xi} + \frac{\partial N_i}{\partial z}\frac{\partial z}{\partial \xi};$$
$$\frac{\partial N_i}{\partial \eta} = \frac{\partial N_i}{\partial x}\frac{\partial x}{\partial \eta} + \frac{\partial N_i}{\partial y}\frac{\partial y}{\partial \eta} + \frac{\partial N_i}{\partial z}\frac{\partial z}{\partial \eta}; \tag{8.5}$$
$$\frac{\partial N_i}{\partial \gamma} = \frac{\partial N_i}{\partial x}\frac{\partial x}{\partial \gamma} + \frac{\partial N_i}{\partial y}\frac{\partial y}{\partial \gamma} + \frac{\partial N_i}{\partial z}\frac{\partial z}{\partial \gamma}.$$

With the gradient of physical properties, e.g., flow potential gradient, the total molar flow rate of component k across the boundaries of an SCV through advection can be evaluated through an integration:

$$F_k = \sum_{l=1}^{n_{ip}} \sum_{j=1}^{n_p} x_{kj}\xi_j \frac{k_{rj}}{\mu_j} \overline{\overline{K}} \cdot \vec{\nabla}\Phi_{jl} \overrightarrow{A_l}, \tag{8.6}$$

where n_{ip} is the number of integration points, n_p is the number of phases, x_{kj} is the mole fraction of component k in phase j, ξ_j is the molar density of phase j, k_{rj} is the relative permeability of phase j, μ_j is the viscosity of phase j, $\overline{\overline{K}}$ is the permeability tensor, $\vec{\nabla}\Phi_{jl}$ is the flow potential gradient in phase j at the lth integration point evaluated by Eq. (8.3), and $\overrightarrow{A_l}$ is the area of the interface. Here, each integration point is the center of the interface between two SCVs. It should be noted that the integration is performed on every interface between two SCVs within the same element. The integration points in 2D elements are shown in Appendix B, Fig. B.1. For 3D elements, the interfaces can also be easily found in Figs. 8.2 through 8.5 (between each pair of SCVs). Because the integration is always calculated within the element, the same permeability value can be used on both sides of the interface, and no interpolation is required for the evaluation of flux. The fluid properties in Eq. (8.6) are evaluated by an upstream scheme depending on the sign of $\overline{\overline{K}} \cdot \vec{\nabla}\Phi_{jl} \overrightarrow{A_l}$.

Ignoring the physical dispersion term, the material balance equation used in the simulator is

$$\frac{\partial N_k}{\partial t} = F_k + q_k, \quad k = 1,\ldots,n_c + 1 \tag{8.7}$$

where N_k is the number of moles of component k, q_k is the injection/production molar rate of component k from wells, and n_c is the number of hydrocarbon components. Component $n_c + 1$ denotes the water component. In the EbFVM, Eq. (8.7) is integrated for every SCV of every element. After that, an assembly process is performed using all SCVs that share the same vertex (within the same CV). Overall, the calculations are performed within each element, and the assembly process is performed to obtain the material balance equation for each CV.

8.3 EDFM in unstructured grids using the EbFVM

The concept used to apply the EDFM to unstructured grids is similar to that used in Cartesian and corner-point grids. Additional CVs are created in the computational domain to represent the fracture segments, and non-neighboring connections (NNCs) are constructed to represent different types of flow associated with fractures and the matrix gridblocks crossed by fractures. The main challenges are in the discretization of fractures and the calculation of transmissibility factors for matrix–fracture connections. In the EbFVM, the matrix permeability is defined on elements. However, the physical properties to evaluate in the simulation (e.g., pressure, saturation) are defined on vertices (or CVs). On the basis of these unique features, a methodology is introduced here to conduct the geometrical calculations of matrix–fracture intersections on SCVs.

Fig. 8.7 illustrates the procedure to obtain fracture segments and calculate matrix–fracture connections. Initially, the fractures are placed inside the grid which is made up of elements (Fig. 8.7A). The elements are then divided into SCVs for geometrical calculation (Fig. 8.7B). The intersections between the fracture polygons and SCVs are calculated. During the calculation, each SCV is treated as a general polyhedron. The geometrical algorithm as described in Chapter 7 for corner-point grids is applied here. After the calculation, the fractures are discretized into a series of fracture segments, and each fracture segment is contained in an SCV, as shown in Fig. 8.7C. The transmissibility factor between an SCV and a fracture segment inside it can be evaluated as

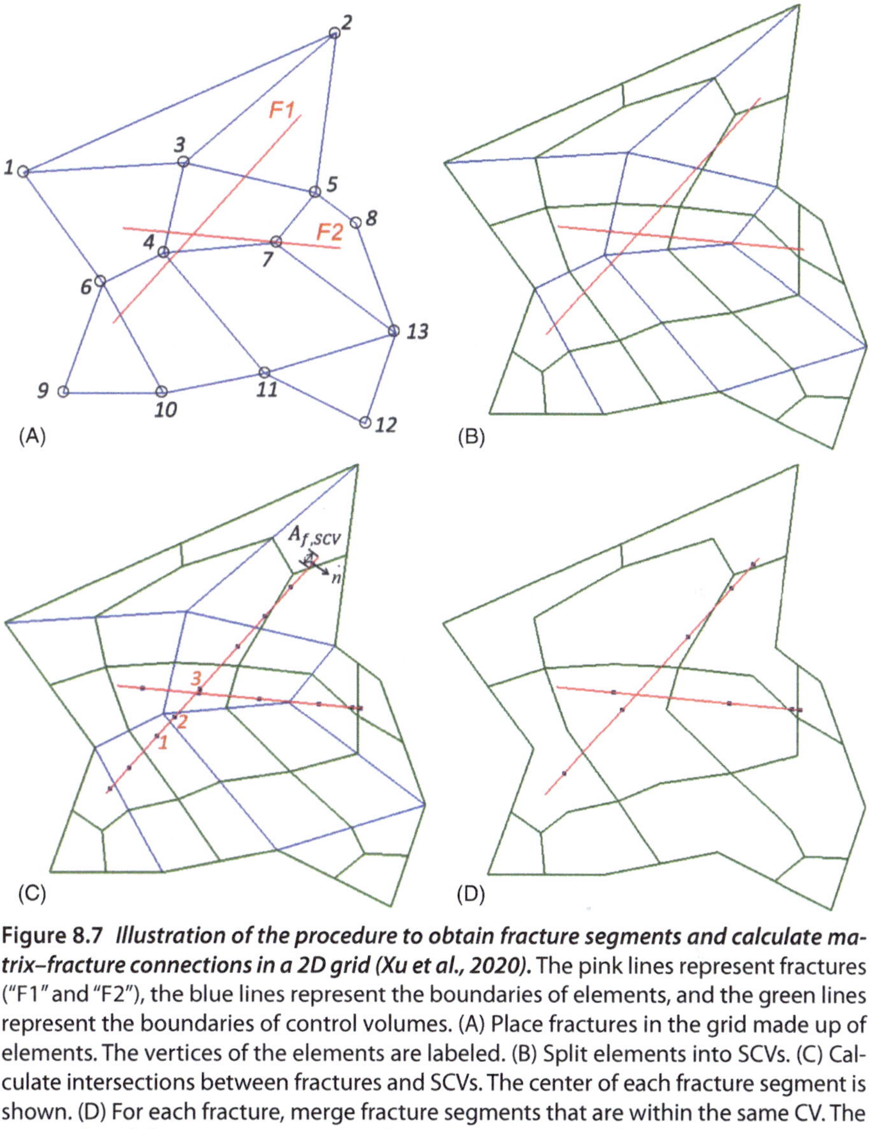

Figure 8.7 ***Illustration of the procedure to obtain fracture segments and calculate matrix–fracture connections in a 2D grid (Xu et al., 2020).*** The pink lines represent fractures ("F1" and "F2"), the blue lines represent the boundaries of elements, and the green lines represent the boundaries of control volumes. (A) Place fractures in the grid made up of elements. The vertices of the elements are labeled. (B) Split elements into SCVs. (C) Calculate intersections between fractures and SCVs. The center of each fracture segment is shown. (D) For each fracture, merge fracture segments that are within the same CV. The center of each fracture segment after the merging process is shown.

$$T_{f\text{-}SCV} = \frac{2A_{f,SCV}\vec{n}\cdot(\overrightarrow{\overrightarrow{K}}_{SCV}\cdot\vec{n})}{d_{f\text{-}CV}}, \tag{8.8}$$

where $A_{f,SCV}$ is the area of the fracture segment in the SCV, $\vec{n}$ is the unit normal vector of the fracture plane, $\overrightarrow{\overrightarrow{K}}_{SCV}$ is the permeability tensor of the

SCV, which is the same as the permeability tensor of the corresponding element, and $d_{f\text{-}CV}$ is the average normal distance from the fracture segment to the CV that the SCV belongs to. For illustration purposes, in Fig. 8.7C, $A_{f,SCV}$ and $\vec{n}$ for a fracture segment is shown. The derivation of Eq. (8.8) assumes a "linear pressure distribution" in the CV, which is suitable for cases where all SCVs in the CV have the same porosity and permeability. For reservoirs with heterogeneous porosity and permeability fields, Eq. (8.8) is a rough approximation, and more studies are needed on the calculation of matrix–fracture transmissibility factors. In the last step, the fracture segments belonging to the same fracture and contained in the same CV are merged if they share a common edge, and the corresponding fracture-SCV transmissibility factors are added together. As an example, in Fig. 8.7C, the three fracture segments ("1," "2," and "3") belonging to Fracture "F1" in the CV around Vertex 4 are merged into one fracture segment, as shown in Fig. 8.7D. The purpose of merging the fracture segments is to reduce the number of fracture segments (equaling the number of CVs) for each fracture. After the merging process, the transmissibility factor between a CV and the fracture segment inside it is calculated as

$$T_{f-CV} = \sum_{i=1}^{n_{merge}} T_{f-SCV,i}, \tag{8.9}$$

where n_{merge} is the number of initial fracture segments (in Fig. 8.7C) that the fracture segment is merged from.

For 2D grids, it should be noted that because some CVs have a concave geometry, not all fracture segments in a CV can be merged into one. It is possible for a fracture to have multiple fracture segments in a single CV. Fig. 8.8 presents an example of this situation.

For 3D grids, similarly, it is not always the case that all fracture segments in a CV that belong to the same fracture can be merged into one fracture segment. In addition, the merging of fracture segments is more complicated compared with that in 2D cases. Fig. 8.9 presents an example where a fracture is placed in the grid shown in Fig. 8.6. The fracture segments contained in the same CV are shown in the same color, and these fracture segments are to be merged. Fig. 8.10 presents a larger example of the merging of fracture segments. In this example, the number of fracture segments decreases from 12,348 to 1,823 after the merging process. It should be noted that because on average each CV contains more SCVs in 3D grids compared with that in 2D grids, the fracture segment merging step is more beneficial in reducing the number of fracture CVs in 3D grids.

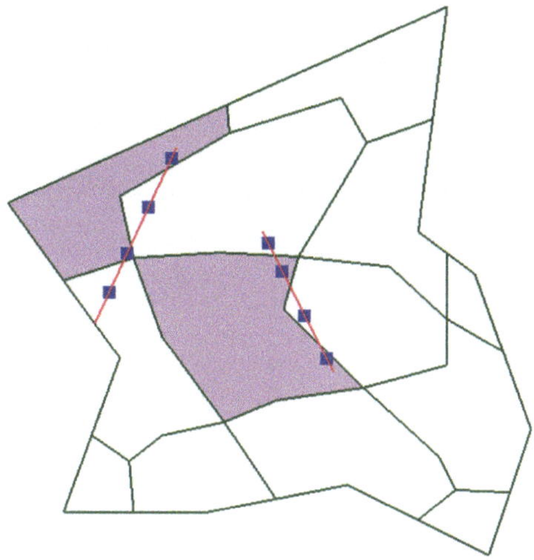

Figure 8.8 ***A 2D example where the fracture segments in a CV are merged into multiple fracture segments (Xu et al., 2020).*** The green lines are the boundaries of CVs. The blue dots represent the centers of fracture segments. The two CVs shown in purple contain multiple fracture segments from a single fracture.

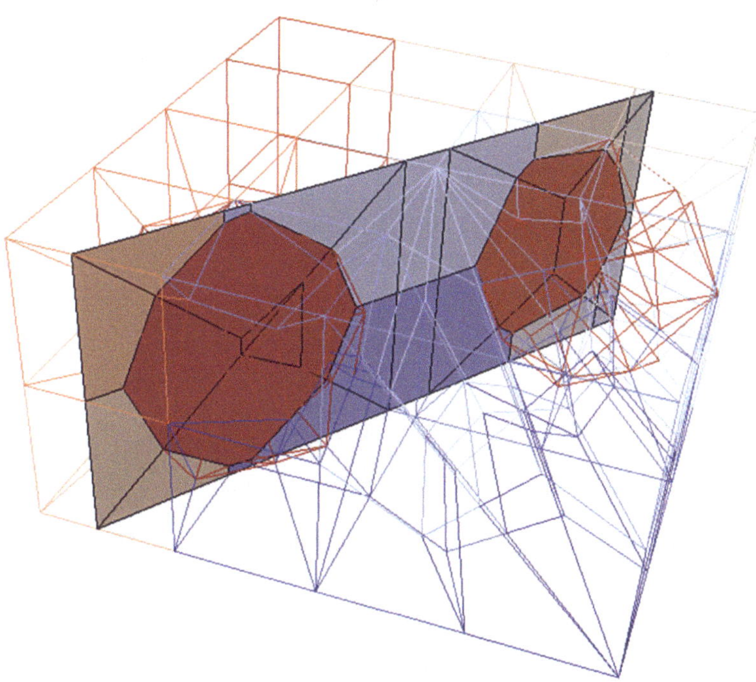

Figure 8.9 ***Illustration of the calculation of matrix-fracture connections in a 3D grid (Xu et al., 2020).*** The fracture segments (before merging) that are contained in the same CV are shown in the same color.

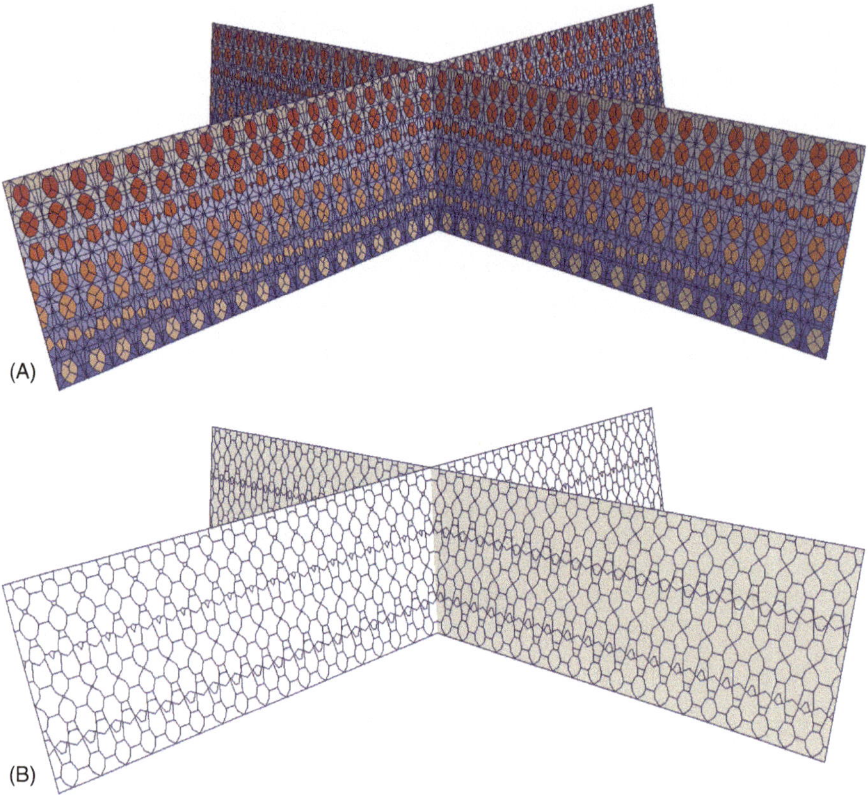

Figure 8.10 ***A 3D example where two fractures are placed in a grid made up of pyramid elements (Xu et al., 2020).*** (A) The fracture segments before the merging process. The fracture segments that are contained in the same CV are shown in the same color. (B) The fracture segments after the merging process.

The transmissibility factor calculations for other types of NNCs and well-fracture intersections in the EDFM are very similar to those in Cartesian and corner-point grids. The same formulations as discussed in Chapter 4 can be used.

8.4 Case studies for 2D unstructured grids using the EbFVM

The methodology discussed in the previous section was implemented in an in-house reservoir simulator UTCOMP-RS and an EDFM preprocessing code. In this and the following sections, case studies are presented using both 2D and 3D grids. The results of the case studies are discussed to demonstrate the convenience and effectiveness of using the methodology presented in this chapter.

8.4.1 Case 1: Homogeneous reservoir with different matrix gridding

In this case study, a waterflooding process in a homogeneous reservoir with a long natural fracture is simulated. The reservoir dimensions are 1,000 × 1,000 × 50 ft. Four different simulation grids with a similar number of vertices (or CVs) are used for result comparison, as shown in Fig. 8.11. The first grid has only square elements, making it like a regular

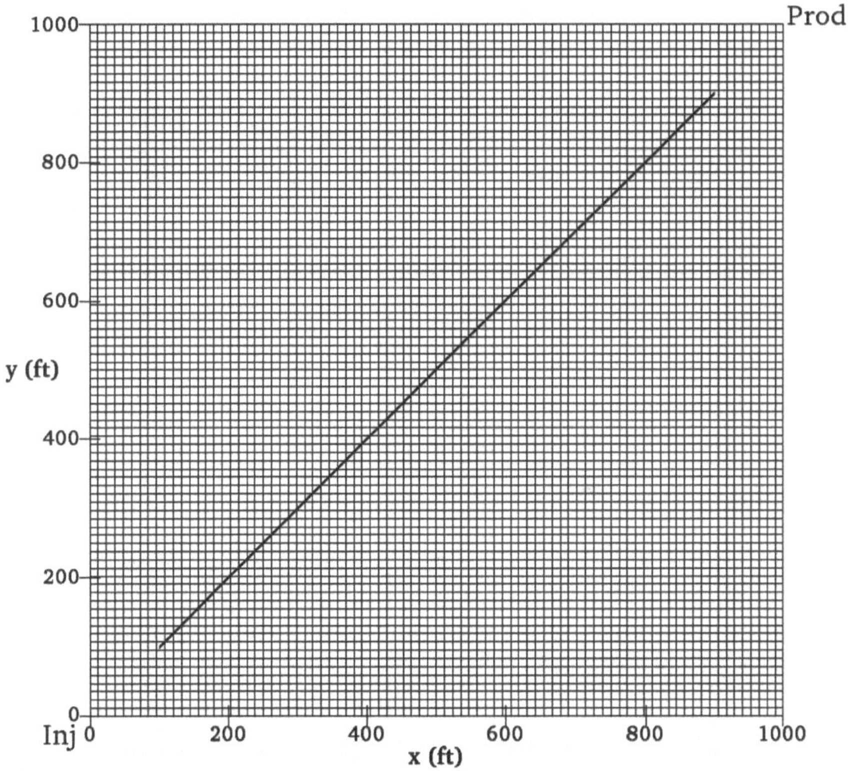

Figure 8.11 ***Four different simulation grids and location of fracture in Case 1 (Xu et al., 2020).*** The water injector ("Inj") is located at (0, 0), and the producer ("Prod") is located at (1000, 1000). The black line represents the fracture. For the matrix grids, the elements are shown. (A) 7,056 square elements (7,225 vertices). (B) 13,650 triangular elements (6,970 vertices). (C) 6,868 quadrilateral elements (7,021 vertices). (D) 9,904 mixed triangular and quadrilateral elements (7,053 vertices).

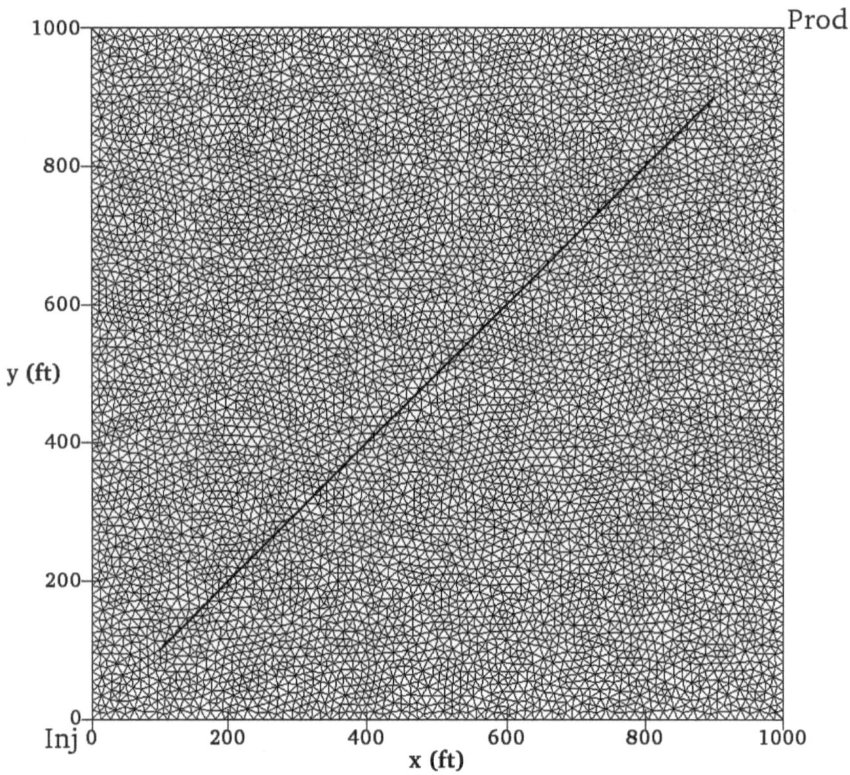

(B)

Figure 8.11 **(*Cont.*)**

Cartesian grid; the second grid is made up of triangular elements; the third grid has only quadrilateral elements; the last grid is a mixture of triangular and quadrilateral elements. A long fracture is placed inside the reservoir, as shown in Fig. 8.11. The fracture width is 0.01 ft, and the fracture permeability is 3×10^6 md. A water injector with a constant injection rate and an oil producer with a constant bottomhole pressure are placed at two corners of the reservoir, as shown in Fig. 8.11. Table 8.1 summarizes the basic parameters for the simulation. In this case study, the oil phase is composed of only one hydrocarbon component (C_{20}), and a two-phase (water and oil) flow is simulated. Corey-type relative permeability curves are used, and the parameters for the relative permeability curves are also provided in Table 8.1. The capillary effect is not considered in this study.

Fig. 8.12 presents the simulated oil flow rate on the four different grids without considering the fracture. It can be seen that similar results were

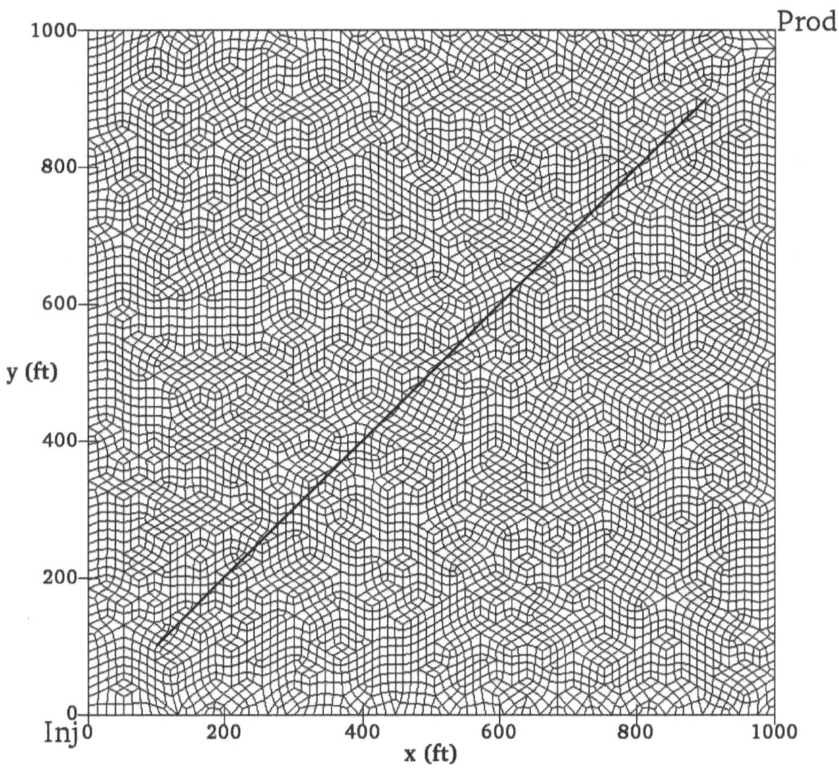

(C)

Figure 8.11 (*Cont.*)

obtained using all four grids, indicating that the EbFVM formulation provides reliable results on all the grids shown in Fig. 8.11.

Fig. 8.13 presents the predicted oil flow rate for simulations that consider the impact of the fracture. Fig. 8.14 shows the water saturation profiles at 500 days, where the impact of the long fracture can be clearly observed. Similar results were obtained for both the oil rate and the water saturation from all four grids, which verifies the effectiveness of the methodology to apply the EDFM to 2D unstructured grids.

8.4.2 Case 2: Homogeneous reservoir with 41 natural fractures

A primary advantage of the EDFM is that fractures with arbitrary orientations can be conveniently simulated. Here, the four simulation grids in Case 1 are used, but more fractures with different orientations are placed in the reservoir, as shown in Fig. 8.15. As can be seen, the fractures form a complex network. The reservoir and fracture properties are the same as in Case 1.

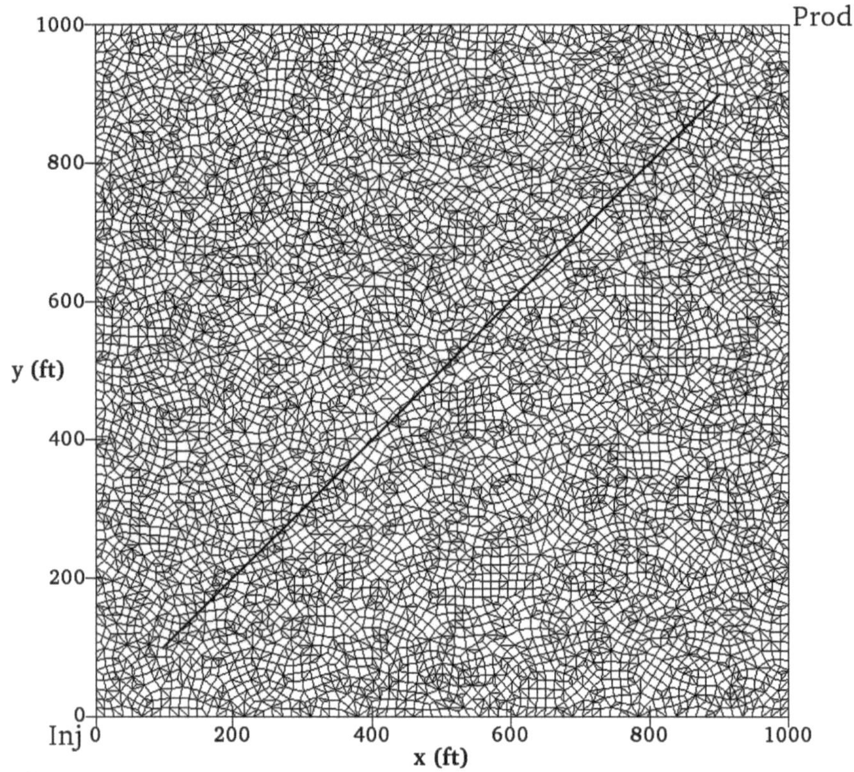

(D)

Figure 8.11 (*Cont.*)

Table 8.1 Key simulation parameters in Case 1 (Xu et al., 2020).

Parameter	Value	Unit	Parameter	Value	Unit
Reservoir permeability	200	md	Residual water saturation	0.24	–
Reservoir porosity	0.25	–	Residual oil saturation	0.2	–
Initial reservoir pressure	4,500	psi	Water rel. perm. endpoint	0.4	–
Reservoir thickness	50	ft	Oil rel. perm. endpoint	0.6	–
Rock compressibility	3×10^{-6}	psi^{-1}	Water rel. perm. exponent	2.0	–
Reservoir temperature	200	°F	Oil rel. perm. exponent	3.0	–
Water viscosity	0.3	cp	Water injection rate	700	STB/day
Initial water saturation	0.3	–	Producer bottomhole pressure	4,000	psi
Wellbore radius	0.5	ft			

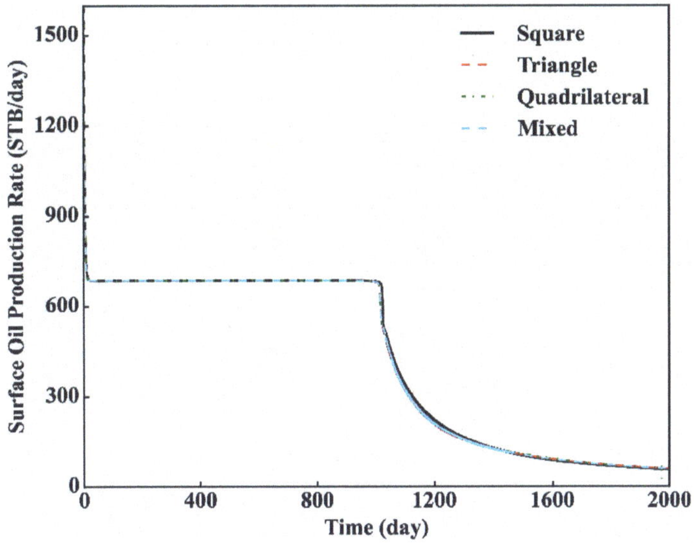

Figure 8.12 ***Comparison of simulation results without fracture (Xu et al., 2020).*** "Square", "Triangle", "Quadrilateral", and "Mixed" represent the simulation grids shown in Fig. 8.11A through D, respectively.

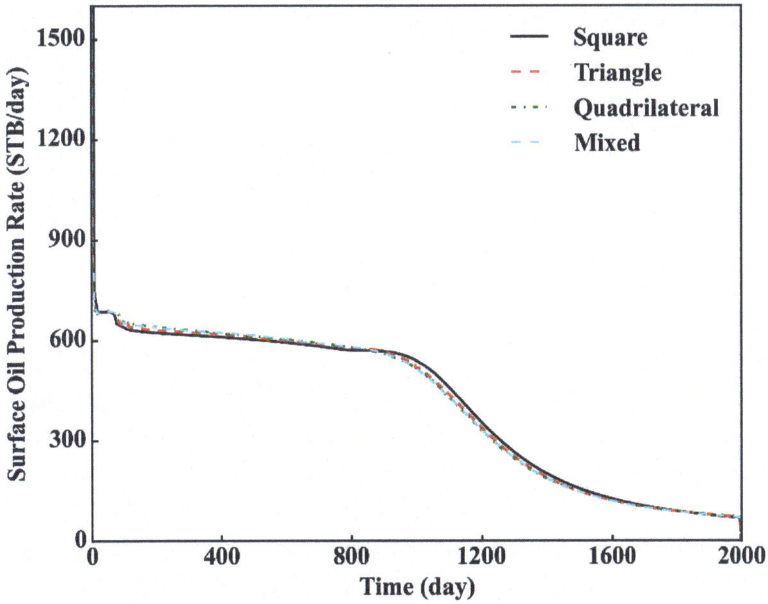

Figure 8.13 ***Comparison of surface oil production rate for simulations with the fracture (Xu et al., 2020).***

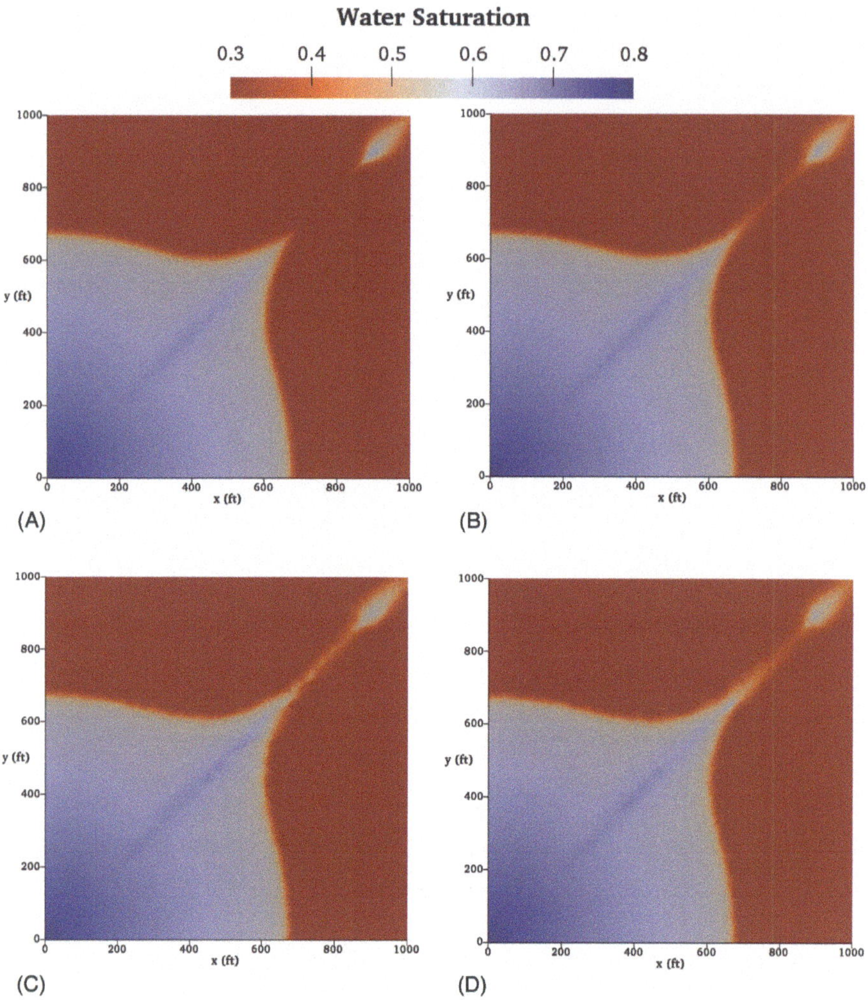

Figure 8.14 ***Water saturation profiles at 500 days (Xu et al., 2020).*** (A) Square elements. (B) Triangular elements. (C) Quadrilateral elements. (D) Mixed elements.

Fig. 8.16 compares the oil rate curves obtained from the simulations on all four grids. Fig. 8.17 presents the water saturation profiles in the matrix at 500 days. The results obtained with the four different grids are very similar, showing the effectiveness of the methodology for simulating complex fracture networks with arbitrary fracture orientations. The physical properties in the fracture network were also calculated during the simulation. Fig. 8.18 shows the predicted water saturation in the fracture network at 500 days using the matrix grid in Fig. 8.11D.

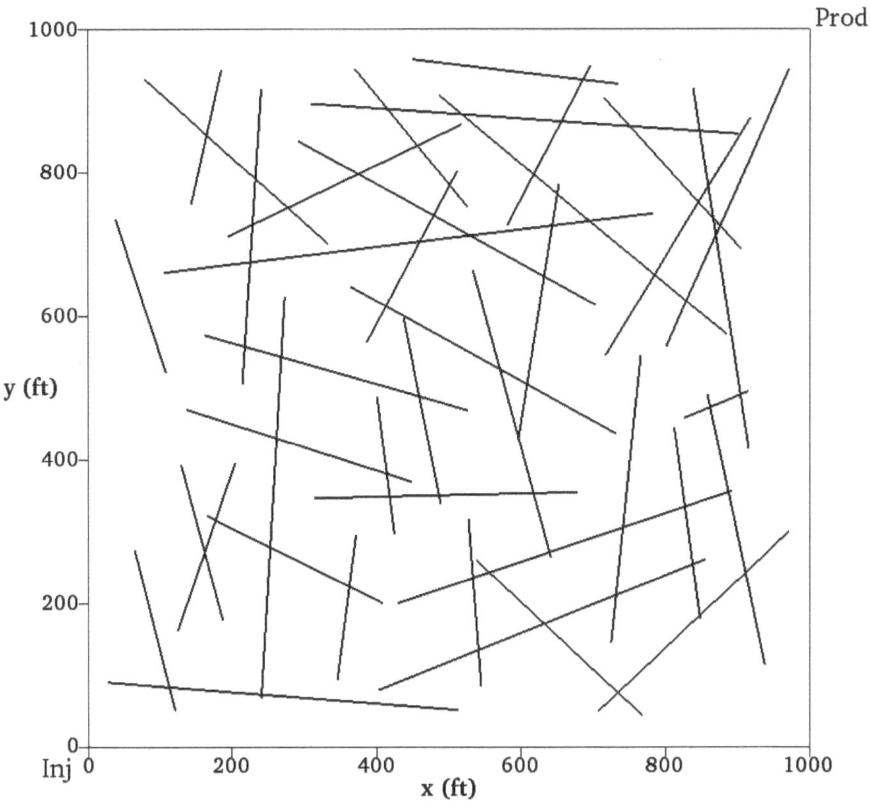

Figure 8.15 ***Locations of the 41 fractures in Case 2 (Xu et al., 2020).***

Comparing Cases 1 and 2, it can be seen that the same matrix grid could be used to simulate different fractures. This is a unique advantage of using non-conforming meshes. With the EDFM, minimum adjustment of the original matrix grid is required, which provides a highly convenient approach when performing history matching and sensitivity studies with different realizations of fracture geometries.

8.4.3 Case 3: Reservoir with irregular geometry

In this case study, a 2D reservoir model with irregular geometry is used, as shown in Fig. 8.19. This reservoir is represented by an unstructured grid with mixed elements. The total number of vertices is 11,857, and the total number of elements is 13,080. The reservoir thickness is 100 ft. A water injector ("Inj") and two producers ("Prod1" and "Prod2") are placed inside the reservoir. The injector has a constant injection rate, and the producers

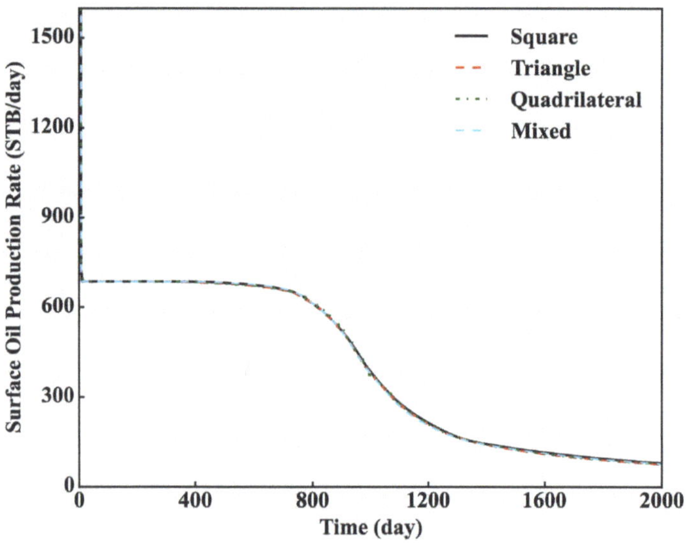

Figure 8.16 ***Predicted surface oil production rate using the four grids (Xu et al., 2020).***

have a constant well bottomhole pressure. Twelve long fractures are placed in the reservoir. The fracture width and fracture permeability are 0.01 ft and 1.0×10^6 md, respectively. Other key simulation parameters are summarized in Table 8.2. In this compositional simulation, three non-water components are considered. Their original composition and properties are summarized in Table 8.3. A Corey-type relative permeability model is used, and the parameters of the relative permeability curves are presented in Table 8.4. The capillary pressure effect is not considered here.

Fig. 8.20 presents the predicted water, oil, and gas production rate curves. The water breakthrough occurs at ~1,000 days, with a sharp increase in water production rate and a decrease in oil and gas production rates. Fig. 8.21 shows the water saturation profiles in the matrix at 500 and 1,000 days, where the movement of water through the fracture network can be clearly observed. The influence of the reservoir geometry can also be seen, as the injected water moves along the irregular reservoir boundary.

8.5 Case studies for 3D unstructured grids using the EbFVM

8.5.1 Case 4: Tight gas reservoir with inclined hydraulic fractures

This study focuses on primary production from a hydraulically fractured horizontal well in a tight gas reservoir. The reservoir dimensions are

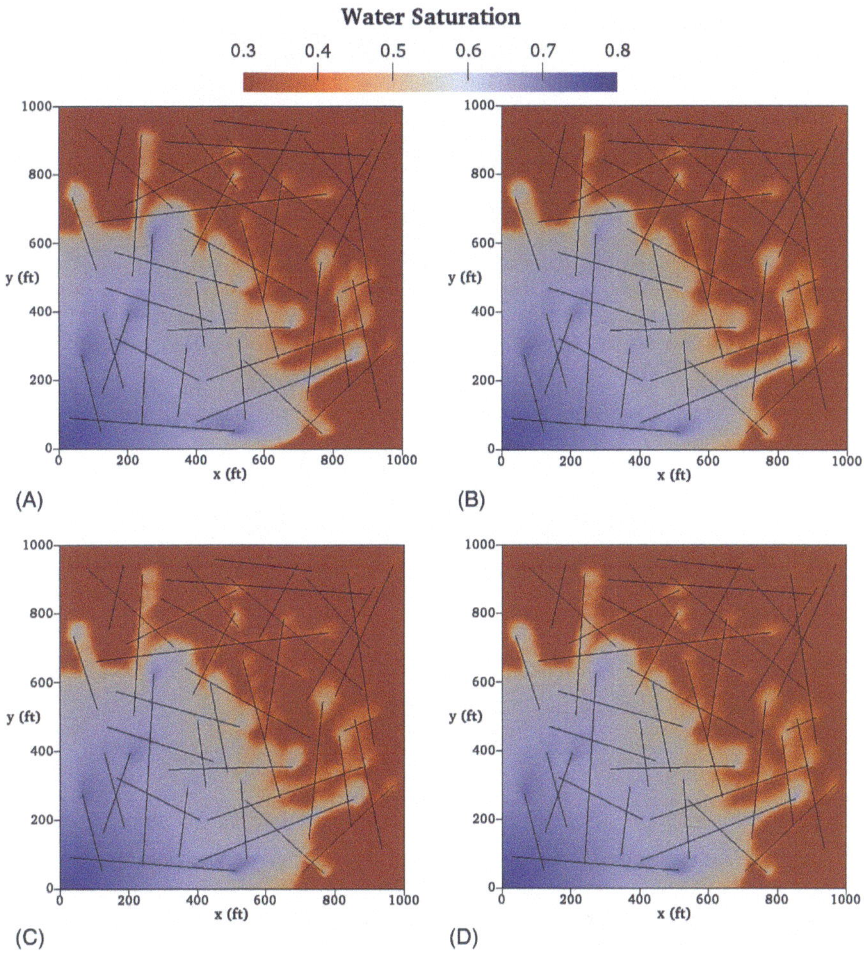

Figure 8.17 ***Water saturation profiles at 500 days (Xu et al., 2020).*** (A) Square elements. (B) Triangular elements. (C) Quadrilateral elements. (D) Mixed elements. The locations of fractures are also shown.

2,000 × 1,600 × 100 ft. A horizontal well and four inclined hydraulic fractures are placed at the center of the reservoir, as shown in Fig. 8.22. The dip angle of the fractures is 70 degrees, and the fracture width and fracture permeability are 0.01 ft and 10,000 md, respectively. All fractures fully penetrate the reservoir in the vertical direction. For verifying the implementation of the EDFM in grids with different types of elements, four matrix grids are created. The numbers of vertices and elements in the four grids are summarized in Table 8.5. As can be seen, all four grids have a similar number of vertices (or CVs), but they are made up of different types of elements.

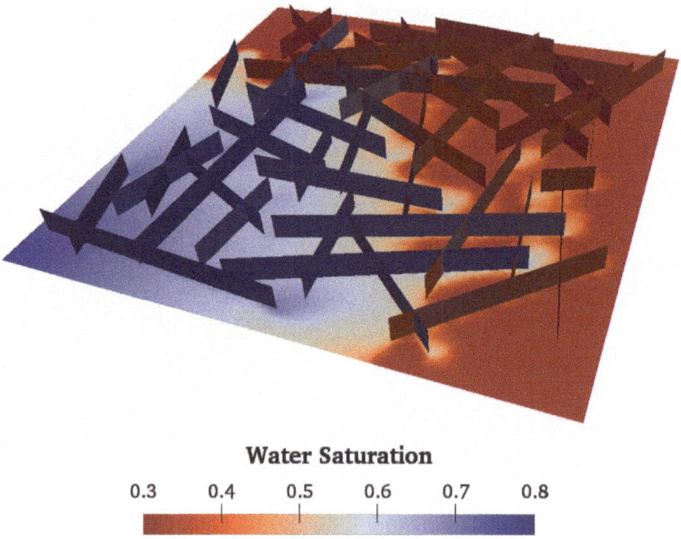

Figure 8.18 ***Water saturation in the fracture network at 500 days using the matrix grid with mixed elements (Xu et al., 2020).*** The water saturation in the matrix is also shown.

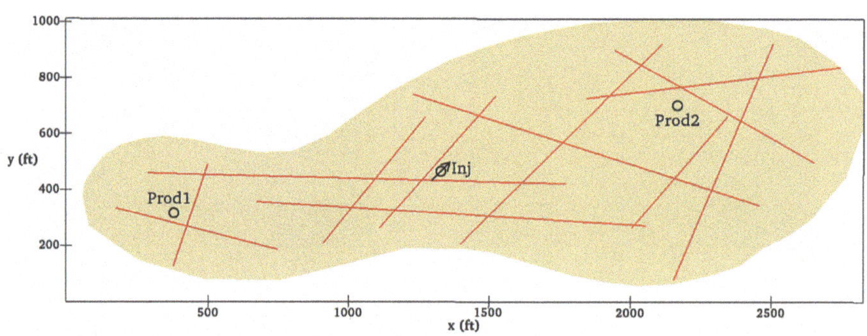

Figure 8.19 ***Reservoir geometry, locations of fractures (shown as red lines), and locations of wells in Case 3 (Xu et al., 2020).***

The grid with mixed elements (Grid D in Table 8.5) is shown in Fig. 8.23. Initially, water is at irreducible saturation, and the hydrocarbon component is methane. Other key simulation parameters are summarized in Table 8.6.

Fig. 8.24 compares the predicted gas production rates using the four grids. Fig. 8.25 presents the pressure profiles in the matrix after 1,000 days of production. As can be observed, an excellent match was obtained between the results using different grids. This verifies the methodology to apply the EDFM to 3D unstructured grids with different types of elements.

Table 8.2 Summary of simulation parameters for Case 3 (Xu et al., 2020).

Parameter	Value	Unit	Parameter	Value	Unit
Reservoir permeability	200	md	Initial water saturation	0.3	–
Reservoir porosity	0.3	–	Water compressibility	1×10^{-6}	psi^{-1}
Initial reservoir pressure	2,000	psi	Well bottomhole pressure	1,500	psi
Reservoir temperature	160	°F	Gas injection rate	3,000	STB/day
Rock compressibility	3×10^{-6}	psi^{-1}	Wellbore radius	0.5	ft
Water viscosity	0.8	cp			

Table 8.3 Properties of hydrocarbon components in Case 3 (Xu et al., 2020).

Component	CO_2	CH_4	NC16
Molar fraction	0.01	0.19	0.80
Molecular weight (lb/lb-mol)	44.0	16.0	222.0
Critical pressure (psi)	1071.6	667.2	252.1
Critical volume (ft^3/lb-mol)	1.505	1.586	13.087
Critical temperature (K)	304.2	190.6	734.7
Acentric factor	0.225	0.008	0.684
Parachor	49.0	77.0	831.9

Table 8.4 Parameters of water, oil, and gas relative permeability curves in Case 3 (Xu et al., 2020).

Parameter	Value	Parameter	Value
Residual water saturation	0.2	Gas rel. perm. endpoint	0.9
Residual oil saturation	0.1	Water rel. perm. exponent	3.0
Residual gas saturation	0	Oil rel. perm. exponent	2.0
Water rel. perm. endpoint	0.4	Gas rel. perm. exponent	2.0
Oil rel. perm. endpoint	0.9		

Together with the results of Cases 1 and 2, it shows the high flexibility of the EDFM. Different grids with different types of elements (2D or 3D) could be conveniently used in conjunction with the EDFM to simulate fractured reservoirs. The accuracy of the long-term prediction using the EDFM is not significantly affected by the type of elements used in the grid.

8.5.2 Case 5: 3D Reservoir with complex geometry

In this case, a 3D reservoir model with irregular geometry is used for the simulation, as shown in Fig. 8.26. The reservoir is represented by an unstructured grid with 42,232 vertices and 44,115 elements, including 4,200 tetrahedron elements, 35,715 hexahedron elements, and 4,200 pyramid elements. This grid was used by Araújo et al. (2016) for a gas-flooding simulation. As can be seen, a "hole" was added in this grid to represent a low-permeability zone. Unlike the case of structured grids, no non-active

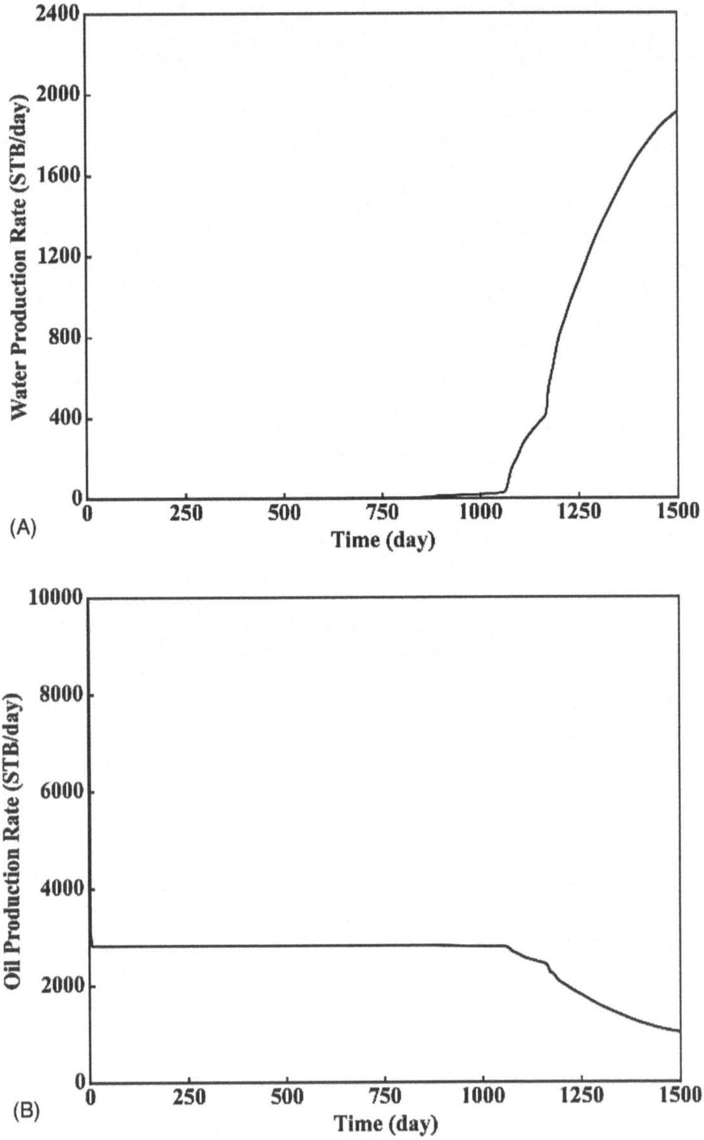

Figure 8.20 Predicted (A) water, (B) oil, and (C) gas production rate (*Xu et al., 2020*).

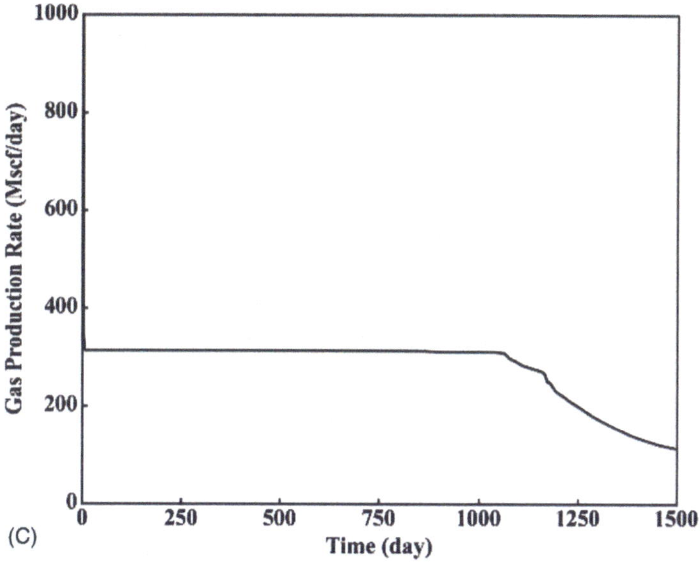

Figure 8.20 (*Cont.*)

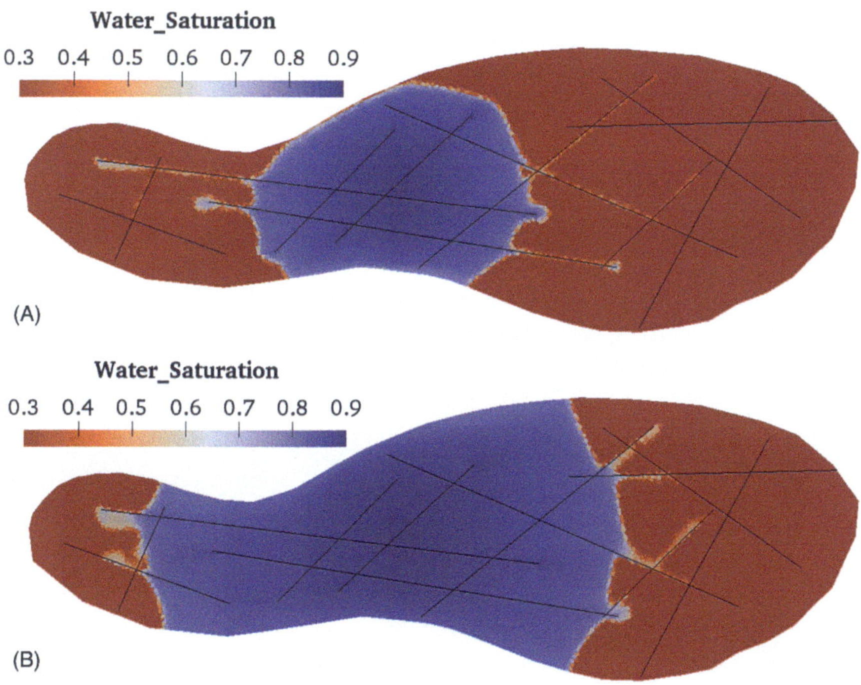

Figure 8.21 Water saturation profiles after (A) 500 days and (B) 1,000 days of injection (*Xu et al., 2020*).

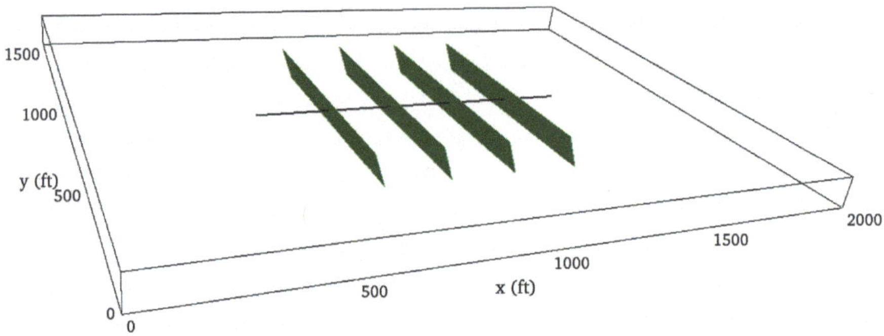

Figure 8.22 ***Reservoir dimensions and locations of hydraulic fractures in Case 4 (Xu et al., 2020).*** The horizontal well (represented by the black line) is in the middle of the reservoir. The four green polygons represent the hydraulic fractures.

Table 8.5 Summary of the numbers of vertices and elements in the four grids in Case 4 (Xu et al., 2020).

Grid	Number of vertices	Number of elements
A	30,250	157, 464 (tetrahedron)
B	33,275	58,320 (prism)
C	30,250	26,244 (hexahedron)
D	31,006	13,392 (tetrahedron), 23,508 (hexahedron), and 3,780 (pyramid)

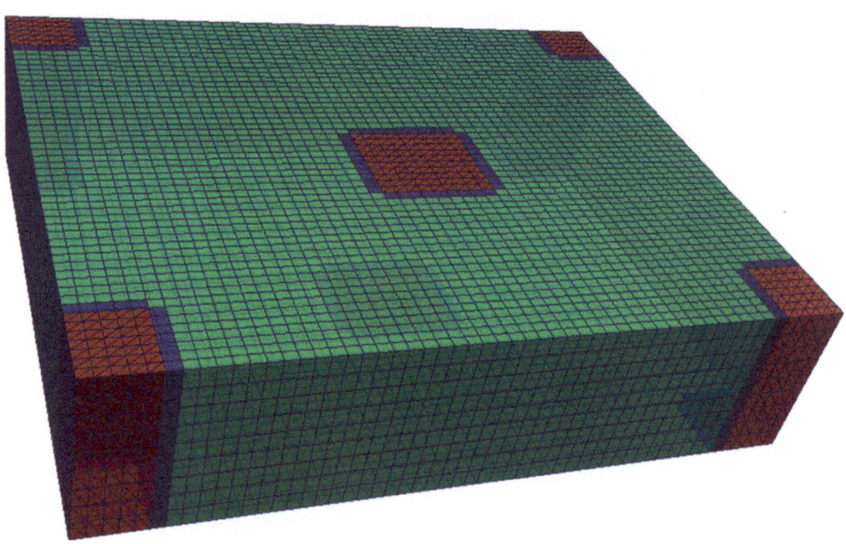

Figure 8.23 ***Grid with mixed elements (Xu et al., 2020). The tetrahedron, hexahedron, and pyramid elements are shown in red, green, and blue, respectively.*** The length in the vertical direction is shown five times the real size.

Table 8.6 Key simulation parameters used in Case 4 (Xu et al., 2020).

Parameter	Value	Unit	Parameter	Value	Unit
Reservoir permeability	0.0003	md	Initial water saturation	0.2	–
Reservoir porosity	0.1	–	Initial gas saturation	0.8	–
Initial reservoir pressure	4,000	psi	Initial gas rel. perm.	0.8	–
Reservoir temperature	200	°F	Well bottomhole pressure	2,000	psi
Rock compressibility	4×10^{-6}	psi^{-1}	Wellbore radius	0.25	ft
Water compressibility	3×10^{-6}	psi^{-1}			

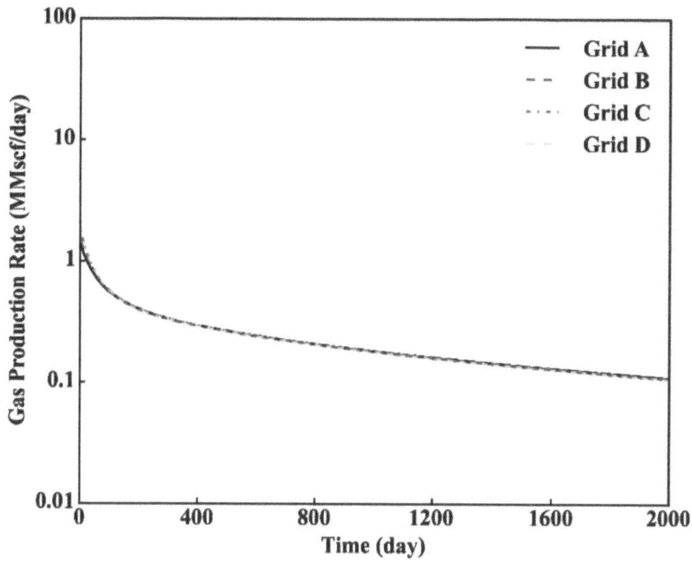

Figure 8.24 ***Predicted surface gas production rates on four different grids in Case 4 (Xu et al., 2020).***

cells need to be used for the representation of the hole in this grid. A horizontal well and 17 vertical fractures are placed in the reservoir, as shown in Fig. 8.26. The six "brown" fractures are directly connected to the wellbore, and their conductivity is 1,000 md-ft; the "grey" fractures are not directly connected to the wellbore, and their conductivity is 10 md-ft. Other simulation parameters are summarized in Table 8.7. Only one hydrocarbon component is used in the simulation (C_{20}). For the water–oil two-phase flow, the same relative permeability curves as in Case 3 are used.

Fig. 8.27 shows the simulated oil flow rate. Fig. 8.28 presents the pressure profile after 10 days of production. The depleted zone around fractures can

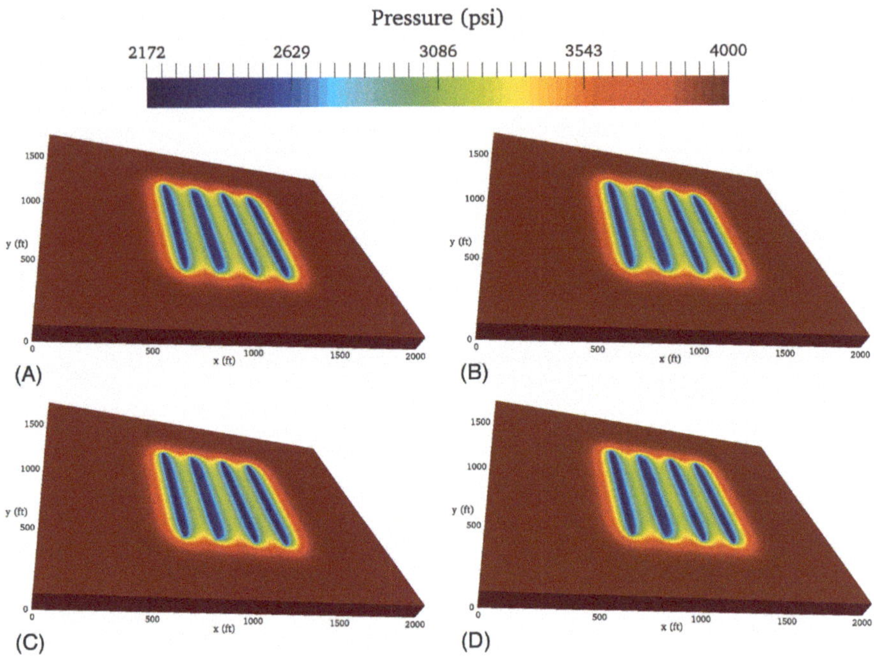

Figure 8.25 ***Predicted pressure profiles after 1,000 days of production (Xu et al., 2020).*** (A) through (D) show the results using Grids A through D in Table 8.5.

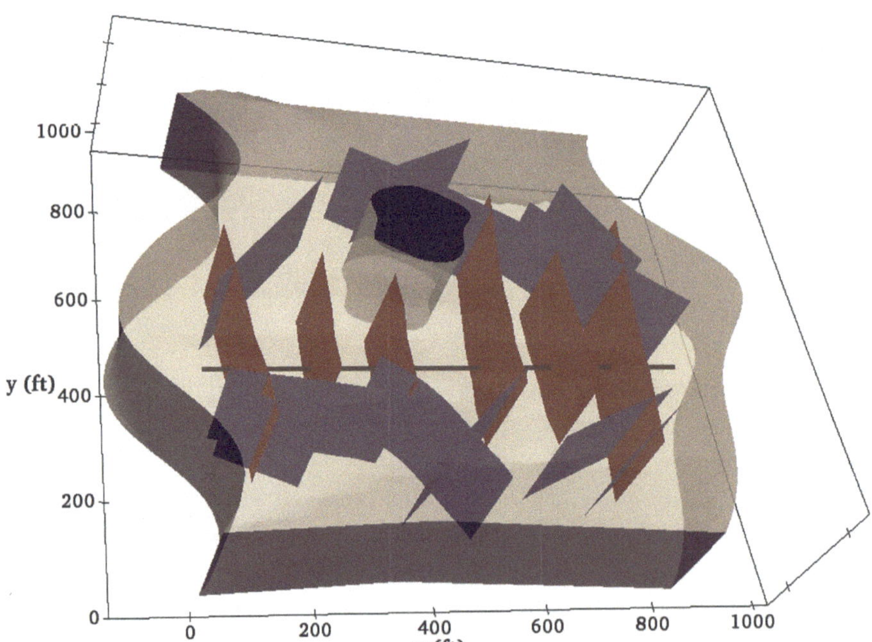

Figure 8.26 ***Reservoir geometry, location of fractures, and location of well in Case 5 (Xu et al., 2020).*** The grey line represents the horizonal well, and the brown and grey polygons represent fractures. The "brown" fractures are directly connected to the wellbore and the "grey" fractures are not directly connected to the wellbore.

Table 8.7 Reservoir parameters used in the simulation of Case 5 (Xu et al., 2020).

Parameter	Value	Unit	Parameter	Value	Unit
Horizontal permeability	0.05	md	Rock compressibility	1×10^{-6}	psi^{-1}
Vertical permeability	0.005	md	Water compressibility	1×10^{-6}	psi^{-1}
Reservoir porosity	0.3	–	Initial water saturation	0.3	–
Initial reservoir pressure	3,000	psi	Well bottomhole pressure	2,000	psi
Reservoir temperature	160	°F	Wellbore radius	0.5	ft

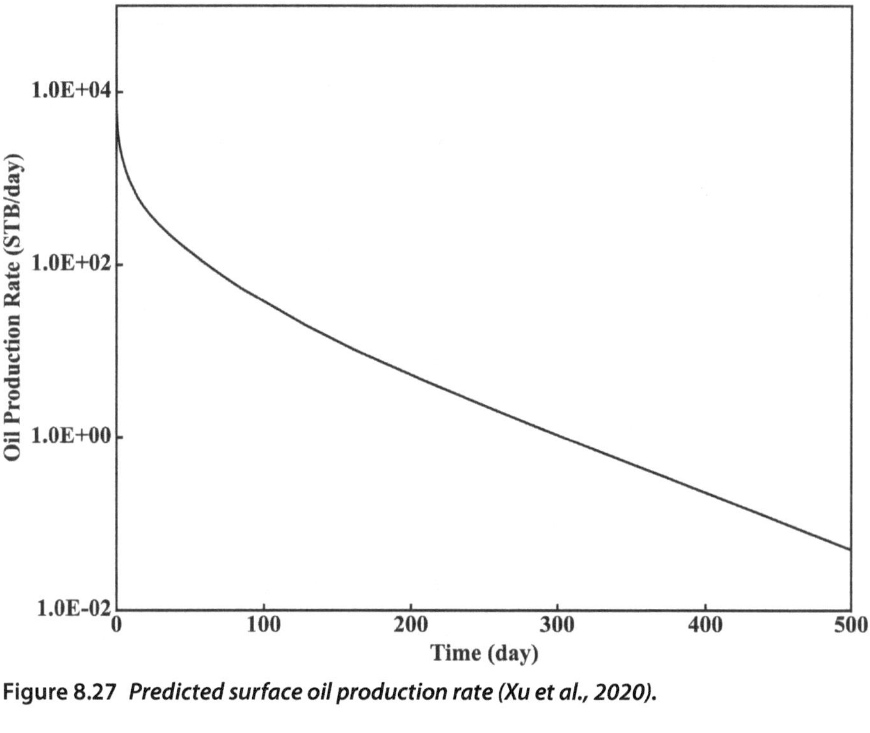

Figure 8.27 ***Predicted surface oil production rate (Xu et al., 2020).***

be observed in the pressure profile. Being connected to the well and having high fracture conductivity, the "brown" fractures have a major impact on oil production and pressure distribution. In contrast, the "grey" fractures have much less impact on pressure distribution compared with the "brown" fractures. The pressure interference between fractures can also be observed from the pressure profile. Furthermore, the impact of the "hole" can also be

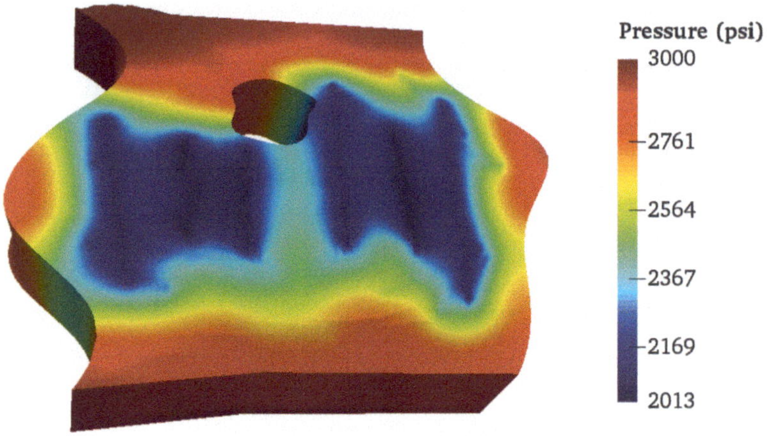

Figure 8.28 ***Predicted pressure profile at 10 days (Xu et al., 2020).***

clearly seen from the pressure distribution around it. Altogether with Case 3, this case study shows the high flexibility of unstructured grids in representing various types of reservoir geometries, such as irregular boundaries and "holes". Because EDFM makes it easy to simulate complex fractures in such grids, the combination of unstructured gridding with the EDFM provides a convenient solution for simulating highly fractured reservoirs with complex geometries.

Nomenclature

$A_{f,SCV}$ = area of fracture, ft^2
$\overrightarrow{A_l}$ = area of interface between sub-control volumes, ft^2
d = distance, ft
F_k = molar flow rate of component k, mole/day
k_{rj} = relative permeability of phase j
$\overrightarrow{\overrightarrow{K}}$ = permeability tensor, md
$\vec{n}$ = unit normal vector
n_c = number of hydrocarbon components
n_{ip} = number of integration points
n_{merge} = number of initial fracture segments to merge
n_p = number of phases
N_i = shape function of vertex i
N_k = number of moles of component k, mole
N_v = number of vertices

t = time, day
T = transmissibility factor, md-ft
q_k = injection/production molar rate of component k from wells, mole/day
x = Cartesian coordinate, ft
x_{kj} = mole fraction of component k in phase j
y = Cartesian coordinate, ft
z = Cartesian coordinate, ft
Φ = physical property value at a point inside an element
$\vec{\nabla}\Phi_{jl}$ = flow potential gradient in phase j at the lth integration point, psi/ft
ξ = local coordinate
ξ_j = the molar density of phase j
η = local coordinate
γ = local coordinate
μ_j = viscosity of phase j, cp

Appendix B: Shape functions for 2D and 3D elements

The shape function N_i is defined for each type of element. In 2D grids, triangular and quadrilateral elements are used. Figs. B.1 and B.2 show the definition of local coordinates (ξ, η) in triangular and quadrilateral elements, respectively. The shape functions for a triangular element are

$$\begin{aligned} N_1(\xi,\eta) &= 1-\xi-\eta; \\ N_2(\xi,\eta) &= \xi; \\ N_3(\xi,\eta) &= \eta. \end{aligned} \tag{B.1}$$

The shape functions for a quadrilateral element are

$$\begin{aligned} N_1(\xi,\eta) &= \frac{1}{4}(1-\xi)(1-\eta); \\ N_2(\xi,\eta) &= \frac{1}{4}(1+\xi)(1-\eta); \\ N_3(\xi,\eta) &= \frac{1}{4}(1+\xi)(1+\eta); \\ N_4(\xi,\eta) &= \frac{1}{4}(1-\xi)(1+\eta). \end{aligned} \tag{B.2}$$

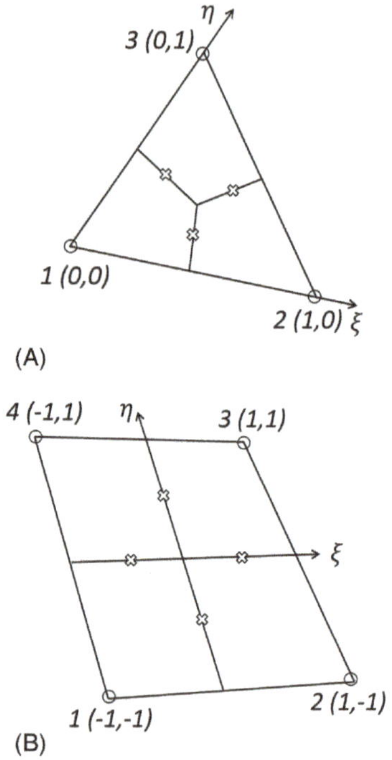

Figure B.1 ***Definition of local coordinates in 2D elements (Xu et al., 2020).*** (A) Triangular element. (B) Quadrilateral element. For each vertex, its index and (ξ,η) coordinates are shown. The integration points in the elements are also shown.

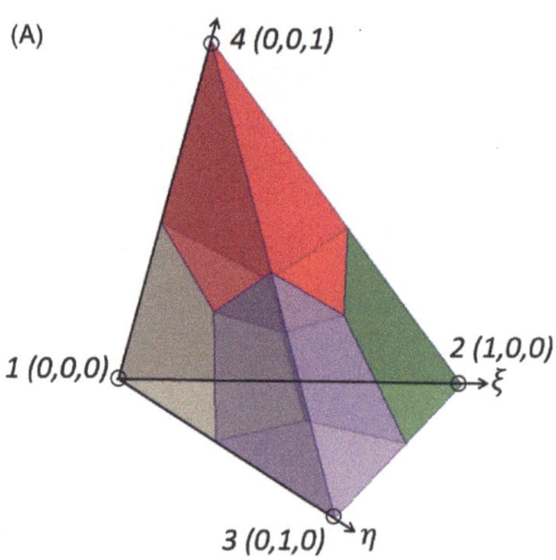

Figure B.2 ***Definition of local coordinates in 3D elements (Xu et al., 2020).*** (A) Tetrahedron element. (B) Prism element. (C) Hexahedron element. (D) Pyramid element. For each vertex, its index and (ξ,η,γ) coordinates are shown.

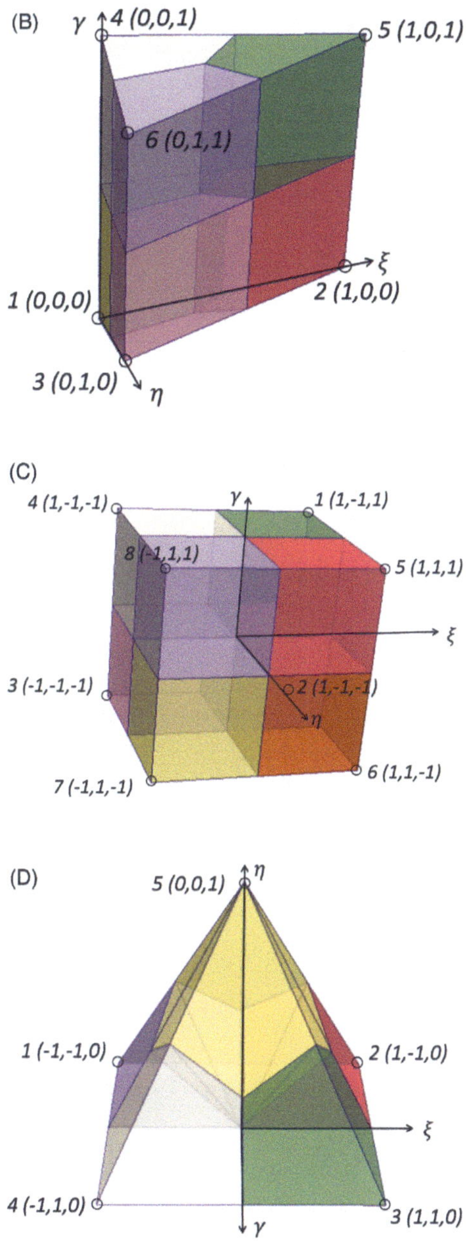

Figure B.2 (*Cont.*)

In 3D grids, four types of elements can be used: tetrahedron, prism, hexahedron, and pyramid. The definition of local coordinates (ξ,η,γ) in 3D elements is presented in Fig. B.2. The shape functions for a tetrahedron element are

$$\begin{aligned}
N_1(\xi,\eta,\gamma) &= 1-\xi-\eta-\gamma; \\
N_2(\xi,\eta,\gamma) &= \xi; \\
N_3(\xi,\eta,\gamma) &= \eta; \\
N_4(\xi,\eta,\gamma) &= \gamma.
\end{aligned} \tag{B.3}$$

The shape functions for a prism element are

$$\begin{aligned}
N_1(\xi,\eta,\gamma) &= (1-\xi-\eta)(1-\gamma); \\
N_2(\xi,\eta,\gamma) &= \xi(1-\gamma); \\
N_3(\xi,\eta,\gamma) &= \eta(1-\gamma); \\
N_4(\xi,\eta,\gamma) &= \gamma(1-\xi-\eta); \\
N_5(\xi,\eta,\gamma) &= \xi\gamma; \\
N_6(\xi,\eta,\gamma) &= \eta\gamma.
\end{aligned} \tag{B.4}$$

The shape functions for a hexahedron element are

$$\begin{aligned}
N_1(\xi,\eta,\gamma) &= \frac{1}{8}(1+\xi)(1-\eta)(1+\gamma); \\
N_2(\xi,\eta,\gamma) &= \frac{1}{8}(1+\xi)(1-\eta)(1-\gamma); \\
N_3(\xi,\eta,\gamma) &= \frac{1}{8}(1-\xi)(1-\eta)(1-\gamma); \\
N_4(\xi,\eta,\gamma) &= \frac{1}{8}(1-\xi)(1-\eta)(1+\gamma); \\
N_5(\xi,\eta,\gamma) &= \frac{1}{8}(1+\xi)(1+\eta)(1+\gamma); \\
N_6(\xi,\eta,\gamma) &= \frac{1}{8}(1+\xi)(1+\eta)(1-\gamma); \\
N_7(\xi,\eta,\gamma) &= \frac{1}{8}(1-\xi)(1+\eta)(1-\gamma); \\
N_8(\xi,\eta,\gamma) &= \frac{1}{8}(1-\xi)(1+\eta)(1+\gamma).
\end{aligned} \tag{B.5}$$

The shape functions for a pyramid element are

$$
\begin{aligned}
N_1(\xi,\eta,\gamma) &= \frac{1}{4}[(1-\xi)(1-\eta)-\gamma+\xi\eta\gamma/(1-\gamma)];\\
N_2(\xi,\eta,\gamma) &= \frac{1}{4}[(1+\xi)(1-\eta)-\gamma-\xi\eta\gamma/(1-\gamma)];\\
N_3(\xi,\eta,\gamma) &= \frac{1}{4}[(1+\xi)(1+\eta)-\gamma+\xi\eta\gamma/(1-\gamma)];\\
N_4(\xi,\eta,\gamma) &= \frac{1}{4}[(1-\xi)(1+\eta)-\gamma-\xi\eta\gamma/(1-\gamma)];\\
N_5(\xi,\eta,\gamma) &= \gamma.
\end{aligned}
\tag{B.6}
$$

References

Araújo, A.L.S., Fernandes, B.R.B., Drumond Filho, E.P., Araujo, R.M., Lima, I.C.M., Gonçalves, A.D.R., Marcondes, F., Sepehrnoori, K., 2016. 3D compositional reservoir simulation in conjunction with unstructured grids. Braz. J. Chem. Eng. 33 (2), 347–360.

Baliga, B.R., Patankar, S.V., 1980. A new finite-element formulation for convection-diffusion problems. Numer. Heat Transfer 3 (4), 393–409.

Cordazzo, J., Maliska, C., da Silva, A.F.C., Hurtado, F.S.V., 2005. An element based conservative scheme using unstructured grids for reservoir simulation. 18th World Petroleum Congress. Johannesburg, South Africa.

Durlofsky, L.J., Chien, M.C.H., 1993. Development of a mixed finite-element-based compositional reservoir simulator. Paper SPE 25253, SPE Symposium on Reservoir Simulation. New Orleans, Louisiana.

Edwards, M.G., 2002. Unstructured, control-volume distributed, full-tensor finite-volume schemes with flow based grids. Comput. Geosci. 6, 433–452.

Fang, S., Cheng, L., Huang, S., Renyi, C., 2016. Laplace domain coupled dual continuum and embedded discrete fracture model for rate transient analysis in low permeability reservoirs. Paper SPE 180265, SPE Low Perm Symposium. Denver, Colorado.

Fernandes, B.R.B., Lima, I., da, C.M., Drumond Filho, E.P., de Oliveira, J.C.B., Marcondes, F., Sepehrnoori, K., 2017. A TVD scheme for 3D unstructured grids applied to compositional reservoir simulation. Braz. J. Chem. Eng. 34 (4), 1161–1174.

Fernandes, B.R.B., Marcondes, F., Sepehrnoori, K., 2016. Comparison of two volume balance fully implicit approaches in conjunction with unstructured grids for compositional reservoir simulation. Appl. Math. Modell. 40, 5153–5170.

Forsyth, P.A., 1990. A control-volume, finite-element method for local mesh refinement in thermal reservoir simulation. SPE Reservoir Eng. 5 (4), 561–566.

Fung, L.S.-K., Hiebert, A.D., Nghiem, L.X., 1992. Reservoir simulation with a control-volume finite-element method. SPE Reservoir Eng. 7 (3), 349–357.

Hoteit, H., Firoozabadi, A., 2006. Compositional modeling by the combined discontinuous Galerkin and mixed methods. SPE J. 11 (1), 19–34.

Karimi-Fard, M., Durlofsky, L.J., Aziz, K., 2004. An efficient discrete-fracture model applicable for general-purpose reservoir simulators. SPE J. 9 (2), 227–236.

Maliska, C.R., 2004. Computational Heat Transfer and Fluid Mechanics, second ed. Livros Tecnicos e Cientificos, Rio de Janeiro.

Marcondes, F., Santos, L.O.S., Varavei, A., Sepehrnoori, K., 2013. A 3D hybrid element-based finite-volume method for heterogeneous and anisotropic compositional reservoir simulation. J. Pet. Sci. Eng. 108, 342–351.
Marcondes, F., Sepehrnoori, K., 2007. Unstructured grids and an element-based conservative approach for compositional reservoir simulation. 19th International Congress of Mechanical Engineering. Brasilia, Brazil.
Marcondes, F., Sepehrnoori, K., 2010. An element-based finite-volume method approach for heterogeneous and anisotropic compositional reservoir simulation. J. Pet. Sci. Eng. 73, 99–106.
Norbeck, J.H., McClure, M.W., Lo, J.W., Horne, R.N., 2016. An embedded fracture modeling framework for simulation of hydraulic fracturing and shear stimulation. Comput. Geosci. 20, 1–18.
Prévost, M., Edwards, M.G., Blunt, M.J., 2002. Streamline tracing on curvilinear structured and unstructured grids. SPE J. 7 (2), 139–148.
Prévost, M., Lepage, F., Durlofsky, L.J., Mallet, J.-L., 2005. Unstructured 3D gridding and upscaling for coarse modelling of geometrically complex reservoirs. Pet. Geosci. 11 (4), 339–345.
Samier, P., Masson, R., 2017. Implementation of a vertex-centered method inside an industrial reservoir simulator: practical issues and comprehensive comparison with corner-point grids and perpendicular-bisector-grid models on a field case. SPE J. 22 (2), 660–678.
Santos, L.O.S., Marcondes, F., Sepehrnoori, K., 2013. A 3D compositional miscible gas flooding simulator with dispersion using element-based finite-volume method. J. Pet. Sci. Eng. 112, 61–68.
Verma, S., Aziz, K., 1997. A control volume scheme for flexible grids in reservoir simulation. Paper SPE 37999, SPE Reservoir Simulation Symposium. Dallas, Texas.
Vitel, S., Souche, L., 2007. Unstructured upgridding and transmissibility upscaling for preferential flow paths in 3D fractured reservoirs. Paper SPE 106483, SPE Reservoir Simulation Symposium. Houston, Texas.
Xu, Y., Lima, I.C.M., Marcondes, F., Sepehrnoori, K., 2020. Development of an embedded discrete fracture model for 2D and 3D unstructured grids using an element-based finite volume method. Accepted for publication.

CHAPTER NINE

Concluding remarks

9.1 Key conclusions

The key conclusions of this book are summarized as follows:

- The results of fracture diagnostics, core analysis, lab experiments, and fracture propagation modeling show the complexity of fracture geometries, which makes it necessary to model complex fracture geometries in reservoir simulation.
- Dual-continuum models and discrete fracture models (DFMs) are two widely used types of models for the simulation of fractured reservoirs. Dual-continuum models are effective for modeling naturally fractured reservoirs with high fracture density, but limitations exist for these models when simulating disconnected fractures, a small number of large-scale fractures, or highly localized anisotropy caused by natural fractures (Karimi-Fard et al., 2004; Moinfar et al., 2011). Accurate evaluation of the transfer function is also challenging. DFMs using conforming meshes are able to capture the detailed geometry of fractures accurately, but it is challenging to generate conforming meshes around complex fractures, and the resulting computational efficiency is usually low.
- The EDFM was developed to explicitly simulate fractures in traditional reservoir simulators without using conforming meshes. Because the EDFM keeps the original matrix grids used by traditional simulators and simulates the fractures through different types of connection factors, it can be applied to different reservoir simulators in a non-intrusive manner. This takes advantage of the existing functionalities offered by these simulators to simulate complex recovery processes involving different physical mechanisms. Fig. 9.1 summarizes the key steps of a typical EDFM simulation. These steps include performing the geometrical calculations, calculating the transmissibilities, and modifying the input data of traditional simulators. Three types of non-neighboring connections (NNCs) are calculated in this procedure, including NNCs between matrix blocks and fracture segments (Type I), between fracture segments within an individual fracture (Type II), and between intersecting fracture segments (Type III).

Embedded Discrete Fracture Modeling and Application in Reservoir Simulation.
http://dx.doi.org/10.1016/B978-0-12-821872-3.00009-5

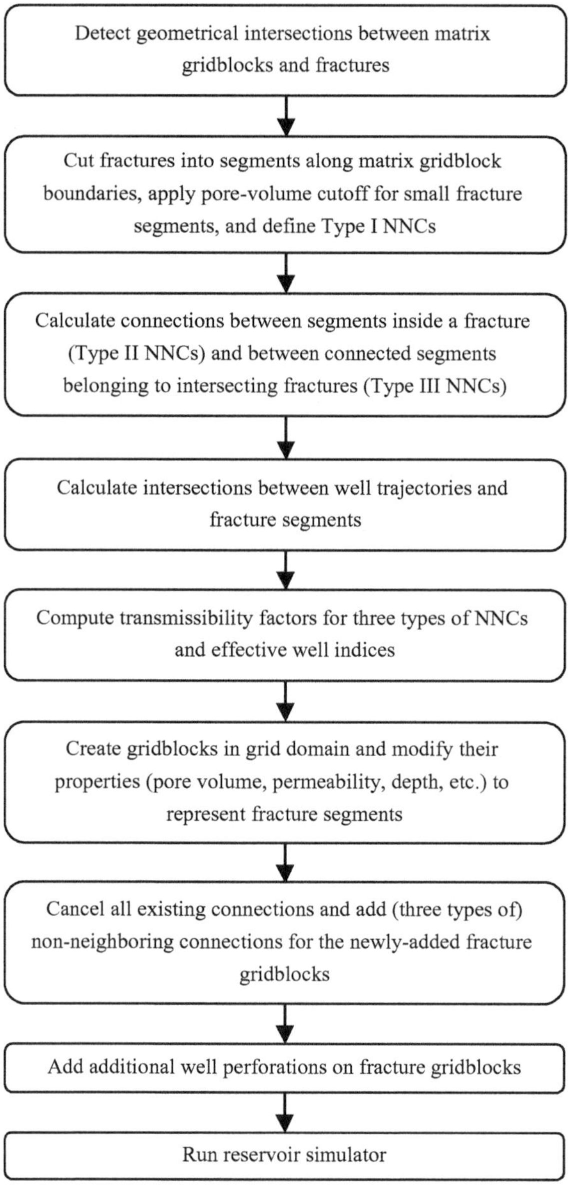

Figure 9.1 ***Flowchart of a typical EDFM simulation (Xu and Sepehrnoori, 2019).*** This flowchart shows the key steps to create fracture control volumes, evaluate transport indices, and set up connections inside reservoir simulators.

- The EDFM is an accurate method for simulating various types of fracture geometries. Good agreements in simulation results were obtained between the EDFM and other approaches.
- The EDFM has an advantage in computational efficiency over the local grid refinement (LGR) model for simulating complex fracture geometries.
- With the EDFM formulations in the half-transmissibility form, the transmissibility factors are linked with gridblock permeabilities, and the dynamic behaviors of complex fractures can be conveniently and effectively simulated by modifying the permeabilities of fracture gridblocks with time. These dynamic fracture behaviors include the creation of new fractures, growth of existing fractures, fracture conductivity variation during simulation, and fracture activation.
- The EDFM is a practical approach for use with history matching and sensitivity studies for complex fracture systems. Engineers can easily alter the number, location, geometry, and properties of fractures. These parameters need to be frequently modified in history matching and sensitivity studies.
- The use of EDFM in corner-point grids and unstructured grids makes it convenient to simulate fractures in geologically complex reservoirs with irregular boundaries. Complicated gridding around fractures is avoided, and minimum adjustment is required for the original grid.
- The implementation of the geometrical calculations of the EDFM in different types of grids is an important step in applying the EDFM. The geometrical algorithm to find the intersection between a polyhedron and a polygon, which covers the situations of both convex and concave polyhedra and polygons, was found to be general and effective in computing the matrix–fracture intersections in different types of grids, including Cartesian grids, corner-point grids, and unstructured grids with various types of elements.
- In corner-point grids, it was found that the accuracy of the EDFM for long-term production prediction is not sensitive to matrix gridding when the average size of gridblocks is similar.
- For the application of the EDFM in unstructured grids using the element-based finite-volume method (EbFVM), when the number of control volumes is similar, the EDFM obtains similar results on different grids with various types of elements. The types of elements tested include triangular and quadrilateral elements in two-dimensional (2D)

grids and tetrahedron, prism, hexahedron, and pyramid elements in three-dimensional (3D) grids.

9.2 Summary of the advantages of the EDFM

With the case studies presented in this book, we have found several advantages of the EDFM in terms of ease of gridding, flexibility, computational efficiency, and extensibility.

First, the gridding in the EDFM is relatively easy. The EDFM uses separate gridding for the matrix and fractures. This is convenient for matrix gridding and avoids the challenges involving gridding around fractures which are faced by other approaches using conforming meshes. It also makes it convenient to place fractures with complex geometries. In addition, the preprocessing complexity of the EDFM increases almost linearly with the increase of fracture number, which makes it easy to handle a large number of fractures.

Second, the EDFM provides high flexibility when modeling fractures. Because of the separate gridding of the matrix and fractures, it is not necessary to re-grid the matrix after changing the fracture locations or geometries. This greatly simplifies the procedures of sensitivity studies and history matching with fractures. In addition, the simulation results are also insensitive to matrix gridding for long-term production forecasting. This allows the possibility of modifying matrix grids without considering the existence of fractures.

Third, the EDFM has high computational efficiency. This is extremely important for sensitivity studies and history matching. Besides, the simulation computational cost using the EDFM increases almost linearly with the increase in the number of fractures, making it computationally feasible to include a relatively large number of fractures in the simulation.

Lastly, the EDFM has high extensibility. It can be extended to traditional simulators with NNC functionalities. It is also compatible with different types of matrix gridding and numerical formulations. Overall, it is a general method to model fractures in reservoir simulation.

9.3 Discussion on the disadvantages of the EDFM

As with other numerical models, the EDFM also has some disadvantages. One such disadvantage is related to the "linear pressure distribution" assumption used in the derivation of the matrix–fracture transmissibility formulation. A large error could be introduced if the pressure distribution around fractures deviates significantly from this assumption. The calculations of fracture–fracture connection/intersection transmissibility factors also involve approximations, which might also introduce error. In addition, because the matrix blocks intersected by fractures are not split into two parts, the EDFM has difficulty simulating situations where the matrix has different properties on either side of the fracture. For the same reason, an error will be introduced in the simulation of fluid flow from one side of the fracture to the other side. Also, compared with DFMs using conforming meshes, the EDFM is not suitable for simulating fractures with a lower permeability than the matrix (flow barriers).

Although many case studies with different types of reservoirs and recovery mechanisms have been presented in this book to demonstrate the effectiveness and efficiency of the EDFM, it should be noted that the EDFM is a special type of DFM. Attention needs to be paid during the selection of suitable models (homogenization, dual-continuum model, or DFM) for different types of fractured reservoirs. Factors such as the scale of fractures, the connectivity between fractures, the connectivity within the matrix, and the kinetics of matrix–fracture interactions need to be considered when selecting the model (Bourbiaux, 2010). Furthermore, it usually requires much more information to build a DFM than in building other fracture models, which should also be taken into account in the engineering practice. It is not recommended to use the EDFM when the nature of the problem does not require an explicit representation of each fracture. In such cases, other models might achieve a better balance between accuracy and computational efficiency.

9.4 Recommendations

In this book, we demonstrated that the EDFM is a promising approach to the simulation of fractured reservoirs. However, there are still many aspects of the EDFM that require future research. Here, based on our experience, we are providing recommendations for future work related to the EDFM.

There are several variations of the EDFM in the literature. Chai et al. (2018) presented a method to split the matrix gridblock whenever a discrete fracture crosses a matrix gridblock. Yang et al. (2018) presented their work on splitting the fracture segments at fracture intersections. There are also other variations of the EDFM that modify the transmissibility formulations

(Tene et al., 2017; Jiang and Younis, 2017). More studies are needed to investigate the advantages and disadvantages of these variations in terms of accuracy, computational efficiency, extensibility, and applicability. Also, in the area of reservoir simulation, multi-point flux approximation has been used for the evaluation of fluid flow within the matrix when a non-K-orthogonal grid is used. It is recommended to develop the EDFM formulations based on multi-point flux approximation to achieve higher accuracy for the evaluation of matrix–fracture flow. In addition, transient flow around hydraulic fractures has long been an important topic in tight reservoirs. Panfili et al. (2015), Yang et al. (2018), and Xu (2018) presented examples where the accuracy of the EDFM for simulating fracture transient flow could be improved by applying LGR to the matrix blocks around fractures. More detailed research is recommended to demonstrate the effectiveness of this methodology for complex fractures.

Further improvement of the simulation performance as the number of fractures increases is also a priority. The increase in the number of fracture segments in the EDFM will undermine the simulation efficiency, and some upscaling techniques, such as static upscaling and flow-based local upscaling, have been used in the literature to solve this issue (Moinfar et al., 2013; Fumagalli et al., 2016). Future research is recommended on applying other types of upscaling procedures to reduce the number of fracture segments in the EDFM and speed up simulations for cases with a large number of fractures.

A great advantage of DFMs over other methods for fracture simulation is their capability of accurately simulating fracture geometries. It is recommended to further develop the EDFM to simulate different types of 2D and 3D hydraulic and natural fracture geometries observed in outcrops and predicted by fracture propagation modeling tools.

In Chapter 8, the EDFM was applied to homogeneous reservoirs using unstructured grids with the EbFVM. The evaluation of matrix–fracture transmissibility for heterogeneous reservoirs is more challenging, as the pressure distribution in a control volume intersected by the fractures might deviate from the "linear pressure distribution" assumption if the control volume is made up of sub-control volumes with different permeabilities. More studies need to be conducted on the matrix–fracture transmissibility calculation for heterogeneous reservoirs using these types of grids. In addition, in recent years, several commercial simulators using unstructured grids have been developed. It would be of great interest to extend the EDFM to these simulators and study the combination of the EDFM with different types of numerical schemes for unstructured grids.

Finally, more work is needed to compare DFMs using conforming and non-conforming meshes. In recent years, both types of DFMs have gained much attention from both academia and industry. It is recommended to perform a detailed quantitative comparison between the EDFM and DFMs using conforming meshes in terms of accuracy and computational efficiency. The comparison should be performed with varying numbers and complexity of fractures.

The development of the EDFM has greatly impacted the field of fractured reservoir simulation. With the increasing interest in unconventional reservoirs, the improvement of fracture characterization and diagnostic techniques, and the increase in computing power, there is a trend toward including more and more fractures explicitly when building the reservoir model. The EDFM will make the model building and simulation simple, fast, and reliable. We hope to see further improvement of the EDFM and more applications of this model to different types of engineering problems in fractured reservoirs in the near future.

References

Bourbiaux, B., 2010. Fractured reservoir simulation: a challenging and rewarding issue. Oil Gas Sci. Technol. – Rev. IFP 65 (2), 227–238.

Chai, Z., Tang, H., He, Y., Killough, J., Wang, Y., 2018. Uncertainty quantification of the fracture network with a novel fractured reservoir forward model. Paper SPE 191395, SPE Annual Technical Conference and Exhibition. Dallas, Texas.

Fumagalli, A., Pasquale, L., Zonca, S., Micheletti, S., 2016. An upscaling procedure for fractured reservoirs with embedded grids. Water Resour. Res. 52 (8), 6506–6525.

Jiang, J., Younis, R.M., 2017. An improved projection-based embedded discrete fracture model (pEDFM) for multiphase flow in fractured reservoirs. Adv. Water Resour. 109, 267–289.

Karimi-Fard, M., Durlofsky, L.J., Aziz, K., 2004. An efficient discrete-fracture model applicable for general-purpose reservoir simulators. SPE J. 9 (2), 227–236.

Moinfar, A., Narr, W., Hui, M.-H., Mallison, B.T., Lee, S.H., 2011. Comparison of discrete-fracture and dual-permeability models for multiphase flow in naturally fractured reservoirs. Paper SPE 142295, SPE Reservoir Simulation Symposium. The Woodlands, Texas.

Moinfar, A., Varavei, A., Sepehrnoori, K., John, R.T., 2013. Development of a coupled dual continuum and discrete fracture model for the simulation of unconventional reservoirs. Paper SPE 163647, SPE Reservoir Simulation Symposium. The Woodlands, Texas.

Panfili, P., Colin, R., Cominelli, A., Giamminonni, D., Guerra, L., 2015. Efficient and effective field scale simulation of hydraulic fractured wells: methodology and application. Paper SPE 175542, SPE Reservoir Characterisation and Simulation Conference and Exhibition. Abu Dhabi, UAE.

Tene, M., Bosma, S.B.M., Al Kobaisi, M.S., Hajibeygi, H., 2017. Projection-based embedded discrete fracture model (pEDFM). Adv. Water Resour. 105, 205–216.

Xu, Y., 2018. Development and application of embedded discrete fracture model for conventional and unconventional reservoir simulation. PhD Dissertation. University of Texas at Austin.

Xu, Y., Sepehrnoori, K., 2019. Development of an embedded discrete fracture model for field-scale reservoir simulation with complex corner-point grids. SPE J. 24 (4), 1552–1575.

Yang, D., Xue, X., Chen, J., 2018. High resolution hydraulic fracture network modeling using flexible dual porosity dual permeability framework. Paper SPE 190096, SPE Western Regional Meeting. Garden Grove, California.

Index

Note: Page numbers followed by "f" indicate figures, "t" indicate tables.

F

FSC
www.fsc.org
MIX
Papier | Fördert gute Waldnutzung
FSC® C083411

Kolibri 360 GmbH
Ferdinand-Jühlke-Straße 7
99095 Erfurt, Deutschland
produktsicherheit@kolibri360.de